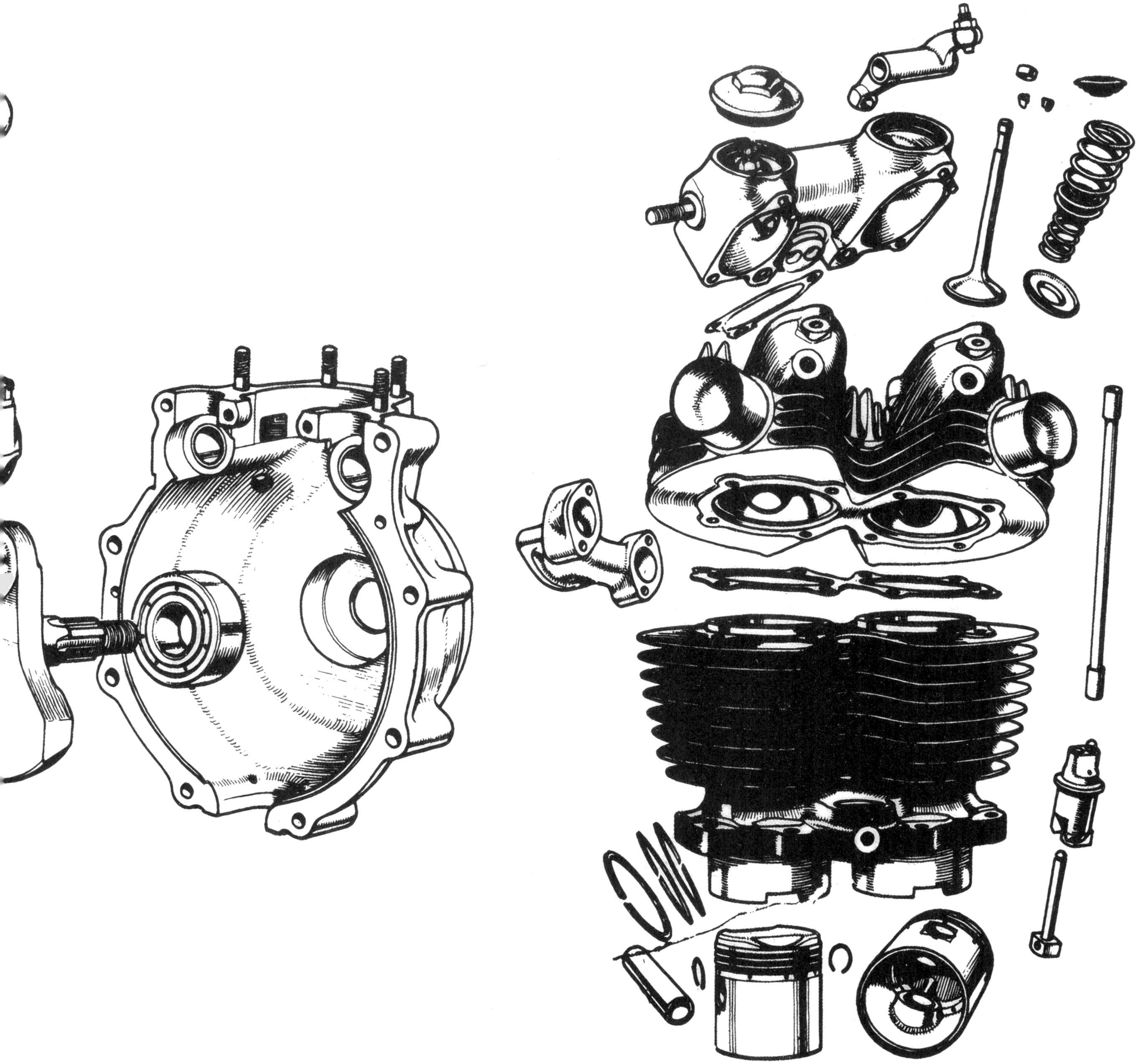

THE COMPLETE BOOK OF
CLASSIC AND MODERN
TRIUMPH
MOTORCYCLES
1937 – TODAY

THIRD EDITION

IAN FALLOON

motorbooks

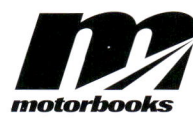

Quarto.com

© 2024 Quarto Publishing Group USA Inc.
Text © 2015–2024 Ian Falloon

Second Edition published in 2019.

First Published in 2015 by Motorbooks, an imprint of The Quarto Group,
100 Cummings Center, Suite 265-D, Beverly, MA 01915, USA.
T (978) 282-9590 F (978) 283-2742

All rights reserved. No part of this book may be reproduced in any form without written permission of the copyright owners. All images in this book have been reproduced with the knowledge and prior consent of the artists concerned, and no responsibility is accepted by producer, publisher, or printer for any infringement of copyright or otherwise, arising from the contents of this publication. Every effort has been made to ensure that credits accurately comply with information supplied. We apologize for any inaccuracies that may have occurred and will resolve inaccurate or missing information in a subsequent reprinting of the book.

Motorbooks titles are also available at discount for retail, wholesale, promotional, and bulk purchase. For details, contact the Special Sales Manager by email at specialsales@quarto.com or by mail at The Quarto Group, Attn: Special Sales Manager, 100 Cummings Center, Suite 265-D, Beverly, MA 01915, USA.

28 27 26 25 24   1 2 3 4 5

ISBN: 978-07603-9217-1

Digital edition published in 2024
eISBN: 978-07603-9218-8

Originally found under the following Library of Congress Cataloging-in-Publication Data

Falloon, Ian.
  The complete book of classic and modern Triumph motorcycles, 1937-today / by Ian Falloon.
    pages cm
  ISBN 978-0-7603-4545-0 (hardback)
 1. Triumph motorcycle--History. I. Title.
 TL448.T7F35 2015
 629.227'5--dc23
                              2014039497

Page Design: Rebecca Pagel
Front cover: 1961 Bonneville T120/James Mann
Back cover: 2024 Thruxton Final Editon/Triumph Motorcycles; Guy Allen (author photo)

Printed in China

Contents

Acknowledgments — 4

Introduction — 6
Where It All Began

1 **1937–1949: Before and After the War** — 10
The Speed Twin, Tiger 100, 3T, Grand Prix, and TR5 Trophy

2 **1950–1955: America and the World's Fastest Motorcycle** — 28
The Thunderbird, T110 and T100 Tiger, Speed Twin, and TR5 Trophy

3 **1956–1962: Birth of Iconic 650s** — 50
The Trophy and Bonneville

4 **1963–1970: Unit Construction 650s** — 82
Bonnevilles, Trophys, and Tigers

5 **1971–1975: Turmoil Times** — 126
Norton Villiers Triumph and the Meriden Blockade

6 **1976–1987: The Last Hurrah** — 152
Meriden Cooperative and Les Harris Bonnevilles

7 **1990–1996: Resurrection at Hinckley** — 172
The Modular Approach; Spine-Frame Triples and Fours

8 **1997–2002: Consolidation** — 188
Second Generation: New Frames and Fuel Injection

9 **2003–2009: After the Fire—Bigger Cruisers and Classics** — 208
The Rocket III, Thruxton, Daytona 675, Scrambler, and Street Triple

10 **2010–2024: Building on Tradition** — 232
Larger Thunderbirds, Tigers, Trophys, Bonnevilles, Scramblers, and new Singles

Index — 301

Acknowledgments

In a career of writing historical books on motorcycles, this is the first I have ever undertaken on a British marque. This was not intentional. Owning both a Triumph T160 Trident and Norton Commando back in the day when they were the hot ticket, I have always had a strong interest in and fondness for British motorcycles. After missing an opportunity to write the history of Triumph motorcycles 20 years ago, the time came when my friend at Motorbooks, Senior Editor Darwin Holmstrom, offered me this project. To cover all Triumph motorcycles over an 80-year period is a significant undertaking, and when it comes to putting something together as big as this you can't do it alone.

The biggest support comes from my wife, Miriam, and my sons Ben and Tim. They often wonder what it is about the attraction of old motorcycles, but they are always there, whether it is visiting collections, attending motorcycle events for photo shoots, or putting up with incessant motorcycle conversation.

Triumph Motorcycles, in America and Australia, have also been extremely forthcoming with material and photographs. Special thanks must go to Mark Berger of Peter Stevens Importers for providing access to hundreds of slides from the

Although the Speed Twin provided the impetus for Triumph's postwar success, arguably the most significant model of the 1950s was the TR6. Ostensibly a 650cc Tiger engine in an off-road TR5 chassis, the TR6 was targeted at American riders. The first "street scrambler," it was perfectly suited to desert racing and established Triumph's competition reputation.

early Hinckley era, while Matt O'Connor and Keith May in Atlanta were exceptionally forthcoming with more recent press images. Ex-Triumph Australia press officer and good friend Guy Allen has always been most supportive, as was Kellie Buckley, editor of *Australian Motorcycle News*, who allowed access to photo archives.

Many enthusiasts provided bikes for photography, but particular thanks must go to Jon Munn, John Grauduszus, Greg Lawn, and Colin Osborne, who went out of their way to help with multiple classic bikes. Others who allowed their bikes to appear are Dave Carlson, Kevin Grant, Don Kotchoff, Brian McInnes, Bob Nesbitt, Allen Smith, Lorin Smith, and Mike Walker. John Jessop, Tim White, and John Walker also supplied official Triumph Motorcycles images, while Gary McDonnell provided a valuable insight into how Triumph had changed over the years. Gary joined Meriden as an apprentice fitter in 1969 and, but for a short spell with Jaguar, has worked his entire life with Triumph. Ex-Britalia Motors dealer John McCoy also read the text of relevant chapters, providing a useful view of the later Meriden era from an American dealer's perspective.

Edward Turner, the father of the Speed Twin, believed 500cc to be the optimum size for a parallel twin. He would probably find even the idea of the 2014 1,700cc Thunderbird LT inconceivable. When released for 1950, Turner's 650cc Thunderbird was the largest capacity production parallel twin, as is the latest Thunderbird. *Triumph Motorcycles America*

5

Introduction
Where It All Began

It is somewhat ironic that Britain's most successful motorcycle manufacturer, Triumph, owes its origins to a German immigrant, Siegfried Bettmann. Bettmann moved to England from Nuremburg in 1884, but he was by no means the penniless European immigrant of legend. From a wealthy background, he was university educated, and his linguistic skills initially landed him a job as a translator with a publishing company in London. As a bicycle craze swept Europe, in 1885 he began exporting Birmingham-made bicycles with a "Bettmann" label. Sensing that this didn't sound British enough, in 1886 he changed the name to "Triumph." Business prospered, and a year later Bettmann took on another German immigrant, Mauritz Schulte. Schulte was a trained engineer, and in 1888 the company moved to Coventry, where they established their own manufacturing facility.

Schulte saw a future in the developing motor industry and in 1902 fitted a Belgian Minerva 2.5-horsepower engine to a Triumph bicycle. The first Triumph motorcycle was born, but Schulte wasn't content with simply using proprietary engines and in 1905 designed and built the first Triumph engine. Claimed to be the first successful engine with the mainshaft running in ball main bearings, this 363cc (78x76mm) side-valve single produced 3 horsepower at a modest 1,500 rpm. It was also so well made that it soon provided Triumph an enviable reputation for reliability and performance. Over the years the engine was gradually enlarged, becoming 453cc in 1907 and 550cc by 1914. In the meantime the company moved to new premises in Priory Street, and following Jack Marshall's victory in the 1908 Isle of Man single-cylinder Tourist Trophy (TT) Race, business prospered. Three thousand machines left the Priory Street factory in 1909. In 1913 Bettmann became the first non-British subject to be mayor of

The first engine made entirely in the Triumph works appeared in 1905.

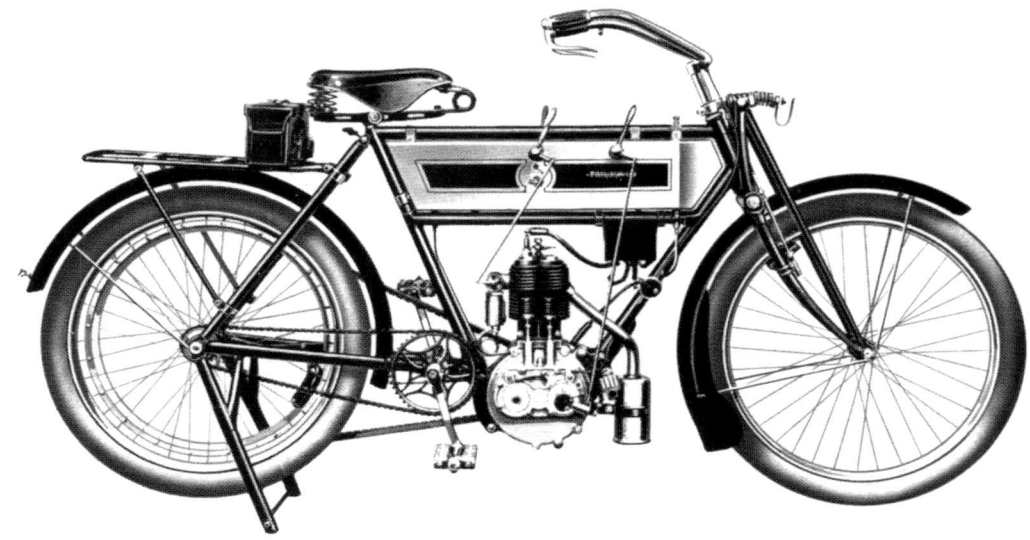

The 1906 Triumph three-horsepower model with accumulator ignition. Magneto ignition was an option.

Triumph produced a booklet to commemorate the service of their motorcycles during the Great War. Most of these were the Model H with a three-speed Sturmey-Archer gearbox and used for dispatch-rider duties.

The Triumph works at Priory Street, Coventry, in 1915. Most of these are Model Hs for war duties. 57,000 Model Hs were produced between 1915 and 1923.

Coventry, a position he lost at the outbreak of World War I, but Triumph still supplied 30,000 Model H singles to the British and Allied forces.

After the war, Triumph's model range remained centered on the side-valve Model H, but this simply wasn't fast enough for racing. In 1921 Triumph commissioned Harry Ricardo to develop a replacement cylinder head for the Model H. Ricardo was a pioneer in combustion chamber design and eventually settled on an overhead-valve four-valve design.

Ricardo's design was actually quite conservative, and from the head down the engine was virtually a Model H, retaining the Model H 80.5x98mm bore and stroke. But the four-valve head provided an immediate improvement, and the Ricardo Triumph set a new standard for 500cc motorcycles. The engine produced 20 horsepower at 4,600 rpm, virtually the same as a 1,500cc car at the time, and the initial Model R Fast Roadster proved very popular. Frank Halford, on a racing version, set a flying mile record of 83.91 miles per hour, but although three Ricardos were entered for the 1921 TT, handling deficiencies saw only one finish. A year later, the Triumph Ricardo was back with the new lubrication system, a stronger valve gear, and a new front fork. This time it finished second in the TT, but from 1923 Triumph decided on a different path. Schulte left, and the new manager, Claude Holbrook, favored car and increased motorcycle production. The result was the immensely popular Model P. Released in 1925, this 494cc side-valve machine was the simplest possible design, providing reasonable performance while still capable of being both made and sold cheaply. At £42 17s. 6p., the Model P was the cheapest 500 ever offered for sale, and production often exceeded 1,000 a week. With Triumph also expanding car production, by 1927 Priory Street was the largest factory in Coventry, employing 3,000 workers and producing 30,000 vehicles a year.

The Ricardo was the first overhead-valve Triumph. Not renowned for its excellent handling, the Model H chassis was tested by the more powerful four-valve engine. The wheels were 26 inches front and rear and the braking rudimentary. In production form it weighed 250 pounds.

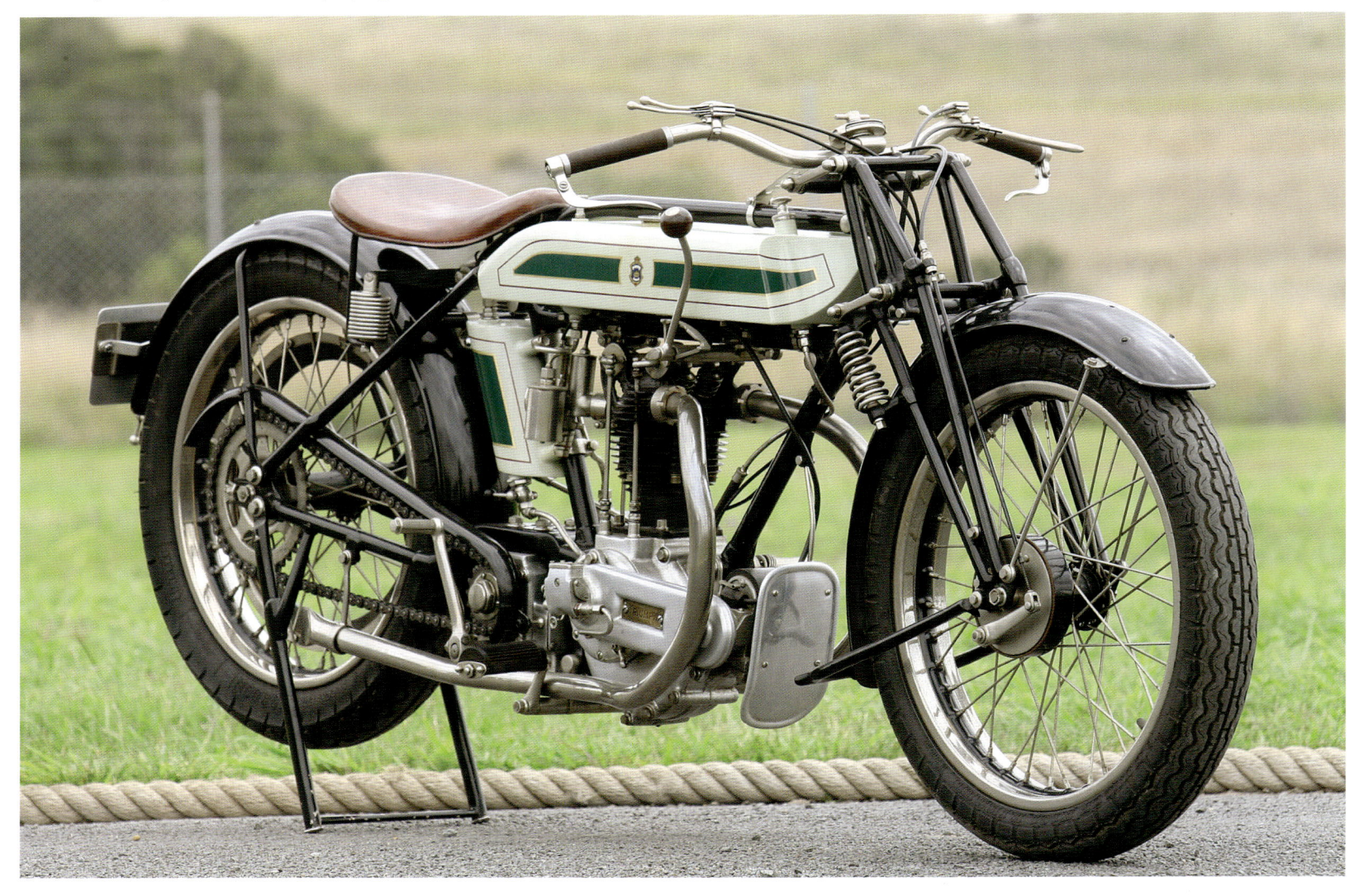

Apart from the four-valve cylinder head, much of the Ricardo was similar to the Model H. The ports were small, with the parallel valves recessed and set at 90 degrees to each other. Smaller flywheels improved revving, and carburetion was by a Triumph twin-barrel carburetor.

Although 1936 was the year Turner arrived at Triumph, Val Page's singles still initially headed the lineup.

But the 1930s slump was just over the horizon, and soon Triumph faced enormous difficulties keeping their large works in operation. By 1932 most of the engineering emphasis was on cars, and as sales of larger motorcycles crashed, designer Val Page was hired from Ariel. Page was allowed to create an autonomous motorcycle design unit and set about producing a completely new range of motorcycles. A talented and meticulous engineer, Page had immediate influence in the form of new, more modern-looking, side-valve and overhead-valve singles and an entirely new semi-unit construction 650cc vertical twin. This twin, the 6/1, featured a double helical gear primary drive, dry sump lubrication, one-piece crankshaft, single rear-mounted camshaft, and gear-driven Magdyno. The four-speed gearbox was in unit with the engine, but although a sound design, the 6/1 was heavy and more suited for sidecar work. Unfortunately, the 6/1 was not successful commercially and early in 1936 Triumph was put into receivership. Triumph and the receiver then decided to close the motorcycle manufacturing operation to concentrate on cars.

Hearing this rumor, Ariel chief Jack Sangster acted quickly, negotiating a deal with the bank receiver to allow motorcycle production to continue at Coventry, but under new management. By that stage Val Page had already left for BSA, and Sangster placed Edward Turner in charge of the new company, Triumph Engineering Co. Ltd. Siegfried Bettmann was initially chairman, but he soon retired and Jack Sangster took his place. The real power lay with Edward Turner, managing director and chief designer. After struggling for most of a decade, Triumph was about to experience the Turner revolution, and it wouldn't look back.

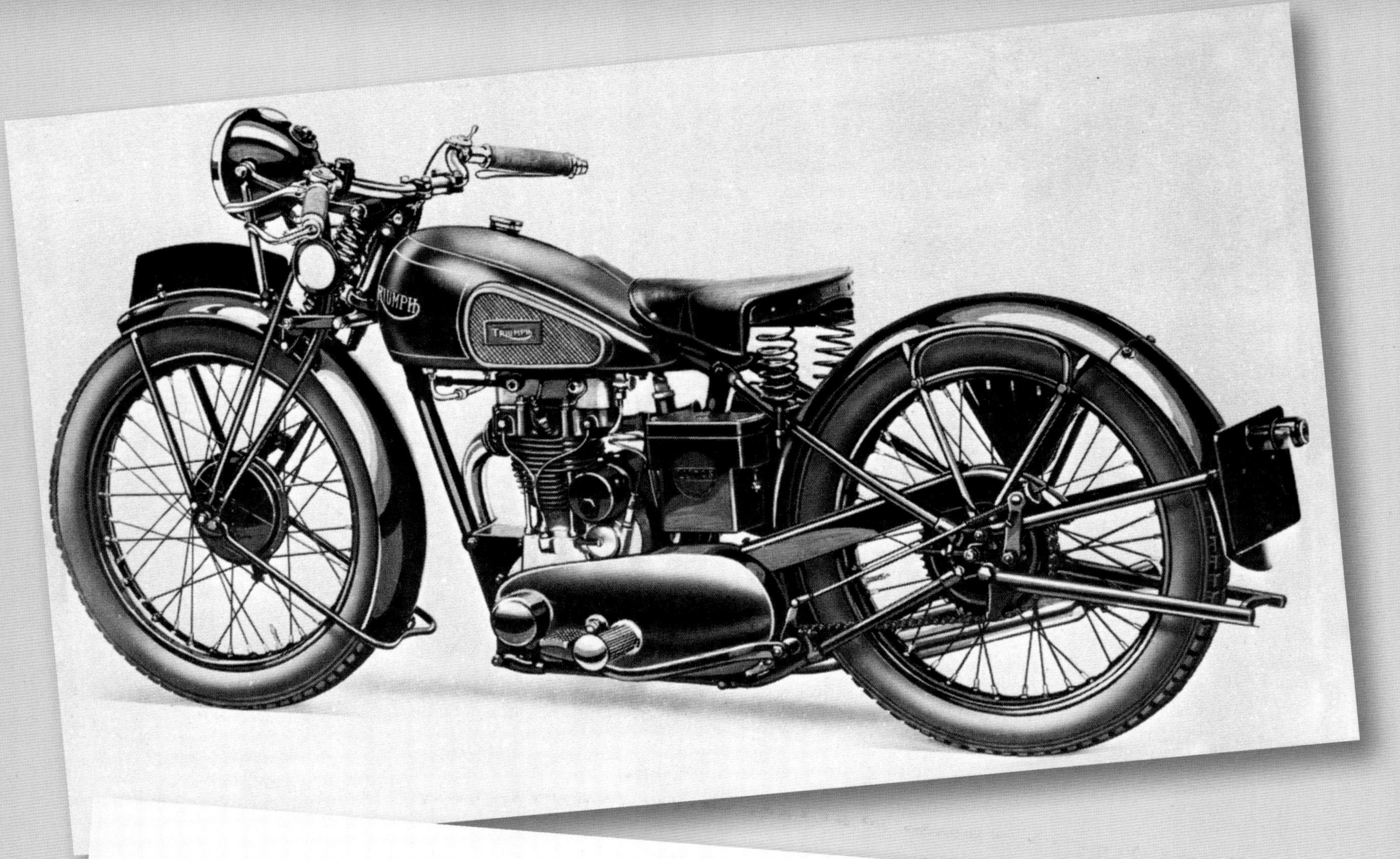
TRIUMPH
TRIUMPH

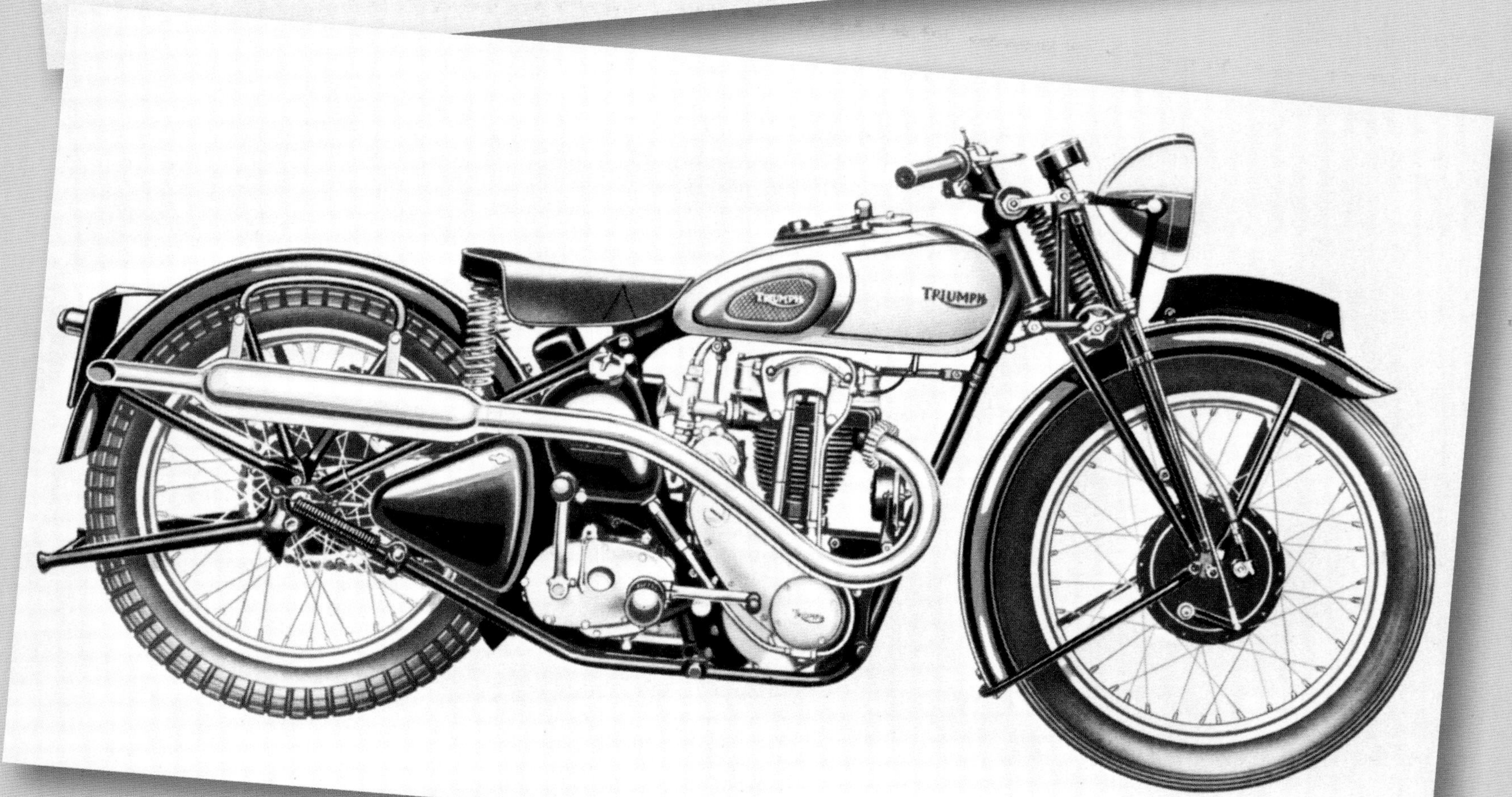
TRIUMPH
TRIUMPH

1
1937–1949
Before and After the War

The Speed Twin, Tiger 100, 3T, Grand Prix, and TR5 Trophy

By 1936 Triumph was already a significant motorcycle manufacturer, but as Triumph Engineering Co. Ltd., they were about to experience a new level of prosperity. Edward Turner may have come with the titles of managing director and chief designer, but it was soon quite obvious that he was also the chief of every other department, from sales to engineering and styling. As Triumph's publicity man Ivor Davies wrote, "Turner was in short the Great White Chief, full stop." Not only was Turner totally in charge, he was also a confident and fearless operator. Turner's initial plan involved assessing the available resources and reducing the multiplicity of component parts. Rationalizing manufacture involved streamlining the existing range of Val Page–designed Triumph motorcycles. First Turner selected one model from each capacity, 250, 350, and 500cc, and through judicious facelifting and a sporty name created an updated range. Although technical improvements were minimal, Turner's new Tigers were an overnight success, going from also-rans to world-beaters.

As an engineer, Turner was unique at the time in that he saw the importance of styling as a sales tool. Most engineers considered styling unimportant, but Turner thought otherwise. He also had the ability to anticipate what the public would accept—and buy. Without enduring the expense of a redesign, to the existing 5/5, 3/2, and L2/1 he fitted new chrome-plated gas tanks, finished in silver and highlighted with dark blue. With molded rubber knee grips, the new tank made the Tiger look lighter and sportier. He also replaced the traditional inverted clutch and brake levers with a racier-looking, longer, TT-style type, and fitted high-level exhaust pipes. Bling extended to polished aluminum timing covers and chain cases, along with chrome-plated saddle springs and headlamp and fork links. However, Turner's ace was to call them the Tiger 70, 80, and 90, indicating their top speeds. Turner was proved right as customers lined up at dealers' doors for the new Tigers.

OPPOSITE TOP: Val Page's 250cc overhead valve L2/1 was a fine design, but as it was expensive to manufacture, Turner dropped it from the range. All the Page-designed singles had a utilitarian appearance that limited their appeal.

OPPOSITE BOTTOM: Turner transformation. In 1936 Turner took the L2/1 and applied his cosmetic touch to create the Tiger 70. Even though the style was more modern, manufacturing costs saw it replaced shortly afterwards by the 2/1, which shared more with other models in the range of Page singles.

CHAPTER ONE

Edward Turner

Only a few individuals have been able to create a motorcycling genetic code and change the direction of motorcycle design, and one was Edward Turner. In a world dominated by the large-capacity, four-stroke single, Turner managed to overturn tradition and rewrite the rules by introducing a vertical twin engine that would set a design trend for the next 40 years. But Turner was much more than a designer of the parallel twin. He was a giant of the British motorcycle industry, dictating the style and engineering of motorcycles for decades.

Hailing from a family with engineering and coach-building traditions, Turner was born on January 24, 1901, and showed an early mechanical aptitude before joining the Merchant Navy at the age of 17. He fancied himself as a baritone singer before finding a more realistic occupation by purchasing a small motorcycle shop with his war gratuity. Benefitting from an explosive growth in the motorcycle industry, Turner decided to build his own motorcycle, a 350cc overhead camshaft single. Here his abilities began to shine. Turner produced all the detail drawings and his own cutting tools, and he machined all the components on his own lathe, milling machine, and bench grinder. In 1928, with his 350 completed, Turner then headed to the heart of the British motorcycle industry, the Midlands, to find someone to make it commercially. Although he wasn't successful in getting his home-brewed 350 into production, it gave him an immediate entry into the upper echelons of the motorcycle industry. Ariel appointed him chief development engineer alongside one of the industry greats, Val Page. And when Page departed for Triumph in 1932, Turner inherited the chief designer's chair.

Following Page's move to Triumph in 1932, Edward Turner became Ariel's chief designer. He was only 32 years old but managed to restore Ariel's fortunes.

Not only did Turner hawk his 350 single around the motorcycle factories of the Midlands, he also had a cigarette packet with the germ of an idea for a four-cylinder motorcycle sketched on the back. Apologizing to Ariel's boss Jack Sangster that the packet was only a cheap brand (Wild Woodbine), Turner got the go-ahead to complete a full set of drawings. After 18 months of development, Ariel unveiled the Square Four at the 1930 Olympia Show in London. In 1935 Turner undertook a major redesign of the Square Four, enlarging the engine to 997cc. So advanced was Turner's 1,000cc Square Four that it remained in production until 1959.

The departure of Val Page also provided Edward Turner an opportunity to revive the fortunes of the ailing Ariel concern. Ariel was facing bankruptcy, times were tough, and with no money for brand-new designs, Turner decided to revamp the existing Val Page–designed singles. No one lusted over the workmanlike Ariel singles, but Turner wove his magic and, with deft styling and brilliant engineering streamlining, managed to turn an uninspiring run-of-the-mill motorcycle into highly desirable property. All Red Hunters had distinctive dark red petrol and oil tanks, and their engines were run for two hours on a test bench. Turner knew that lots of sparkle and glossy paint was the styling most sporting Englishmen hoped to own, and the Red Hunter restored Ariel's fortunes. Not only was the Red Hunter good looking, it was also fast, rugged, and reliable, successful in trials, scrambles, and even on the banking of Brooklands.

Turner may have been an engineer, but he had a greater flair for commerce. His dynamic approach was far reaching, and he met and talked with dealers, supervised the production of sales literature and advertising, and even indulged in road testing. He was known to distrust the opinions of professional testers, and he was quite happy to hop on a new bike and ride down the road, still wearing his trilby hat. But perhaps most importantly, Turner was a master at cost cutting and metal pairing, providing Jack Sangster with a handsome profit on each machine. This business acumen, setting him apart from other engineers, was something Sangster appreciated, and it would later solidify Turner's position at Triumph.

Above: One of Turner's most enduring and individual designs was the Ariel Square Four. After beginning life as a 500, it grew to 600 and eventually 1,000cc. Ariel claimed the new Square Four could accelerate from "10 to 100 miles per hour" in top gear and had rider Freddie Clarke do just that at Brooklands in 1936. The 1,000cc Square Four continued after the war; this is a 1948 Mark 1.

RIGHT: Another Turner success story was the Ariel Red Hunter. Either 500 or 350cc, the 350 (a 1934 version shown here) was ostensibly a smaller-bore version of the 500. Ariel wanted performance to accompany the sporting looks, so the 500 was provided with two pistons: a 7:1 compression ratio for road use (producing 28 horsepower), and a higher-compression racing piston said to propel the Red Hunter to a top speed of around 90 miles per hour. This was unsubstantiated because the Red Hunter didn't achieve any notable prewar racing success. The closest the Red Hunter came to a TT victory was in the 1935 film *No Limit*, where star George Formby rode a Red Hunter, thinly disguised as a "Rainbow," to a win in the Senior TT.

But while the new singles sold well, Turner was continually looking at ways to cut costs. With its unique integrally forged flywheel and mainshaft, the Tiger 70, initially based on the L2/1, was one of the finest British prewar 250s, but in Turner's eyes its days were numbered. Turner claimed the company was placing £5 in the toolbox of every one sold at the manufacturer's price of £38, so after only one year it made way for a more economical version, the 2/1. This shared more parts with the 350cc Tiger 80 and was another example of Turner's economic pragmatism.

1938 5T Speed Twin 500

Edward Turner's intention was always to produce a vertical twin, a design he had considered at Ariel after creating an experimental twin out of half the Square Four. Turner removed the front crankshaft to allow the engine to function as a vertical twin, and both Val Page and Bert Hopwood were watching this experiment. Both would go on to design vertical twins of their own, and Turner inherited Val Page's 650cc 6/1 vertical twin when he took over Triumph in 1936. While he considered it extremely sound from an engineering point of view, Turner didn't see it having wide appeal. So he immediately embarked on creating the Speed Twin, which appeared in July 1937.

One of the reasons the Speed Twin appeared so quickly was that while Turner was not really an innovator, he was extremely adept at incorporating existing components and design features. He knew the conservatism of the motorcycling public dictated that his new twin should look similar to a traditional twin-port single. And his eclecticism was wide ranging: some of his design features shared with Val Page's 250 single, others were borrowed from Turner's Riley Nine car, and the crankshaft layout came from Vauxhall. Back in 1932, Page's hemispherical cylinder head included two valves set at 90 degrees, as in the Riley, and this design had carried through to the Tiger 70. An even stronger association with the Riley was evident from the short pushrods and rockers, and the gear-driven fore and aft camshafts, although Turner couldn't justify the expense of the Riley's helical timing gears. Turner claimed the mechanical layout was entirely new, but the inside flywheel supported midway between the crank journals by a stiff web had been used by Vauxhall in their 1922 3-liter racing engine. A quest for strength and reliability saw Turner choose light RR 56 Hiduminum forged aircraft-quality alloy con-rods with manganese-molybdenum steel rod caps. The cylinder head was a one-piece iron casting, and a Magdyno was gear-driven behind the cylinders off the inlet camshaft.

Turner was committed to maintaining commonality with the existing 250cc Tiger 70 single, and thus it shared the same 63mm bore and 80mm stroke,

One of the first things Turner did when he arrived at Triumph was to standardize the name logo. He wanted it as recognizable as the London Underground sign. The Speed Twin was also as intentionally narrow as a single.

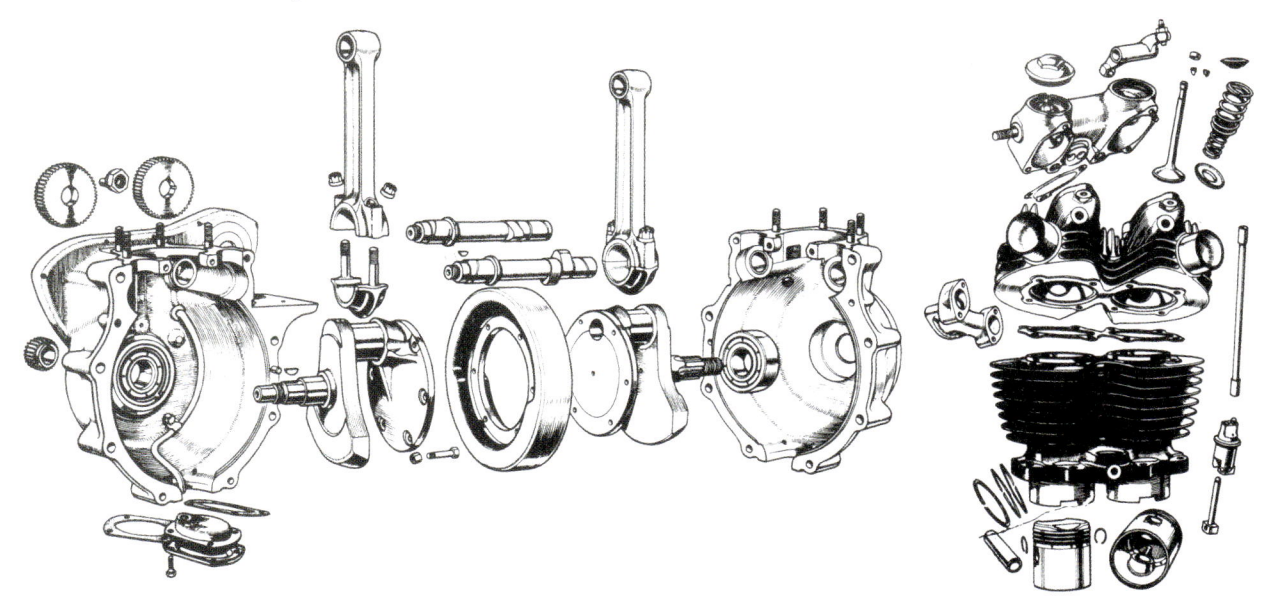

Exploded view of the prewar Speed Twin engine showing the crankshaft with plain big-end bearings and the simple construction that carried through for several decades. The cylinder head and cylinders were cast iron.

CHAPTER ONE

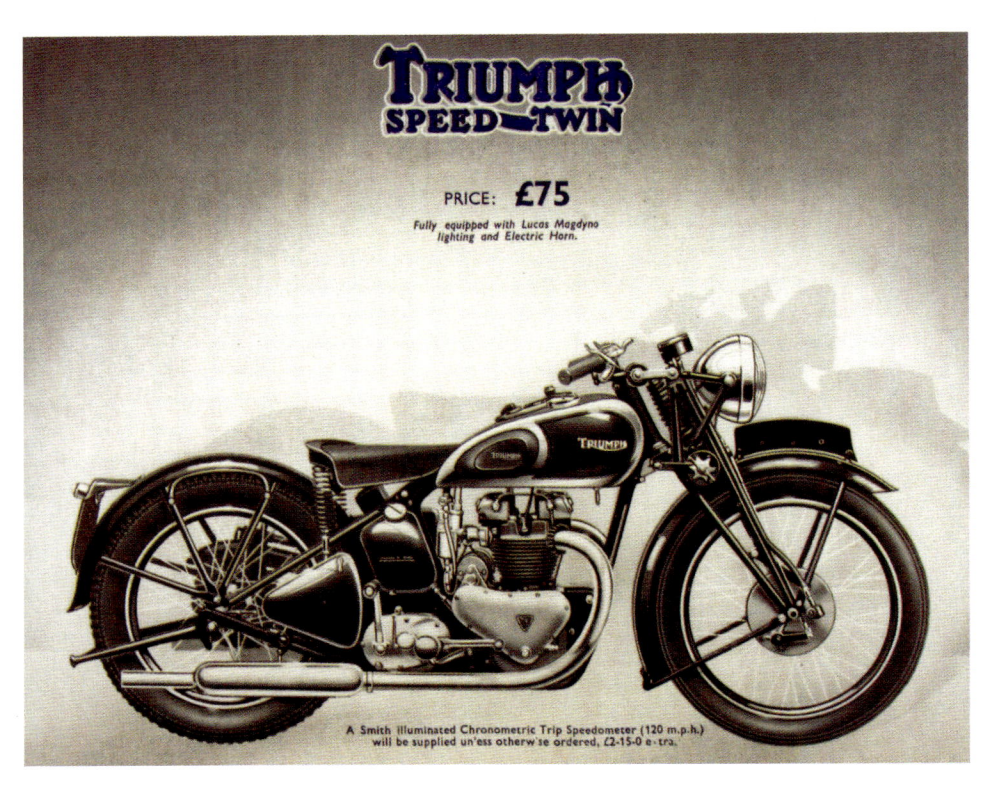

In 1938 the Speed Twin set a new standard of performance and smoothness for the day.

The Speed Twin was as compact as a single and unusual in that no pushrods were visible on the timing side. External pipes drained oil from the rocker box to the push-rod tubes.

as well as a variety of engine components. As he favored simplicity and lightness, the new engine was slightly narrower than the 500cc Tiger 90 single and could share the same chassis. All this contributed to affordability; at £75, the Speed Twin was only £5 more than the Tiger 90 and barely 4 pounds heavier. And while the Speed Twin may have looked similar to a single, even with Turner's favored 360-degree crankshaft (with the pistons rising and falling together) the Speed Twin was smoother, started easier, and provided superior acceleration. Considering the Triumph twin's later reputation for vibration, this smoothness was a surprise, but keep in mind that these early twins had a very low compression ratio and the pre-unit design (with separate gearbox) provided a smoother transmission of the modest power. Another significant feature of the Speed Twin was the quietness of the engine and exhaust compared to the bark of a big single. As expected, the Speed Twin was finished with plenty of chrome, right down to the tank-top instrument panel, but the Amaranth Red bodywork color, extended to the frame and forks, set the twin apart from the singles.

The Speed Twin was first unveiled to the press at the end of July 1937, initially in a description in *The Motor Cycle*, and then with an official unveiling at the Earls Court Motorcycle Show in August. As Turner had envisaged, the public liked what they saw and the Speed Twin was an instantaneous success. The first road tests appeared in October 1937, with *The Motor Cycle* achieving an impressive average top speed of 93.75 miles per hour over four runs. Their best-timed run, albeit with a strong tailwind, of 107 miles per hour was quite astonishing for a 500 at the time. After commenting on "the exceptional smoothness at all speeds apart from a slight period around 60 miles per hour in top gear," *The Motor Cycle* summed up the Speed Twin by commenting, "after hundreds of miles of really hard driving, the Triumph was as clean and smart as at the beginning, and apart from a very slight seep of oil from the rear end of the primary chain case, not a spot of oil had leaked from any of the joints of the power unit."

For a completely new design it had few initial weaknesses, but some problems occurred with the six-studs fastening the cylinders to the vertically split crankcase. These could pull off or break the cylinder casting under prolonged high-speed use and were updated with a more robust eight-stud cylinder flange during 1938.

It wasn't long before the Speed Twin attracted the interest of tuners attempting speed records. Marius Winslow developed a supercharged Speed Twin, and in October 1938, Ivan Wicksteed set a new flying lap of Brooklands of 118.02 miles per hour. As the Brooklands circuit closed soon afterwards, this record still stands. The Speed Twin also found favor for speed-record attempts as far away as Australia. In March 1938 Les Fredericks set a new 12-hour distance record on the clay Coorong Circuit in South Australia. Overnight, and in dense sea spray, he covered an astonishing 8,055 miles at an average speed of 65.7 miles per hour. Visibility was so poor he couldn't see the lamps marking the circuit and at 3:00 a.m. tore right through a stake and lamp, nearly ending up in a rocky outcrop. Frederick wanted to set a 24-hour record, but the conditions were so bad that the attempt was aborted after 14 hours, when water and sand got into the engine and the Speed Twin began to run on one cylinder. This record proved the Speed Twin's toughness in difficult conditions.

Another significant occurrence transpired in early 1938 when the Metropolitan (London) Police Force decided to enlarge their motorcycle force. They selected the Speed Twin over models from seven manufacturers, and although only around 20 were initially purchased, this eventually led to Triumph producing specific police models, both for the United Kingdom and abroad. Police models may not have been the most profitable, but they kept Triumph's profile high, and their high mileages were legendary. In 1946 the average mileage of many of the 42 Metropolitan Police Speed Twins was 98,000 miles, with six covering 150,000 miles and one 161,000 miles. This last machine was still capable of reaching 75 miles per hour.

1939 T100 Tiger 500

Only a year after the launch of the Speed Twin came the Tiger 100, an £80 sports version. With its chrome and silver finish, the Tiger 100 was extremely attractive, and in typical Turner fashion, the "100" indicated the potential top speed. Although the top speed claim could only be achieved with the removal of the silencer end caps, it again demonstrated Turner's marketing genius. The Tiger's formula was tried and tested: introduce a basic model to iron out the problems and test the market and release a sport version soon afterwards.

Powering the Tiger 100 was the same iron engine that drove the 5T, but with the new eight-stud cylinders, polished ports, strengthened crankshaft and con-rods, new flywheel, and higher-compression forged-aluminum alloy slipper pistons. The Amal carburetor was slightly larger and fitted with a bell mouth, and for an additional £5 there was an optional aluminum-bronze cylinder head. This provided improved cooling at higher rpm, but the valves sat directly in the bronze, without cast-iron valve seats. Also setting the Tiger 100 apart was the polished aluminum primary chain case and larger (1.2-gallon instead of 6-pint) oil tank with quick-release cap.

To cope with the extra power the Tiger 100 received an updated full cradle frame, the 29¼-degree steering head angle providing more trail for increased stability. The front fork was also new, with 4-inch-longer bottom links, and the 7-inch cast-iron front brake was ribbed for additional cooling. Several other components set the Tiger apart from the 5T, including silencers with detachable end caps and baffles, a narrow, rubber-mounted handlebar, chrome-plated brake and clutch levers, and a friction-damped quick-action twistgrip. The beautiful chrome and silver gas tank was larger (at 4.8 gallons as opposed to the 5T's 4.2 gallons) and for this year only included a Bakelite panel incorporating ammeter and oil pressure gauges. The gas tank featured diecast badges instead of the earlier 5T's embossed type, and fuel pipes were a metal braided flexible type. Also new this year was a

RIGHT ABOVE: Triumph Twin assembly line at Priory Street in 1939. This was heavily bombed in 1940.

RIGHT BELOW: The 1939 Tiger 100, higher performance for very little extra money. The "100" indicated the model's potential top speed.

1938 5T Speed Twin

Bore	63mm
Stroke	80mm
Capacity	498cc
Horsepower	26 at 6,000 rpm
Compression Ratio	7.0:1
Carburetor	Amal 276/132 LH 15/16-inch
Ignition	Lucas Magdyno
Gearbox	Four-speed
Front Fork	Girder
Brakes	7x1⅛-inch
Wheels	WM2x20-inch front, WM2x19-inch rear
Tires	3.25x20-inch front, 3.50x19-inch rear
Wheelbase	54 inches
Dry Weight	355 pounds
Color	Amaranth Red
Engine Numbers	8-T-12345 (1938) 9-T-xxxxx [actual numbers lost in Coventry Blitz] (1939) 40-5T-xxxxx (from 1940)
Frame Numbers	TH or TF prefix, number didn't correspond with engine

CHAPTER ONE

1939 T100 Tiger

Bore	63mm
Stroke	80mm
Capacity	498cc
Horsepower	34 at 7,000 rpm
Compression Ratio	7.8:1
Carburetor	Amal 76 1-inch
Ignition	Lucas Magdyno
Gearbox	Four-speed
Front Fork	Girder
Brakes	7x1⅛-inch
Wheels	WM2x20-inch front, WM2x19-inch rear
Tires	3.00x20-inch front, 3.50x19-inch rear
Wheelbase	54 inches
Dry Weight	355 pounds
Color	Silver and chrome
Engine Numbers	9-T100-xxxxx (1939) 40-T100-xxxxx (1940) actual numbers lost in Coventry Blitz
Frame Numbers	TF prefix, number didn't correspond with engine

ABOVE: The Tiger 100 had distinctive silencers with removable baffles and endcaps. The front fork was new, as was the front brake.

TOP: Inside each Tiger 100 engine were polished ports, a strengthened crankshaft, and higher compression pistons.

RIGHT: The gas tank instrument surround was Bakelite only for 1939. After the war it was painted metal.

FAR RIGHT: For 1939 the gas tank received diecast badges.

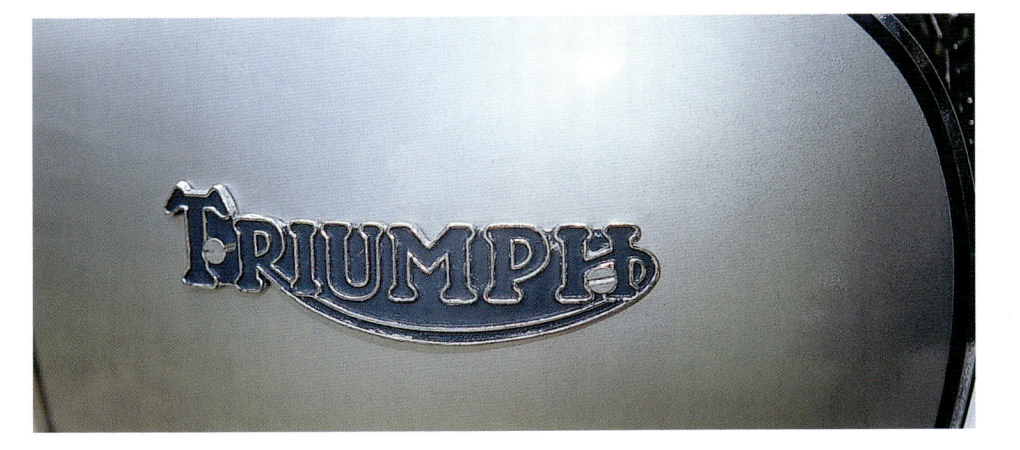

16

1937–1949

The Tiger 100 provided a lot of motorcycle for £80 in 1939. The speedometer driven from the front wheel was optional.

chrome-plated front license plate surround, and a leak valve at the rear of the primary chain case provided positive lubrication to the drive chain. The T100's top speed of 96 miles per hour in standard form was impressive for a 500 in 1939, though, despite the new frame, the handling at this speed left something to be desired. One of Turner's weaknesses was refusing to accept the machine's limitations and implement suggested improvements, and despite the modifications suggested by tester Freddie Clarke, the handling deficiencies remained unaddressed. The taciturn Clarke was often at loggerheads with the cantankerous Turner, but his developmental skills were impressive and he was a formidable rider, setting an all-time 750cc lap record at Brooklands on a 503ccc Speed Twin at 118.02 miles per hour.

According to the 1939 brochure, each Tiger 100 engine "was individually tested on a Heenan and Froude brake, then stripped and reassembled by skilled mechanics. A Test Card, signed by the Chief Tester was supplied to each machine." This was a great selling point, and despite handling deficiencies, the Tiger 100 was extremely popular.

1939 5T Speed Twin 500

As the new T100 Tiger usurped the Speed Twin as the range leader, there were only minor updates to the 5T for 1939. Most of these were shared with the T100, all engines now with the eight-stud cylinder flange and rear chain lubrication. Other new features included revised valve timing and a new engine-shaft shock absorber cam contour altered to provide a smoother operation. As on the T100, the front license plate surround was chrome-plated, the gas tank incorporated a Bakelite instrument panel, and the tank badges were diecast. The handlebar this year was narrower and more back-swept to provide a more comfortable riding position.

Eager to promote their twins as reliable touring machines, Triumph decided to stage an attempt on the Maudes Trophy. This trophy was awarded each year to the manufacturer whose product performed best in a reliability test observed by the Auto-Cycle Union (ACU). Two machines, a Speed Twin and a T100, were selected by the ACU from random dealer's stock and in February 1939 were ridden from Coventry to John o'Groats in Scotland and Land's End in Cornwall, finishing at Brooklands in England. The 1,806 miles were covered at an average speed

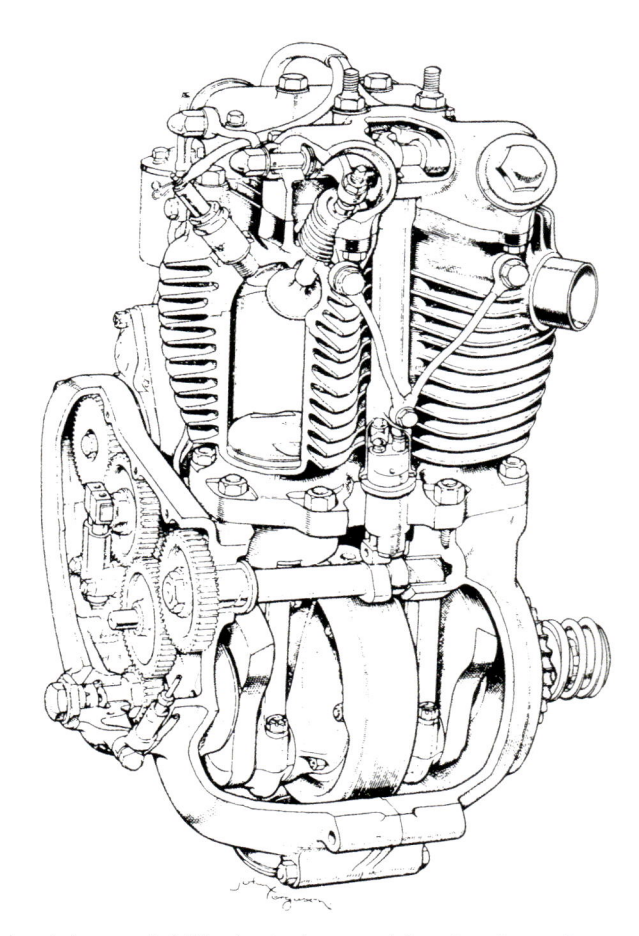

One of the few updates to the Speed Twin engine for 1939 was the eight-stud cylinder flange.

The 1939 5T Speed Twin. This publicity picture doesn't show the diecast gas tank badges.

17

CHAPTER ONE

of 42 miles per hour, and at Brooklands the Tiger 100 averaged 78.5 miles per hour over six hours and the Speed Twin, 75.02 miles per hour. The Tiger 100's best lap was 88.46 miles per hour and the Speed Twin's was 84.41 miles per hour. As both machines performed the test without any major mishap apart from punctures and falling off a stand, Triumph won the Maudes Trophy. But by the time it was announced in November 1939 it meant little, as Britain was now at war. Also in November 1939, Triumph's publicity received a boost with the public endorsement of the Speed Twin by three-time holder of the world's land speed record, Sir Malcolm Campbell. A Triumph owner since 1908, Campbell's opinion was "the Triumph Speed Twin has no equal."

Apart from the war, other events occurred in 1939 that profoundly affected Triumph and its future. In July, Turner's wife, Edith, was killed in a head-on car crash and Turner went into a decline. Jack Sangster understood Turner's situation and in the summer of 1939 sent him to the West Coast of the United States, where he began to forge links with Bill Johnson. Turner had been corresponding with Johnson, a lawyer, since 1937 and encouraged him to purchase British and American Motors in Pasadena in 1938. Turner gave Johnson direct sales rights for Triumph in Southern California, and in 1940 Johnson established Johnson Motors, Inc., moving his shop to West Pico Boulevard in Los Angeles. So, as the skies looked increasingly gloomy in Europe, the stage was set for Triumph to prosper a decade later.

1940

Edward Turner was committed to producing twins at the expense of singles, and soon after the release of the Speed Twin set about designing a 350 twin, the 3T, or Tiger 85. This initially didn't prove as satisfactory as the 500, and it was released just as war broke out in September 1939. Most of Triumph's existing production was then requisitioned for military service, and further development of the 3T had to wait until 1946. Disaster struck Triumph when German bombing destroyed the Priory Street works in the Coventry Blitz of the night of November 14, 1940. The company was forced to relocate, initially into an old foundry at Warwick, and while Jack Sangster wanted to rebuild Priory Street, Turner negotiated with the government to purchase a 22-acre plot in the village of Meriden, five miles outside Coventry. Work commenced in July 1941, and the new factory was up and running in an astonishingly short time given the difficult wartime circumstances. By July 1942, 3HWs (based on the prewar Tiger 80) were in production, and soon afterwards, following a disagreement with Jack Sangster, Edward Turner departed unexpectedly for BSA.

1940 T100 Tiger 500

With most production requisitioned by the military, only a small number of T100s were produced before the Priory Street factory was bombed. These were either export versions for the United States and Canada or versions intended for special dispatch duties for dignitaries such as British Prime Minister Winston Churchill and James Bond's creator, Commander Ian Fleming.

Although production was limited, a number of updates were incorporated this year. Lower-compression (6:1) full-skirt pistons replaced the earlier slipper type, and to prevent oil pressure loss, a bronze piston with a tubular extension was fitted between the timing cover bush and the end of the crankshaft. The big-end bearing clearance was increased, improving crankshaft oil flow and piston and cylinder wall

Valanced fenders and a rear rack were optional extras on the 1940 Tiger 100. Note the check spring above the fork's friction damper.

lubrication. As gas was rationed, a larger, 23-tooth engine sprocket replaced the 22-tooth. Other updates saw the Bakelite switch panel replaced by a steel type with a crinkle black finish, a check spring above the fork friction damper to smooth fork deflection, and a slimmer speedometer cable for those models with a speedometer. Options now also included fully valanced fenders front and rear.

1940 5T Speed Twin 500

The 1940 Speed Twin received the same engine updates as the Tiger 100 and now shared many of the Tiger's chassis components. The frame incorporated the T100's increased steering head angle, and the front fork included a lighter main spring with two small check springs on either side. Other changes saw the adoption of the larger T100 4.8-gallon gas tank with a black, pressed-steel instrument panel. Although Triumph had patented a spring wheel in February 1939 with the intention of introducing it on the 1940 range, it wouldn't be introduced until 1947.

Most 1940 Speed Twins (and Tiger 100s) were shipped to the United States, where they ended up in Johnson Motors' Pico Boulevard showroom. Amazingly, considering the difficulties, Johnson sold around 300 Triumph twins before the supply ended with the bombing of the factory at Priory Street. In the meantime, Johnson was intent on building a public image for Triumph in America by entering Tiger 100s in as many competition events in California as he could. It paid dividends, with Bruce "Boo-Boo" Pearson winning 32 of 36 events in the 1940 season. Triumph finally began to get some serious recognition in the US motorcycle press. At this time Triumph motorcycles began their association with celebrities and film stars, with MGM actor Robert Taylor taking delivery of a 1940 Speed Twin that somehow ended up in a 1942 shipment of spares from the Triumph factory. The story has it that, as new vehicles were unavailable to the public during the war, Bill Johnson swapped the Speed Twin for Taylor's Dodge pickup.

1943 to 1945

In mid-1943 Sangster, with the offer of a shareholding increase, enticed Turner back to Triumph from BSA, and Turner, with this added incentive, was even more intent on maximizing profitability. Already the tide of the war was turning, and Turner was looking ahead, already preparing to turn the entire output of Meriden to twins. In the meantime, the only twin-cylinder engines Triumph was building were auxiliary generator motors for the army and the Royal Air Force (RAF), the airborne auxiliary power plant or AAPP (or "A squared, P squared," to the Experimental Department). These fan-cooled engines had a lighter aluminum cylinder head and barrel and would have an unlikely life after the war as a Grand Prix racer.

The 1940 Speed Twin had a larger gas tank with screw-on knee grips. It also had the fork check springs.

The RAF generator, powered by a Triumph twin-cylinder engine with special aluminum cylinder head and barrel.

1945 to 1946

When peace came, Britain was burdened with a large deficit and manufacturers were encouraged to expand export markets. Envisaging this, in 1945 Turner announced that Meriden would be concentrating solely on twins and later introduced the 3T 350 alongside the Speed Twin and Tiger 100. In early 1945 Johnson was appointed the official US distributor for Triumph, and six months before the war ended he moved back to Pasadena and built a large auto-style dealership. From now on Triumph's fortunes would be totally linked to the United States, and demand for Triumph twins during the 1950s always exceeded supply.

1945–1946 T100 Tiger 500

Production resumed at Meriden in November 1945, and as in 1940, the Tiger 100 remained the range leader. The engine was updated with new crankcases, now with a front-mounted dynamo, and a rear flange-mounted British Thompson & Houston (BTH) magneto. This featured an automatic advance and retard instead of the earlier manual type. Other updates saw the replacement of the external oil drainpipes with internal oil ways and a redesigned rocker oil feed, now taken from a T-junction in the return pipe at the oil tank. The engine breather was now a timed rotary valve driven off the inlet camshaft, and the piston-type oil pressure release valve replaced the earlier ball and spring. Unlike the Speed Twin, the Tiger 100 retained its higher, prewar compression ratio, and the carburetor remained a 1-inch Amal, but now with a spring-loaded plunger for the choke. The silencers were now shared with the 5T Speed Twin.

CHAPTER ONE

1945–1947 T100 Tiger

Bore	63mm
Stroke	80mm
Capacity	498cc
Horsepower	30 at 6,500 rpm
Compression Ratio	7.8:1
Carburetor	Amal 276 1-inch
Ignition	BTH Magneto
Gearbox	Four-speed
Front Fork	Telescopic
Brakes	7x1⅛-inch
Wheels	WM2x19-inch front and rear
Tires	3.25x19-inch front, 3.50x19-inch rear
Wheelbase	54 inches
Dry Weight	361 pounds
Color	Silver and chrome (less chrome in 1947 due to supply difficulty)
Engine Numbers	From 46-T100-72000 (1946) 47-T100-79046 (1947)
Frame Numbers	TF prefix

ABOVE: The 1946 brochure. The Tiger 100 received a telescopic fork and now shared its silencers with the Speed Twin. LEFT: A sectioned Tiger 100 engine specially prepared for the Earls Court Motorcycle Show.

Apart from the engine updates, the main development was to the front suspension, with a telescopic fork replacing the girder type. This featured one-way hydraulic damping and 6½ inches of travel. These long, early Triumph forks were pretty rudimentary, and as the springs were inside the tubes, they were attractively slim. Unfortunately, as they also suffered considerable flex and were prone to leakage, they did little to quell the Tiger 100's reputation for questionable handling. Along with the new fork came a smaller, 19-inch front wheel, but the rear end at this stage remained rigid. Other new features included a new handlebar to suit the telescopic fork, a speedometer drive from the rear wheel, and a smaller, 7-inch Lucas headlamp. The new front end tidied the styling, but some of the elemental simplicity of the prewar model was lost in the process, and the Tiger 100 lost the pure beauty of the 1939 and 1940 models.

The 1946 5T Speed Twin, now with a 4.8-gallon gas tank.

1945–1946 5T Speed Twin 500

As it made sense to increase model uniformity, for 1945 and 1946 the 5T Speed Twin was closer in specification to the Tiger 100 than the previous version. The engine featured the new crankcases with a front-mounted dynamo and shared the other updates of the Tiger T100, but to accommodate Britain's low-octane "Pool" gasoline the compression ratio was reduced to 6.5:1. Apart from the distinctive Amaranth Red color, in most respects the Speed Twin was identical to the Tiger 100. This included the 1.2-gallon oil tank (with 2-inch hinged filler cap), telescopic fork, 19-inch front wheel, and 7-inch headlamp, still with a chrome-plated headlamp shell at this stage.

1947

As there was continued strong demand for the Tiger 100 and Speed Twin, it wasn't surprising that both models continued largely unchanged for 1947. The only updates to the 500 twins extended to the carburetor float chamber, now on the left, and the painted headlamp shell. A new-type optional prop stand was now bolted under the primary chain case. Edward Turner was always wary of producing too many motorcycles, and during 1947 production reached around 12,000, with up to 60 percent exported. But Turner believed the real future lay in exports to America, and by 1947 Johnson Motors was advertising heavily and beginning to establish a nationwide dealer network.

While the 500 twins continued much as before, the 3T finally made it into production. Development coincided with the military side-valve 500cc TRW, a machine that didn't initially find favor with the British Army but was eventually popular enough with other branches of the military to last into the 1960s. As the TRW was never sold as a cataloged model, we won't cover it in detail here.

1945–1947 5T Speed Twin

Bore	63mm
Stroke	80mm
Capacity	498cc
Horsepower	28 at 6,000 rpm
Compression Ratio	6.5:1
Carburetor	Amal 276 LH $^{15}/_{16}$-inch (276 BN/1AT for 1947)
Ignition	BTH Magneto
Gearbox	Four-speed
Front Fork	Telescopic
Brakes	7x1⅛-inch
Wheels	WM2x19-inch front and rear
Tires	3.25x19-inch front, 3.50x19-inch rear
Wheelbase	54 inches
Dry Weight	361 pounds
Color	Amaranth Red
Engine Numbers	From 46-5T-72000 (1946) 47-5T-79046 (1947)
Frame Numbers	TF prefix

3T (3T DeLuxe) 350

The 3T epitomized Edward Turner's desire to maintain model uniformity and thus was an amalgam of several designs. Largely it was a Speed Twin with the smaller engine that had evolved from the prewar Tiger 85, wartime 3TW, and postwar TRW. This long-stroke engine design was laid out like a Speed Twin, but with the simpler TRW crankshaft construction utilizing clamped crankpins and one-piece con-rods. The top end of the overhead-valve engine was also similar to the Speed Twin's, but the cylinder head and rocker boxes were now cast in one piece. As the rest of the machine was

continued on page 24

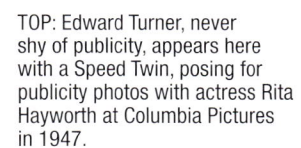

TOP: Edward Turner, never shy of publicity, appears here with a Speed Twin, posing for publicity photos with actress Rita Hayworth at Columbia Pictures in 1947.

ABOVE LEFT: The 5T Speed Twin as depicted in the 1947 brochure. The headlight shell was painted this year. This picture shows a spring wheel, something that wasn't available until 1948.

ABOVE: Turner was a frequent visitor to California and had a close relationship with Bill Johnson (left).

The 3T looked very similar to the Speed Twin. This is the 1949 3T DeLuxe.

CHAPTER ONE

Grand Prix (T100R) 500

The Speed Twin and Tiger 100 were highly regarded, so it was no surprise that many enthusiasts wanted to adapt the Triumph twin for competition. The stumbling block was Edward Turner. Turner believed that, as Grand Prix machinery bore no resemblance to production bikes, there was no value in investing a disproportionate amount of resources in the quest for what he saw as only dubious publicity. But Turner's disapproval of competition didn't stop those at the factory from thinking otherwise, particularly Freddie Clarke, head of the Experimental Department. During 1946 Turner was absent for long periods overseas, which provided Clarke the opportunity to create a hybrid engine using the T100 lower end and the lighter aluminum cylinder head and barrel of the "A squared, P squared" wartime generator units, which featured square barrels and parallel exhaust ports. Developmental work took place at night. The results were extremely satisfying: the power rose to 47 horsepower at 7,000 rpm, so Freddie Clarke installed this engine in a modified Tiger 100 frame with Turner's new spring wheel rear suspension. It was provided to Irish farmer Ernie Lyons, who, after an outing at the Ulster Grand Prix to iron out any teething troubles, entered it in the 1946 Senior Manx Grand Prix. This was the first race held on the mountain circuit since 1939. In appalling conditions, and despite the front frame downtube breaking on the last lap and the uncertain handling provided by the spring wheel, Lyons rode from 12th to finish 1st at an average speed of 76.73 miles per hour. Lyons' heroic victory saw Turner relent, and he even authorized a celebratory dinner for Lyons. He didn't concede in his attitude to Freddie Clarke, however, and instructed his works manager to censure the department head. Clarke subsequently left Triumph for AMC, where he tragically died in 1947 testing a prototype twin.

Ernie Lyons' success led to dealer requests for the production of a limited number of Grand Prix replicas, and in the meantime David Whitworth carried out much of the practical development in races in Europe during 1947. This led to the appearance of the production Grand Prix in February 1948. These over-the-counter racers were not true Grand Prix machines but, as the advertising leaflet proclaimed, were intended "to enable the non-professional rider to compete on level terms in all types of long and short circuit racing." As far as Turner was concerned, the Grand Prix was simply another commercial proposition, and as it shared most of its parts with the T100, it was a suitably profitable venture.

ABOVE: The special Tiger 100 that won the 1946 Manx Grand Prix. This was the prototype for the Grand Prix.

BELOW: The production Grand Prix. This particular bike was rebuilt at the factory in 1949 and provided with a certificate claiming 42.2 horsepower.

It may have been a flawed racer, but the Grand Prix was undeniably good looking.

Ultimately 196 Grand Prix engines were built, with 175 complete motorcycles. The engine featured liberal internal polishing and lightening, no external primary chain case, twin Amal carburetors, and open megaphone exhausts. The valve lifter guides were aluminum, the engine oil filter a cartridge type, the clutch basket lightened, and the four-speed gearbox close ratio. The cylinders on the right side still retained the bosses originally meant for the fastening of the generator's heat guard. Each machine was supplied with a data and horsepower certificate listing around 40 to 42 horsepower at 7,000 rpm, with a safe ceiling of 7,200 rpm, and a quoted top speed of 120 miles per hour running on Pool gasoline with low-compression pistons. The rear suspension included the new spring wheel with Dunlop light alloy racing rims (a 20-inch front), and aluminum fenders contributed to a dry weight of only 314 pounds. Most had Silver Sheen-painted gas tanks, while early tanks were chromed with silver panels. All had a larger capacity than the road-going T100. Also included in the specification were drop handlebars and rear-sets. They differed in many details from the T100 Tiger. A remote canister oil-filter unit was fitted between the gearbox and timing chest, and generally the gearshift lever was reversed, providing a different shifting pattern. The front engine plates included a hole for lubrication of the tachometer drive, the brakes were a larger diameter and wider than on the T100 Tiger, and the fork included a heavy brace and one-piece, dual-attach-point fork clamps. Both the engine and frame numbers carried an "R" designation.

In the wake of Lyons' earlier success, Turner was persuaded to allow the entry of a semi-works team of six bikes in the 1948 Senior TT at the Isle of Man. On paper it looked promising, with well-known Norton entrant Nigel Spring sponsoring some big-name riders such as Freddie Frith and Bob Foster, but all the bikes retired with engine problems. This public relations disaster only cemented Turner's negativity toward racing. The Grand Prix may have failed at the 1948 TT, but later in the year it achieved more success. Don Crossley won the Manx (at 80.62 miles per hour), and in 1949 New Zealander Syd Jensen finished fifth in the Senior TT (at 83.17 miles per hour). The Grand Prix was also very popular in the United States, its biggest success being Ron Coates' victory in the 1950 Daytona 100-mile amateur race at a record speed of 81.26 miles per hour. Coates was service manager for the East Coast distributor TriCor until 1970. The Grand Prix, or T100R, continued to be listed until 1950. It suffered from rather severe vibration, so it was more suited to short circuit racing, but the Grand Prix earned the reputation for being a fast, if somewhat fragile, factory racer.

ABOVE LEFT: All Grand Prix had the new spring wheel, but this was of dubious effectiveness.

ABOVE MIDDLE: The Grand Prix's Amal 1-inch carburetors with remote float bowl.

ABOVE RIGHT: The Grand Prix' exhaust ports were parallel and its cylinder head finning distinctive. The bosses on the cylinder were remnants of the generator heat guard mount.

BELOW: The production Grand Prix was one of the most attractive Triumphs. Missing here are the saddle springs.

CHAPTER ONE

1947–1951 3T (3T DeLuxe)

Bore	55mm
Stroke	73.4mm
Capacity	349cc
Horsepower	19 at 6,500 rpm (17 at 6,000 rpm for 1951)
Compression Ratio	6.3:1
Carburetor	Amal 275 ⅞-inch
Ignition	Magneto
Gearbox	Four-speed
Front Fork	Telescopic
Brakes	7-inch
Wheels	WM2x19-inch front and rear
Tires	3.25x19-inch front and rear
Wheelbase	52.2-inch (53.25-inch from 1950)
Dry Weight	325 pounds
Color	Black with chrome
Engine Numbers	From 47-3T-79046
Frame Numbers	3T prefix

1948–1950 T100R Grand Prix

Bore	63mm
Stroke	80mm
Capacity	498cc
Horsepower	40–42 at 7,200 rpm
Compression Ratio	8.3:1 (optional 8.8 and 12.5:1)
Carburetor	Twin Amal Type 6 1-inch
Ignition	BTH TT Magneto
Gearbox	Four-speed
Front Fork	Telescopic
Rear Suspension	Spring wheel
Brakes	8x1 ⅜-inch front and rear
Wheels	WM2x20-inch front and WM2x19-inch rear
Tires	3.00x19-inch front, 3.50x19-inch rear
Wheelbase	55 inches
Dry Weight	314 pounds
Color	Silver Sheen or chrome
Engine Numbers	T100 xxxxx R (1948–1950 number series)
Frame Numbers	TF xxxxx R (1948–1950 number series)

The 1948 Tiger 100 with the optional spring wheel.

The main features distinguishing the 1948 Speed Twin were the rear fender mounts without handles.

Continued from page 21

essentially a Speed Twin, the 3T was no rocketship. *Motor Cycling*, testing an early model in 1946, managed a top speed of 74 miles per hour.

Johnson Motors imported the 3T through until 1951, but they were never as popular as the larger twins in the United States. They were also updated over the years, their styling mirroring the larger 5T with the incorporation of the instrument nacelle in 1949 and restyled gas tank in 1950. The DeLuxe featured an all-black finish and smaller gas tank, but it was still not very successful. It was reported only three 3T DeLuxes were imported into the United States by the Triumph Corporation (TriCor) in 1951, so it was deleted from the 1952 lineup.

1948 T100 Tiger 500 and 5T Speed Twin 500

The big news this year was the option of the patented spring wheel on both the T100 Tiger and 5T Speed Twin. This now included an 8-inch brake (instead of 7-inch) and, with two compression and one rebound spring inside spring boxes in the hub, provided 2 inches of undamped movement. Relatively complex and difficult to assemble, the spring wheel looked good on paper but in reality did little for the Triumph twin's already suspect high-speed handling. With the spring wheel came a revised speedometer drive, which moved from the rear wheel to the gearbox.

Other updates for the year included new fenders, the front with detachable front stays and the rear without side handles. A new rear number plate and taillight bracket accompanied these, reshaped to provide a handhold while placing the bike on the rear stand. Due to uncertain component supply, the implementation of these updates occurred gradually throughout the year.

1949 T100 Tiger 500 and 5T Speed Twin 500

Gradual evolution of the T100 and Speed Twin saw the replacement of the tank-top instrument panel by a headlamp nacelle, a feature that would distinguish Triumphs through the mid-1960s. An optional parcel grid fitted in place of the previous instrument panel. The most significant engine update was an increase in the 5T compression ratio and the introduction of a Vokes air cleaner. The dynamo was increased to 60 watts, and as the oil pressure gauge was deleted the oil pressure release valve was fitted with an indicator

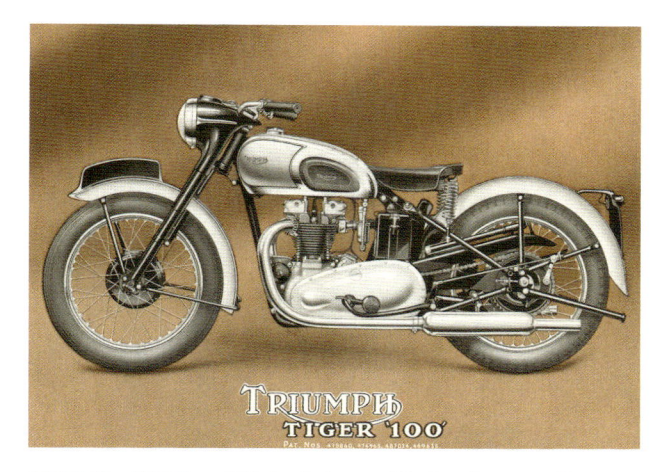

1937–1949

ABOVE: The 1949 Tiger 100, the first year of the headlamp nacelle.

BELOW: Apart from the instrument nacelle, the 1949 Speed Twin still looked similar to the previous year's model. More significant gas tank styling updates would appear in 1950.

button. The T100's oil tank capacity was reduced from 9.6 to 7.2 pints, and an aluminum threaded filler cap replaced the earlier hinged type. Also new on the T100 (but not the 5T) was a single lipped roller main bearing replacing the ball bearing on the timing side.

1949 and 1950 TR5 Trophy 500

During 1948 Triumph decided to enter the International Six Days Trial (ISDT) in San Remo, Italy, and prepared three ISDT bikes based on the stock 5T but using the alloy head and barrel of the Grand Prix. Henry Vale built the bikes, and they scored three gold medals

1948 T100 Tiger and 5T Speed Twin

Bore	63mm
Stroke	80mm
Capacity	498cc
Horsepower	30 at 6,500 rpm (28 at 6,000 rpm 5T)
Compression Ratio	7.8:1 (6.5:1 5T)
Carburetor	Amal 276 1-inch (15/16-in 5T)
Ignition	BTH Magneto
Gearbox	Four-speed
Front Fork	Telescopic
Rear Suspension	Spring wheel (from TF 15530)
Brakes	7x1 1/8-inch (8-inch spring wheel)
Wheels	WM2x19-inch front and rear
Tires	3.25x19-inch front, 3.50x19-inch rear
Wheelbase	54 inches
Dry Weight	365 pounds (381 pounds with spring wheel)
Color	Silver and chrome (Amaranth Red 5T)
Engine Numbers	From 48-T100-88864 to 102235 (T100) 48-5T 88227 to 102160 (Speed Twin)
Frame Numbers	TF 15001 to TF 24765

1949 T100 Tiger and 5T Speed Twin
(Differing from 1948)

Compression Ratio	7.0:1 (5T)
Carburetor	Amal 276/DK/1AT 15/16-inch (5T)
Engine Numbers	From T100-9-102236 to 113386 (T100) 5T-9-102581 to 113386 (Speed Twin)
Frame Numbers	TF 25115 to TF 33615

CHAPTER ONE

1949–1950 TR5 Trophy

Bore	63mm
Stroke	80mm
Capacity	498cc
Horsepower	24 at 6,000 rpm
Compression Ratio	6.0:1
Carburetor	Amal 276 15/16-inch 5T
Ignition	BTH Magneto
Gearbox	Four-speed
Front Fork	Telescopic
Wheels	WM1x20-inch front and WM3x19-inch rear
Tires	3.00x20-inch front, 4.00x19-inch rear
Wheelbase	53-inch
Dry Weight	295 pounds
Color	Silver and chrome
Engine Numbers	TR5-9106001 to 9112671 (1949) TR5 111N-14384N (1950)
Frame Numbers	TC 11010T to TC 13107T (1949)

and a team prize known as the "Trophy" (to distinguish it from the less-important "Vase"). The results were good, but the riders, Captain Alan Jeffries, Bert Gaymer, and Jim Alves, were less than enthusiastic about the bikes. Although fast and reliable, they were overweight and had poor handling, leading to the new TR5 Trophy in 1949.

The TR5 first appeared at the Earls Court Motorcycle Show in November 1948, with production commencing during January 1949. Still using the alloy generator engine, the TR5 was much shorter and lighter than the 1948 version and formed the basis for ISDT-winning bikes for the next three years. The standard gearbox was wide ratio, and the TR5 received a completely new, shorter frame, unlike any other in the range. Each was supplied with a dynamo and quickly detachable headlamp and could be specified with higher-compression (7.5:1) pistons for ISDT duties or 4.5:1 for low-speed trials work. As on the Grand Prix, bronze threaded rings screwing into the head retained the exhaust pipes instead of the usual Triumph stubs. The TR5 also only had one carburetor, not the Grand Prix' two.

The TR5 was the world's first genuine "dual-purpose" bike, light enough and with good ground clearance to go off-road. Most had a rigid frame, with the spring wheel an option. While the rest of the range would be considerably updated for 1950, the TR5 continued largely unchanged for one more year, still with the alloy generator engine. The 1950 TR5 was also the last production Triumph to feature a chrome-plated gas tank until the 1982 Royal Wedding T140.

In the US the TR5 immediately proved popular with off-road riders and was instrumental in raising Triumph's profile. More than two decades later, Triumph legend Bud Ekins was asked to provide a motorcycle to Paramount Studios for Arthur "The Fonz" Fonzarelli (Henry Winkler) to use in the popular 1970s sitcom *Happy Days*. The Fonz could often be seen tooling around on the bike, and it became an essential part of his character and mystique. This was one of three different TR5s of various years used in the decade-long series (1974–1984), the last being a 1949 TR5 Trophy. Mildly customized, it was never actually ridden by Winkler, whose dyslexia prevented him riding a motorcycle. While two of the later TR5s haven't survived, the 1949 square-barreled model has, languishing in original form in a motorcycle shop in Oakland until it was auctioned in 2011. Did it make its way into the hands of someone as cool as The Fonz?

OPPOSITE: The 1949–1950 TR5 is considered by many to be one of the most perfectly proportioned of all Triumph motorcycles.

BELOW: The left siamesed exhaust system was unique to the TR5.

The Motor Cycle, September 29th, 1949

3 × 90 × 500! INTRODUCES ITSELF!

Thunderbird

Under A.C.U. supervision the first three standard production 650cc Triumph "Thunderbirds," fully equipped, completed 500 miles each at over 90 mph on Montlhery Track, Paris on 20th Sept., 1949. This demonstration of high speed reliability introduces the newest and most exciting Triumph Twin yet!

TRIUMPH

TRIUMPH ENGINEERING CO. LTD., MERIDEN WORKS, ALLESLEY, COVENTRY

1950–1955

America and The World's Fastest Motorcycle

The Thunderbird, T110 and T100 Tiger, Speed Twin, and TR5 Trophy

By 1949 Johnson Motors had marked their first decade with annual sales through 100 dealers of more than 1,000 bikes a year. Edward Turner was now totally convinced Triumph's future lay in the United States, and dealers told him many American riders considered the 500cc capacity of the T100 and 5T too small when compared to Harley-Davidson and Indian V-twins. Turner responded in his typically pragmatic way, creating a 650cc twin out of the existing 500 with very little modification and hence minimal outlay. The name "Thunderbird" was inspired by the Thunderbird Motel Turner passed in South Carolina early in 1949; of American Indian origin, it was the name of a giant eagle-like bird capable of unleashing thunder and rain. In 1955 Ford adopted the name for their Thunderbird sports car, but not before entering into a legal agreement with Triumph. Turner was always intent on maximizing publicity, and he launched the new model in a high-speed demonstration at the Montlhéry banked racetrack near Paris in September 1949. Three machines covered 500 miles each at an average speed of over 90 miles per hour, generating considerable publicity. By the time the Thunderbird was officially announced, 2,500 examples had been built and were already in dealers' showrooms. Alongside the new Thunderbird for 1950 were mildly a updated Tiger 100, Speed Twin, and 3T DeLuxe, while the TR5 Trophy was identical to 1949 but for the inclusion of the stronger 6T Thunderbird gearbox. Examples with the stronger gearbox were distinguished by a speedo cable connection in the right gearbox casing.

OPPOSITE: This advertisement was placed in the weekly motorcycle magazines but was soon withdrawn after Turner was told it bore resemblance to *Riders of the Apocalypse*. Turner wanted the Thunderbird to represent joy, not gloom.

1950 6T Thunderbird

Bore	71mm
Stroke	82mm
Capacity	649cc
Horsepower	34 at 6,300 rpm
Compression Ratio	7:1 (8.5:1 optional for the US)
Carburetor	Amal 276 1-inch (increased to 1 1/16 midseason)
Ignition	BTH Magneto with automatic advance and retard
Gearbox	Four-speed
Front Fork	Telescopic
Rear Suspension	Optional spring wheel
Brakes	7x1 1/8-inch (8x1 1/8-inch rear with spring wheel)
Wheels	WM2x19-inch front and rear
Tires	3.25x19-inch front, 3.50x19-inch rear
Wheelbase	55 inches
Dry Weight	370 pounds (385 pounds with spring wheel)
Color	Thunder Blue (lighter blue from 10166N)
Engine Numbers	6T 1017N to 15000N (Reference to the year of manufacture discontinued)
Frame Numbers	From TF 33616

RIGHT: The 1950 6T was very similar to the 5T.

ABOVE RIGHT: For 1950 the engine again featured external oil drain pipes.

1950 6T Thunderbird 650

As with the Speed Twin, Edward Turner again hit the jackpot with the 6T Thunderbird. When it came to marketing and design he was always ahead of the pack, and just as the Speed Twin's success ensured that all other British manufacturers would follow suit with a parallel twin, the Thunderbird soon saw several competing parallel twins larger than the previous standard 500cc. Turner's Triumph twins may not have been the fastest or best handling, but they were cleverly styled and looked more modern than the competition.

Many engine components were updated for the Thunderbird. The crankshaft was now fully machined, and the wider bores saw a reintroduction of the external rocker oil drainpipes. The cylinder head was now no longer spigoted, but sealed by a copper gasket, and the finning was more generous, with four fins on the cylinder head at the exhaust port. The oil pump was modified to provide increased flow and the gearbox redesigned and strengthened to cope with the increased torque of the larger engine. All the ratios were altered, and the earlier weak point, the separate floating layshaft, was now a stronger, driven type. The clutch was also beefed up with two additional plates and the gearing raised by increasing the engine sprocket to 24 teeth (from the 5T's 22 teeth).

One of the advantages of basing the 6T engine on the 5T was that it bolted directly into the existing 5T frame, with the spring wheel still an option. Following some spring breakages, during the year a Mark II spring wheel replaced the Mark I this year. With conventional ball races instead of cups and cones, this was more reliable but offered no functional improvement and remained a handling impediment. The brazed-lug frame, with its unusual tapered single front downtube, was relatively rigid, but the slim Triumph fork with its long springs remained a weakness.

Also new for the Thunderbird was a painted gas tank with a four-band embellishment, and the entire motorcycle was painted Thunder Blue, including the frame. This rather insipid color met with a mixed reception and was changed midyear. The Thunderbird also instigated the Triumph's new "low chrome" policy, initiated by Turner after experiencing union trouble in the Meriden plating shop. Also for the first time a dual seat was offered as an option. Although it wasn't billed as a high-performance model—it was more of an all-rounder—the Thunderbird's performance didn't disappoint. *The Motor Cycle* tested an early example in September 1949 and achieved a top speed of 97 miles per hour. In the United States, for which the Thunderbird was really intended, *Cycle* magazine found the Thunderbird nearly faultless apart from the color, claiming, "the riding characteristics leave nothing to be desired." It didn't take long for Americans realize the Thunderbird's potential as a record breaker, and at the Rosamond Dry Lake in the Mojave Desert, on the first

1950–1955

The Wild One

Although motorcycles in America were primarily about sport and recreation, in the early 1950s they were increasingly associated with rebelliousness. The controversial 1953 movie *The Wild One* was based on a 1951 short story, "The Cyclists' Raid," by Frank Rooney, loosely based on events occurring in Hollister, California, in July 1947. A small group of 3,000 motorcyclists attending an AMA Gypsy Tour became drunk and disorderly, and the incident received wide national publicity. Stanley Kramer's film portrayed motorcyclists in a negative light, with star lead Marlon Brando as Johnny, riding a 1950 Triumph Thunderbird. Intent on promoting motorcycling as a wholesome activity, Bill Johnson strongly protested and tried to stop the film's production. He didn't succeed, and the Thunderbird featured strongly, still retaining its tank badges. But all publicity is good publicity, and Triumph's sales were unaffected. By modern standards *The Wild One* was tame, but it was still banned in the UK for 14 years.

LEFT: Marlon Brando posing with the 1950 Thunderbird used in the movie *The Wild One. Triumph Motorcycles*

FAR LEFT: Brando with Mary Murphy, who played the daughter of the local sheriff. *Triumph Motorcycles*

Thunderbird in the United States, Bobby Turner set a new American Motorcycle Association (AMA) record for unstreamlined motorcycles at 135.84 miles per hour.

1950 T100 Tiger 500 and 5T Speed Twin 500

Triumph's high-performance model for 1950 was still the T100 Tiger, and it and the 5T Speed Twin incorporated some of the updates first seen on the 6T Thunderbird. These included the external rocker oil drainpipes, but not the new 6T crankshaft. Most of the T100 Tiger's engine was shared with the 5T Speed Twin, but the ports were now polished. The gearbox was now the new Thunderbird type, with two sets of ratios available. The most significant styling update saw a painted tank replace the chrome type, but the colors were as before. As on the Thunderbird, a Mark II spring wheel appeared midyear, and a dual seat was an option. Although the T00 Tiger was still arguably

The 1950 T100 Tiger, now with a silver-painted gas tank.

CHAPTER TWO

1950 T100 Tiger and 5T Speed Twin

Bore	63mm
Stroke	80mm
Capacity	498cc
Horsepower	30 at 6,500 rpm (28 at 6,000 rpm 5T)
Compression Ratio	7.8:1 (7.0:1 5T)
Carburetor	Amal 276 1-inch ($^{15}/_{16}$-inch 5T)
Ignition	BTH Magneto
Gearbox	Four-speed
Front Fork	Telescopic
Rear Suspension	Spring wheel (Mark II from eng. 7439N)
Brakes	7x1⅛-inch (8x1⅛-inch spring wheel)
Wheels	WM2x19-inch front and rear
Tires	3.25x19-inch front, 3.50x19-inch rear
Wheelbase	54 inches
Dry Weight	365 pounds (381 pounds with spring wheel)
Color	Silver Sheen (Amaranth Red 5T)
Engine Numbers	1001N–16160N (T100) 1009N–16084N (Speed Twin)
Frame Numbers	From TF 33616

The 1950 Speed Twin incorporated many updates from the Thunderbird.

OPPOSITE: Extremely purposeful and good looking, the 1951 T100 Tiger.

the fastest standard 500cc bike available, and competition success in the United States was keeping demand high, with all the attention on the Thunderbird this year, the T100 was marking time in preparation for a more significant 1951 update.

1951

Although sales in the United States were strong, Edward Turner believed they were only scratching the surface and Triumph required a wider distribution than the West Coast–based Johnson Motors could provide. So in 1951 Triumph established TriCor, a factory-owned distributorship based in Baltimore, Maryland. Johnson now supplied 19 western states—everything west of Texas and north to the Canadian border—with TriCor serving the rest of the United States, which was the majority of the country. Turner chose expatriate Briton Denis McCormack to run TriCor and this proved an inspired move, with Triumph imports almost tripling during 1951 (from 1,000 to 2,730 motorcycles). This year was also the last for the unsuccessful 3T.

The other significant occurrence this year was Jack Sangster's selling of Triumph to BSA in March, although this sale didn't include TriCor in Baltimore. Turner remained in charge of Triumph, and while he later condemned the sale, it was particularly profitable for him as he sold ET Developments (including the patent for the spring wheel) to BSA for nearly a quarter of a million pounds. Turner was subsequently a wealthy man, later living in style and owning a succession of yachts and large houses.

1951 6T Thunderbird 650

With the Thunderbird proving popular, the main update for 1951 was to the earlier, unloved color scheme, now a lighter and more attractive Polychromatic Blue. A few, particularly for the United States, were also black this year. Engine improvements were minor, with the cam wheels incorporating three keyways to allow more precise vernier cam timing adjustment, and (as on the 1949 T100) the timing side main bearing changed from a ball to twin lipped roller type. A Lucas magneto with automatic advance and retard was also specified alongside the earlier BTH unit. Other changes saw the 7-inch front brake drum now in cast iron to provide more rigidity, a new bayonet-type filler cap, and a standard parcel rack. From this year the engine and frame numbers were shared, with the frame number no longer separate. This applied to all models and included an NA suffix for 1951.

In the United States during 1951 Bobby Turner continued to set records on his early Thunderbird, running 129.24 miles per hour on nitro at Daytona Beach and 132.26 miles per hour miles per hour at Bonneville. This was the fastest C-class AMA record for any engine size on pump gasoline, and the record stood for seven years. The Thunderbird also began to establish itself as a road racer during 1951, with Jimmy Phillips winning the 80-cubic-inch class in the Peoria TT and Walt Fulton the inaugural 100-mile Catalina Grand Prix. It was just the beginning of an incredible racing era for Triumph twins in the United States.

1951 T100 Tiger (T100C) 500

With the demand for a higher-performance T100, particularly in the United States, where it was a very popular competition mount, Triumph updated the T100 Tiger for 1951 with a new aluminum cylinder head and barrel. Johnson Motors (JoMo) in California also requested a racing uprating kit, and this was developed at Meriden and offered as an option.

With its splayed exhaust ports, the new close-finned pressure diecast aluminum cylinder head was

1950–1955

The 1951 Thunderbird, similar to 1950 but with a new color scheme. *Triumph Motorcycles*

33

CHAPTER TWO

1951 6T Thunderbird *(Differing from 1950)*

Carburetor	Amal 276 1 1/16
Ignition	Lucas K2F or BTH Magneto
Color	Polychromatic Blue or black and gold from 11130NA–11166NA
Engine and Frame Numbers	6T 136NA to 15808NA

1951 T100 Tiger (T100C) *(Differing from 1950)*

Horsepower	32 at 6,500 rpm (40 HP T100C)
Compression Ratio	7.6:1 (8.25:1 or 9.5:1 T100C)
Carburetors	Twin Amal Type 6 remote bowl (T100C)
Ignition	BTH KC2 or Lucas K2F Magneto
Wheelbase	55 inches
Dry Weight	355 pounds
Color	Silver Sheen
Engine and Frame Numbers	T100 101NA to 15808NA

RIGHT: The headlight nacelle and parcel rack, first introduced in 1949, would distinguish all 1950s Triumph twins.

FAR RIGHT ABOVE: The attractive alloy cylinder head and barrel of the 1951 T100.

FAR RIGHT BELOW: The T100 still retained a single Amal carburetor for 1951.

an extremely attractive casting. It allowed for larger (1 7/16-inch) inlet valves, while other updates included Duralumin pushrods and the stronger 6T con-rods. The T100 clutch was now the 6T six-plate type. Even with the single Amal carburetor, in standard form the T100 was still good for around 92 miles per hour on the low-quality fuel then available. Several chassis updates were also incorporated this year, in particular the inclusion of the 1948 front frame type allowing an additional cylinder head steady. The front brake was also the more rigid cast-iron type of the 6T, and the dual seat was standard. The T100 also received the new filler cap and parcel rack, along with larger-bored gas taps.

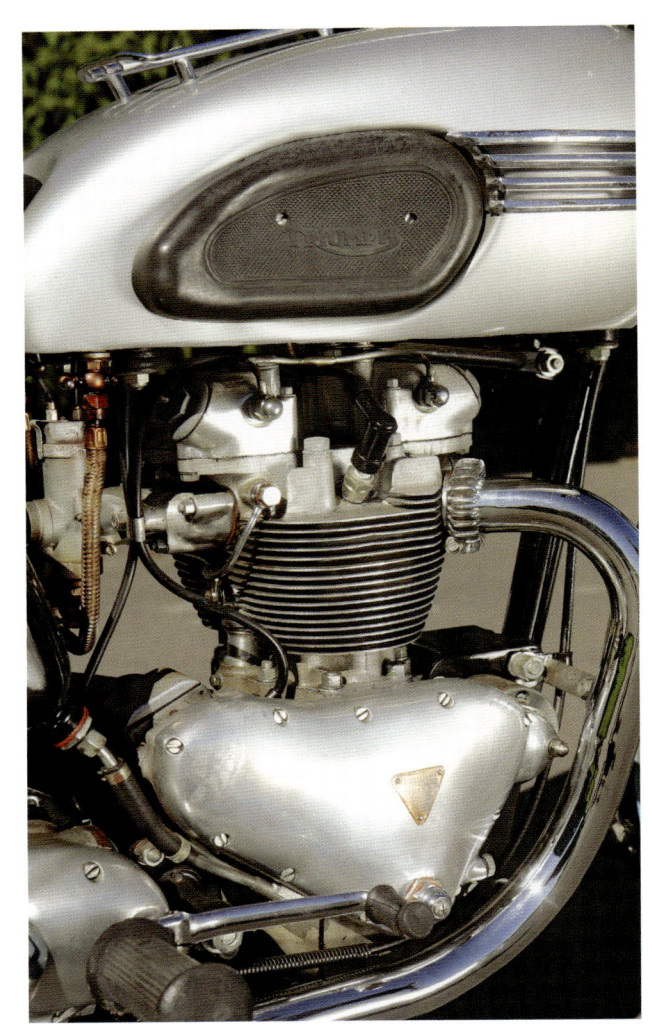

34

For the significant sum of £35, a race kit was available to convert the T100 into the T100C (for "Convertible"). This came with a range of equipment, including a choice of two higher compression ratios (for the CP100 and CP101), racing camshafts and valve springs, twin remote float-bowl Amal racing carburetors, racing braided fuel pipes, 1-gallon oil tank, Smiths 8,000 rpm tachometer, and megaphone exhaust. Also available were a close ratio gearbox and a range of sprockets. This kit transformed the T100 into a 120-mile-per-hour racer, but reliability and handling were questionable.

1951 5T Speed Twin 500

Visually the 1951 5T Speed Twin looked very similar to the previous year, but inside the engine were a number of updates bringing the specification more in line with the 6T Thunderbird. This included the roller timing side main bearing, fully machined crankshaft with heavier bob weights and strengthened con-rods, and three keyways in the cam wheel pinions. The front brake was also cast iron, and the 5T received the new filler cap and parcel rack as on the other models.

1951 5T Speed Twin
(Differing from 1950)

Carburetor
Amal 276 DK/1AT/M 1 1/16
Engine and Frame Numbers
5T 840NA to 15192NA

1951 TR5 Trophy 500

Along with the updated T100, the other significant new model for 1951 was the TR5 Trophy. No longer powered by the square-barreled Grand Prix engine, it now featured the new aluminum T100 unit. Apart from a lower 6:1 compression ratio, the engine was identical to the T100, and the boxed racing kit was also available. The chassis was much as on the previous version, with the Mark II spring wheel still an option. This wasn't popular on the TR5 as it added 20 pounds in weight and couldn't be quickly removed to repair a puncture. Good looking and effective as an off-road motorcycle, the TR5 was a competitive ISDT machine (Triumph won again in 1951), and in the United States it was particularly popular for scrambles and enduro.

ABOVE LEFT: All 1951 models received a new cast-iron front brake.

ABOVE: The Mark II spring wheel was still optional in 1951.

CHAPTER TWO

ABOVE: A dual seat was still an option for the 1951 Speed Twin.

RIGHT; With its distinctive high-level exhaust system, the TR5 Trophy continued the style of the earlier version, but with the new T100 Tiger aluminum engine.

1951 TR5 Trophy
(Differing from 1949–1950)

Horsepower	25 at 6,000 rpm
Carburetor	Amal 276 1-inch
Rear Suspension	Optional spring wheel Mark II
Color	Silver Sheen, Chrome, and Black
Engine and Frame Numbers	TR5 101NA to 14934NA

1952

Although Triumph was now part of the BSA Group, there was considerable antipathy between Meriden and Small Heath (BSA). Edward Turner viewed Triumph as the industry leader, and in the United States this was undoubtedly the case. With more dealers, and Rod Coates' official service schools, demand far exceeded supply. This was despite significant price increases, with the Thunderbird increasing from $712 in 1951 to $837 in 1952. But in the United Kingdom, BSA was outperforming Triumph. The BSA Gold Star single was dominating club racing and scrambles off-road racing, while their Bantam and C11 250 singles provided BSA entry-level machines that increased their market share. Turner had plans to match this, but the Triumph equivalent, the Terrier, wouldn't appear until 1953. In the meantime Britain was still struggling with fuel rationing, and improving fuel economy was the most pressing issue for Triumph in 1952. And with nickel shortages due to the Korean War, some chrome finishes disappeared. Handlebars and wheel rims were painted on the 6T and 5T, and previously chrome-plated parts (such as the kick-start lever) were cadmium-plated.

1952 6T Thunderbird 650

In the interests of improving fuel economy, a single automotive-type SU MC2 vacuum carburetor replaced the Amal. This required a new intake manifold and modified frame. To emphasize their success in achieving exceptional fuel economy, Triumph held a fuel economy run later in 1952. A series of riders, including Edward Turner, managed 129 miles per gallon at an average speed of 30 miles per hour. Not exactly exciting performance figures, but good publicity for the day, particularly in the United Kingdom, where fuel was still expensive despite the end of rationing. Other updates for 1952 (shared with the entire range) included a larger diameter headlamp nacelle with an underneath pilot light and 7-inch Lucas sealed beam headlight. The electrical system also changed from negative to positive earth, and the fork springs were shorter.

1952 T100 Tiger 500 and 5T Speed Twin 500

As the T100 was the performance spearhead it was spared some of the "low chrome" policy that characterized the 6T and 5T. Along with the updated headlight nacelle, both the T100 and 5T received the standardized frame, with a hole in the seat tube for a new air filter. Engine updates included a return to the 1950-type pushrod tube and 5T-type tappet guides. The CP100 racing kit now included higher-compression pistons, as well as a cast-iron cylinder barrel for use with alcohol fuel. This was good enough for Bryan Hargreaves to win the 1952 Clubmans Senior TT, albeit by stressing the handling and bottom end of the engine to its limit. *The Motor Cycle* tested a T100 in

1950–1955

An auto-style SU carb distinguished the 1952 Thunderbird.

The 1952 Tiger 100 with a new headlight nacelle.

August and found it to be a "Jekyll and Hyde," "docile and gentle in heavy traffic, yet possessed with truly tigerish characteristics when given its head on the open road." The mean maximum top speed was 92 miles per hour.

1952 TR5 Trophy 500

While the engine was ostensibly unchanged this year, unlike the T100 still with the 1951-type tappet guides and pushrod tubes, the TR5 chassis received a number of small updates. The steering head angle was increased to provide more trail, and the rear brake pedal pad was reduced in size. The rear brake drum and sprocket were now integral, and the gas tank was painted silver, as were the wheel rims.

1952 TR5 Trophy (Differing from 1951)

Color Silver Sheen with dark blue lining
Engine and Frame Numbers TR5 16000NA to 22000NA, then 26046 to 31625

1952 6T Thunderbird (Differing from 1949–1950)

Carburetor SU MC2I
Ignition Lucas magneto
Engine and Frame Numbers 6T 15809NA to 21999NA 6T 25001 to 32300 (NA suffix deleted after 25001)

1952 T100 Tiger (T100C) and 5T (Differing from 1951)

Compression Ratio 8:1 (US export T100) 8.5:1 or 12.0:1 (CP100)
Ignition Lucas K2F RO Magneto
Engine and Frame Numbers T100 15809NA to 25000NA, 25000 to 32302; 5T 1600NA to 22000NA, 26096 to 31901

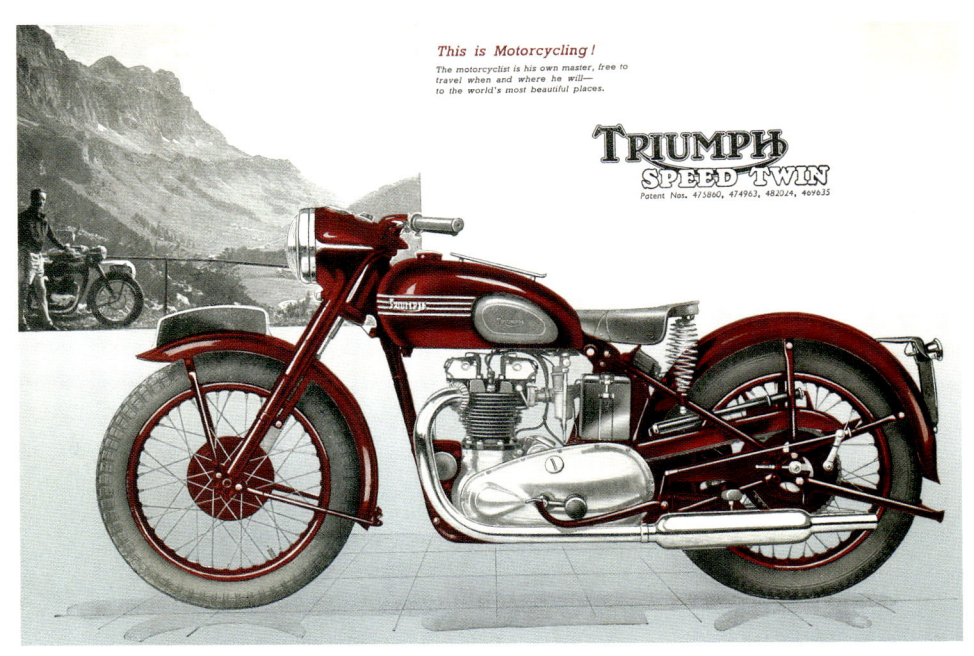

The standard 1952 Speed Twin still had a single seat.

CHAPTER TWO

ABOVE: A painted tank distinguished the 1952 TR5 Trophy. Most were rigid frame, without a spring wheel.

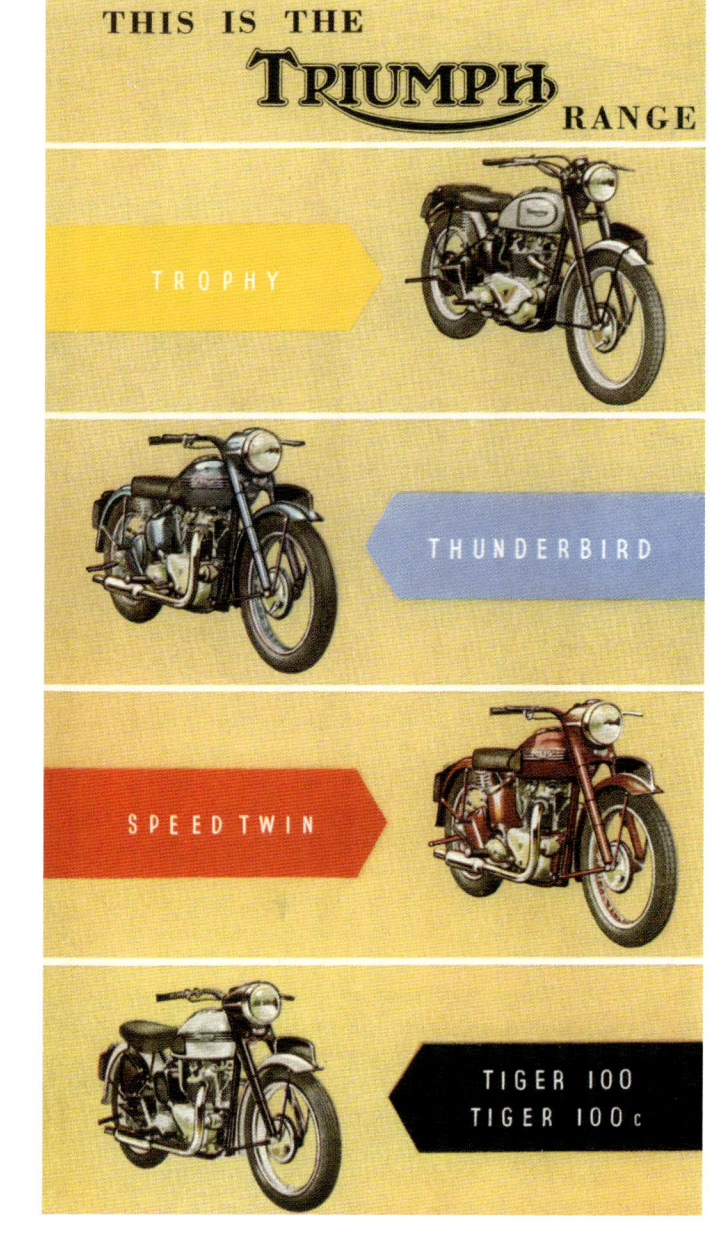

RIGHT: The Triumph range for 1953 consisted only of 500 and 650 twins.

1953

Despite Edward Turner's best efforts, even printing sales brochures, the new single-cylinder 150cc Terrier wasn't ready for production for 1953, and the range continued as before. As expected, there were some technical updates, in particular the dubious introduction of alternator electrics. This was primarily a cost-cutting move, and Triumph was the first major manufacturer to include an alternator and coil ignition. As was usual with Triumph, this first appeared on the most basic model, the 5T Speed Twin. This year not only was the racing kit offered for the T100 Tiger, but a full factory-built racing T100C was also available. New features shared over the range were camshafts with quieting ramps and a rectangular taillight.

1953 6T Thunderbird 650

By now the Thunderbird had evolved into a reliable and long-lived motorcycle, particularly if used at moderate speeds. Early crankshaft breakages were a thing of the past, and the only update was to the engine shock absorber, which moved from the engine shaft to within the clutch assembly. Although the Thunderbird retained the dynamo, this was to keep uniformity with the 5T Speed Twin and was generally considered an inferior update. As the blue Thunderbird still remained unpopular in the United States, dealers demanded black versions, which became known as the "Blackbird." This became the only US export color during 1953.

1953 T100, T100C Tiger 500, and 5T Speed Twin 500

The 5T received coil ignition with a Lucas distributor and a crankshaft-mounted Lucas RM12 alternator. As there was no voltage control, this early system was

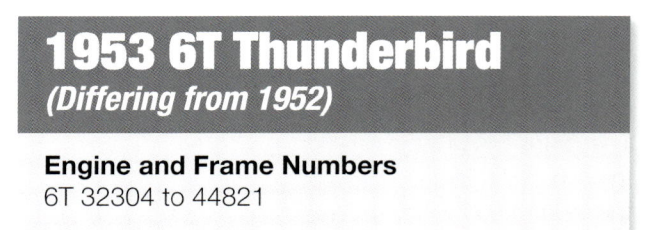

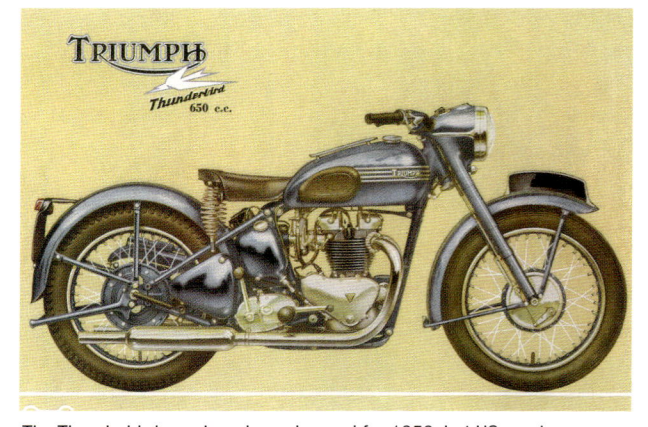

1953 6T Thunderbird
(Differing from 1952)

Engine and Frame Numbers
6T 32304 to 44821

The Thunderbird was largely unchanged for 1953, but US versions were now black.

ABOVE: The Lucas alternator was a first for a motorcycle from a major manufacturer.

LEFT: Underneath the 1953 Speed Twin's new chain case was an alternator.

BELOW: Triumph's 1953 sporting range included two of the most handsome machines of the day, the T100C and TR5 Trophy.

quite rudimentary and provided two switches in the nacelle, one for the lights and the other for ignition. The system was modified twice during the year, but still the unit wasn't trouble-free: the distributor in particular remained problematic. As on the 6T, the engine shock absorber was incorporated in the clutch. These updates required new crankcases and inner and outer chain cases.

With some skepticism surrounding an alternator charging system and coil ignition, it wasn't surprising to see Triumph's top sport model, the T100, retain magneto ignition. The magneto ignition was more reliable, and the dynamo could be removed for competition. The crankcases were not the modified 5T type, but the T100 did include the clutch shock absorber. For this year only, Triumph offered the T100C, a T100 with a factory-fitted kit; this magnificent machine came with twin Amal remote float-bowl carburetors, E3134 inlet and exhaust camshafts, and 8.0:1 pistons. The engine included many polished parts, with the option of a close- or wide-ratio gearbox. The 9.6-pint T100C oil tank incorporated a quick-release filler cap and an anti-froth tower.

1953 TR5 Trophy 500

Apart from the few updates shared with other models in the range, including clutch shock absorber, quieter camshaft ramps, and rectangular taillight, the TR5 continued largely unchanged this year. The TR5 remained very popular as an off-road competition mount in both ISDT and the United States. For the ISDT, held this

1953 T100 Tiger, T100C, and 5T
(Differing from 1952)

Compression Ratio	8:1 (T100C)
Horsepower	42 at 7,000 rpm (T100C with racing exhaust)
Ignition	Lucas DKX2A Distributor (5T)
Weight	364 pounds (T100C)
Engine and Frame Numbers	T100 32303 to 44134; 5T 33868 to 45575

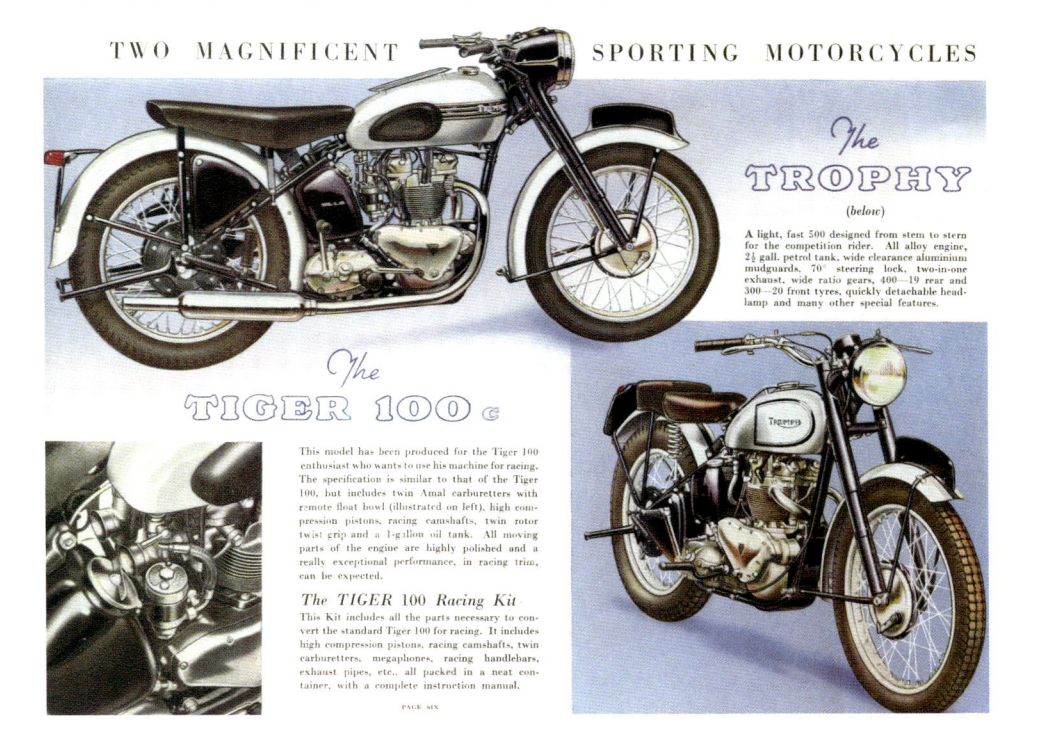

CHAPTER TWO

Jim Alves' 1953 factory ISDT TR5 featured swingarm rear suspension and a 650cc engine. Notice the 5T alternator-type primary chain case.

year in Czechoslovakia, Triumph used a swinging arm frame. A 650cc engine powered Jim Alves' TR5, and Britain again won the Trophy.

1954

This was a big year for Triumph. It was almost as if Edward Turner wanted to take the parent company BSA head-on: Triumph's range nearly doubled from four to seven. Factory extensions allowed for increased production, and new models appeared at both ends of the lineup. Catering to the US demand for higher performance, the 650cc T110 Tiger was offered, and as an antithesis, the entry-level Terrier and 200cc Tiger Cub finally made it into production. Along with these new models came a significant number of detail changes.

1953 TR5 Trophy
(Differing from 1952)

Engine and Frame Numbers
TR5 32854 to 44121

The 1954 T110 set the new standard for a good-looking, high-performance motorcycle.

1954 T110 Tiger 650

Since the release of the Thunderbird three years earlier, West Coast US distributor JoMo and other mechanics had worked hard at developing the 6T for competition. They were extremely successful, and many of these developments appeared on the new T110, or "Ton-Ten." Meriden also worked on the T110's development, and a prototype appeared as Jimmy Alves' 1953 ISDT machine. The T110 Tiger was the highest-performing Triumph twin yet and one of the fastest motorcycles available in 1954, easily capable of the 110 miles per hour indicated by the model description. Many believe the first-year T110 the fastest of them all, and there wasn't much argument regarding the styling. In its Blue Sheen and black finish, the T110 was simply stunning.

Unlike on the T100, the T110 cylinder head and barrels were cast iron, but the cylinder head design was new, with modified ports and larger inlet valves. The E3325 camshafts were based on the renowned "Q" sports cam developed in the United States, and the engine had higher-compression pistons and a larger Amal carburetor with TT-like float chamber. Other updates saw a stiffer crankshaft with larger diameter shafts, new flywheels, and larger diameter big-end journals. A ball bearing replaced the timing side roller bearings, the crank now running in ball bearings on both sides. The oil pump was also uprated and the primary chain cases shortened. Unlike the 6T this year, the T110 retained the earlier Lucas dynamo and magneto ignition.

It wasn't only the engine that received significant updates for the T110 Tiger. After experimenting with swingarm rear suspension in the 1953 ISDT, Triumph finally followed other British manufacturers in adopting it on their premier sport models. The frame featured a single downtube, Triumph's first center stand, and twin Girling shock absorbers. The swingarm was unbraced, so handling was still less than perfect. A clever shortening of the engine managed to keep the wheelbase increase moderate, and completing the new chassis specification was a new 8-inch single leading shoe front brake with polished aluminum anchor plate and chrome wire mesh–covered air scoop. The rear fender was more deeply valanced. A quickly detachable rear wheel was offered as an option, as was a 3.6-gallon tank in the United States. The seat was a new dual-level type.

The T110 Tiger was so popular it quickly supplanted the Thunderbird as Triumph's premier model, relegating the 6T to sidecar and more mundane duties. Americans took to it immediately, and a T110 even won the lightweight class in the 1954 Cambridge (Minnesota) Enduro, an event for which it was never really intended. The T110's speed was never in doubt: *Cycle* magazine managed nearly 110 miles per hour running into a headwind. But the T110 wasn't without

40

1954 T110 Tiger 650

Bore	71mm
Stroke	82mm
Capacity	649cc
Horsepower	42 at 6,500 rpm
Compression Ratio	8.5:1
Carburetor	Amal 289 1⅛-inch
Ignition	Lucas K2F magneto
Gearbox	Four-speed
Front Fork	Telescopic
Rear Suspension	Swingarm
Brakes	8x1⅛-inch front and 7x1⅛-inch rear
Wheels	WM2x19-inch front and rear
Tires	3.25x19-inch front, 3.50x19-inch rear
Wheelbase	55.75 inches
Dry Weight	395 pounds
Color	Blue Sheen and black
Engine and Frame Numbers	T110 44135 to 56699

New for the T100 and T110 this year was an 8-inch single leading shoe cast-iron front drum brake. The scalloped flange only featured for 1954.

The 1954 headlight nacelle for magneto models included a lights-only switch and central kill button.

1954 6T Thunderbird
(Differing from 1953)

Ignition
Lucas DKX2A Distributor
Engine and Frame Numbers
6T 44235 to 55593

The 1954 6T, now with alternator electrics but still a rigid frame.

problems. When used hard, the iron cylinder head could overheat, sometimes distorting and warping and leading to a blown head gasket.

1954 6T Thunderbird 650

For this year the 6T Thunderbird shared the 5T's alternator electrics and distributor ignition. The engine featured the T110 big-bearing crankshaft and timing side ball bearing, with new primary chain cases to accommodate the Lucas RM14 alternator. The new electrical system required an updated headlight nacelle. In other respects the Thunderbird was unchanged, still with a rigid frame and optional spring wheel.

1954 T100 Tiger 500

Sharing many updates with the new T110 Tiger, the T100 Tiger was now able to match any 500 from other British manufacturers. The engine was still the close-finned aluminum unit, but now with the stronger T110 crankshaft and timing side ball bearing. As on the T110 the chain case was shortened, and the electrical system remained a dynamo with magneto ignition. Twin carburetors were an option, and according to factory literature a racing kit was still available (although the T100C was now discontinued). The frame was T110's brazed cradle single downtube unit and swingarm, with the new 8-inch front brake. As on the T110, the

CHAPTER TWO

ABOVE: Many updates were incorporated on the 1954 T100 Tiger, but the style was unchanged.

LEFT: The most notable new feature was swinging arm rear suspension.

BELOW: Inside the aluminum T100 engine was the T110's stronger crank.

42

1954 5T Speed Twin and TR5 Trophy *(Differing from 1953)*

Engine and Frame Numbers
5T 45578 to 55493; TR5 45595 to 52969

1954 T100 Tiger *(Differing from 1953)*

Rear Suspension	Swingarm
Brakes	8x1⅛-inch front and rear
Wheelbase	55.75 inches
Dry Weight	375 pounds
Color	Shell Blue Sheen and black
Engine and Frame Numbers	T100 44135 to 56699

seat was a new dual-level type, with a slight reduction in seat height (from 31.0 inches to 30.5 inches), and a QD wheel was an option. Like the T110, the handling was somewhat scary at higher speeds, but overall the swingarm rear suspension was a marked improvement over the earlier rigid type or spring wheel. And while the power was somewhat less than the T110, the T100 maintained the advantage of the aluminum cylinder head and barrel.

1954 5T Speed Twin and TR5 Trophy 500

The Speed Twin continued with very few changes this year, still with the rigid frame. Visually the most noticeable update was to the silencers, now barrel-shaped as on other models this year, and a circular shaped rectifier. The headlight nacelle now featured a combined ignition and lighting switch. Also largely unchanged was the TR5 Trophy, which now featured the stronger T110 crank. Both these models were biding time for more significant updates in 1955.

1954 T15 Terrier 150 and T20 Tiger Cub 200

After several years of gestation, both the 150cc Terrier and 200cc Tiger Cub finally made it into production for 1954. The Terrier was originally revealed at the 1952 Earls Court Show, and it surprised many pundits with specifications that departed noticeably from usual Triumph practice. Most of the competition was two-stroke, so the Terrier looked extremely advanced with its high specification overhead-valve engine. Unit with the transmission, the compact engine featured a diecast aluminum cylinder head, and early examples had a troublesome roller bearing big end. The electrical system included a Lucas alternator and points and coil ignition. Rear suspension was by a plunger: the Terrier and Tiger Cub were the only Triumphs to feature this type of rear suspension.

Terrier production commenced in July 1953, but actual delivery was delayed nine months due to many problems surfacing in early examples. Most of these were due to Turner's philosophy of saving costs by paring metal, and as a result the Terrier had many faults that should arguably have been eliminated in the design stage. As the Terrier was a relatively low-profit model anyway (compared to the twins), this was a quite shortsighted. Turner boasted that the Terrier was designed in eight weeks, and by the end of 1954 Triumph published a seven-page booklet listing 23

ABOVE: A new chain case contributed to a shorter engine and gearbox unit.

LEFT: Only a few visual clues distinguished the 1954 Speed Twin from the 1953 version.

LEFT BELOW: The TR5 Trophy was also largely unchanged for 1954.

CHAPTER TWO

ABOVE: The 1954 T15 Terrier was the first Triumph single with plunger rear suspension.

BELOW: The high-level exhaust was standard on the 1954 T20 Tiger Cub.

1954 T15 Terrier (T20 Tiger Cub)

Bore	57 (63)mm
Stroke	58.5 (64)mm
Capacity	149 (199)cc
Horsepower	8 (10) at 6,000 rpm
Compression Ratio	7:1
Carburetor	Amal 332 1 1/16 (3/4)-inch
Ignition	Lucas AC/DC
Gearbox	Four-speed
Front Fork	Telescopic
Rear Suspension	Plunger
Brakes	5 1/2-inch front and rear
Wheels	WM1x19-inch front and rear
Tires	2.75 (3.00)x19-inch front and rear
Wheelbase	49 inches
Dry Weight	175 (182) pounds
Color	Amaranth Red (Shell Blue Sheen)
Engine and Frame Numbers	T15 101 to 8517; T20 3000 to 8517

common faults. These included a broken crankpin, oil leaks, transmission problems, and fractures of various parts. Many of the problems were the result of shortcomings in the lubrication system, with both the oil pump and oil tank capacity too small. It was hardly an auspicious beginning, but when it was running the Terrier was quite sprightly. *Cycle* magazine, in June 1954, managed a top speed of 68 miles per hour running into a slight headwind. Gasoline consumption of 84 miles per gallon was also impressive.

Late in 1953 the T20 Tiger Cub was introduced as a sport version of the Terrier. Unlike the Terrier, the Tiger Cub had a plain bearing big end from the outset; the cylinder head was new and the Amal carburetor larger. With a high-level exhaust system, the Tiger Cub was about 8 or 9 miles per hour faster than the Terrier.

To allay dealers' fears about the Terrier's reliability, in October 1953 Edward Turner arranged for three machines to be ridden from Land's End to John O'Groats. With Turner riding one of the Terriers, visiting dealers on the way, they averaged 36.68 miles per hour over the 1,000 miles. Later called "The Gaffers' Gallop," the demonstration was so successful that order books for the new singles were full for the next 12 months. Turner made the most of the publicity, and it was another example of his understanding of the marketplace. Although the problems remained, they were gradually ironed out over the next few years, and tens of thousands of riders worldwide learned to ride on Terriers and Tiger Cubs.

1955

With so many new models and updates incorporated for 1954, it wasn't surprising to see only a few updates for the 1955 season. All twins now received swingarm rear suspension and the T110's stronger crankshaft, but it was developments in America that stole the headlines this year. Johnson signed star Californian off-road racer Bud Ekins to ride a TR5 Trophy, and Texan Johnny Allen embarked on a quest to make his Thunderbird streamliner the fastest motorcycle in the world. Both events would be pivotal to Triumph's later success, particularly in the United States.

1955 T110 Tiger 650 and T100 Tiger 500

As the T110 Tiger's cast-iron cylinder was problematic, Triumph added a fifth cooling fin near the exhaust pipe clamp and, in a rather dubious move, painted the iron head silver to give the impression it was made of aluminum. The T100 Tiger was also largely unchanged this year, but the compression ratio for all versions rose to the 8.0:1 of US export models. The frame and cycle parts also continued much as before, with only small updates. These included sidecar lugs added to the frame. Twin Amal Type-6 carburetors were still optional on the T100, and in the United States a smaller, 3.6-gallon gas tank was also

1950–1955

The T110 Tiger set a new standard for affordable sport motorcycles in 1955.

The 6T Thunderbird gained swingarm suspension for 1955.

1955 T110 and T100 Tiger (Differing from 1954)

Compression Ratio
8:1 (T100)
Engine and Frame Numbers
T110 56700 to 70929; T100 56700 to 70929

available. A feature shared across the range this year was the one-piece oil tank unit incorporating the oil tank, air cleaner, battery, and toolkit. By now the T110 had found its niche in British production racing, finishing first, second, and third in the 750cc class of the 1955 Thruxton Nine Hour race.

1955 6T Thunderbird 650 and 5T Speed Twin 500

This year saw the most significant change to the Speed Twin since 1945, when the telescopic fork was introduced. with both the 6T Thunderbird and 5T Speed Twin receiving swingarm rear suspension. Also introduced were the associated inner and outer primary crankcases required to reduce engine length. The Speed Twin also received the stronger crankshaft and ball timing side main bearing, and it was the first twin to specify the new Amal Monobloc carburetor. US versions of the 6T Thunderbird also received this new carburetor, but the SU continued for other markets. Both the 6T and 5T featured a new Lucas distributor with reduced ignition advance. The frame for both the 6T and 5T was similar to that of the 1954 Tiger, with

1955 6T Thunderbird and 5T Speed Twin (Differing from 1954)

Carburetor	Amal 376/25 $^{15}/_{16}$-inch (5T); Amal 376/40 $1^{1}/_{16}$-in (US 6T)
Rear Suspension	Swingarm
Wheelbase	55.75 inches
Dry Weight	385 pounds (6T); 380 pounds (5T)
Engine and Frame Numbers	6T 56700 to 70874; 5T 55494 to 70196

twin Girling shock absorbers. With the new dual-level seat, the seat height increased to 30.5 inches (from 29.5 inches). The introduction of the swingarm also saw a significant weight increase.

continued on page 48

CHAPTER TWO

James Dean's Triumphs

Few Hollywood stars have become as iconic and revered after their death as James Dean. Dean was only 24 years old when he died in a crash in his Porsche 550 Spyder in September 1955, but he had already made his mark with a lead role in the major film *East of Eden*. Hailing from Fairmount, Indiana, Dean had been riding motorcycles since the age of 15, first a CZ, then a Royal Enfield 500 and, in 1953, an Indian Warrior TT. Coincidentally, Dean stored the Indian in a Greenwich Village garage where budding actor Steve McQueen worked as a part-time motorcycle mechanic.

In April 1954 Dean signed a contract to play Cal Trask in Steinbeck's *East of Eden*. He moved from New York to Los Angeles, and one of the first things he did was buy a new Triumph Tiger T110. Three days after filming wrapped on *East of Eden* he traded the T110 for a shell blue 1955 Triumph TR5 Trophy. This he soon lightly customized, with higher handlebars, an open exhaust pipe with no muffler, knobby tires, and a single 6T-type spring seat with a pillion seat bolted backwards on the rear fender. Dean admired Marlon Brando, and this was a tribute to the similar pillion seat on Brando's Thunderbird in *The Wild One*.

In the meantime Dean filmed *Rebel Without a Cause* and co-starred in the epic *Giant* (with Rock Hudson and Elizabeth Taylor). *Rebel Without a Cause* was released 27 days after Dean's death, with *Giant* appearing over a year later. Although involved in only three major films, he has become an American cultural icon.

Dean's TR5 Trophy was sold back to the original dealer, Ted Evans Motorcycles in Culver City, California, after his death, and it was subsequently modified and raced in Minnesota. In the late 1980s Dean's cousin, Marcus Winslow Jr., bought the bike and had it restored in the same configuration as Dean had customized it. It is currently on display at the Fairmount Historical Museum in Indiana, along with a large assortment of James Dean artifacts.

James Dean and his TR5 Trophy. *Triumph*

Dean posing on the Trophy before it was mildly customized. *Triumph*

The 1955 Speed Twin, still in Amaranth Red.

1950–1955

LEFT: A new primary cover was included with the adoption of swingarm suspension.

FAR LEFT: The Speed Twin retained the smaller, 7-inch front brake.

LEFT: The Speed Twin was the first model to feature the new Amal Monobloc carburetor.

FAR LEFT: Rear suspension was by a pair of Girling shock absorbers.

LEFT: The Speed Twin may have gained some weight by 1955, but it was still a handsome motorcycle.

BELOW: The headlight nacelle for the 1955 Speed Twin.

47

CHAPTER TWO

TRIUMPH TROPHY
Patent Nos. 475860, 474963, 482024

A High Performance Model for the Competition Enthusiast

A new model, of high performance, designed to be readily adaptable to most forms of motorcycle competition. Its specification includes many practical features which will appeal instantly to the really experienced competition rider.

SPECIFICATION

ENGINE. O.H.V. high compression vertical twin with die-cast alloy head and barrel, two gear-driven camshafts, "H" section RR56 alloy connecting rods, plain big ends, and central flywheel. Dry sump lubrication, pressure-fed big ends and valve gear. Air cleaner. Upswept two-in-one exhaust pipe with silencer.

FOUR - SPEED GEARBOX. Positive foot-change, large diameter multi - plate clutch with rubber pad type shock absorber.

FRAME. Brazed cradle type frame with swinging arm rear suspension with hydraulic damping adjustable for varying loads.

FORKS. The famous Triumph telescopic pattern with long supple springs and hydraulic damping.

FUEL TANKS. All-steel welded with quick-release caps and accessible filters.

BRAKES. Large diameter cast iron drums, polished front anchor plate, finger adjustment.

ELECTRICAL EQUIPMENT. Powerful headlamp with quickly detachable harness.

Lucas 6 volt 60 watt dynamo, automatic voltage control. Lucas gear-driven "Wader" type magneto.

OTHER DETAILS. 120 m.p.h. (or 180 km.p.h.) Smiths Speedometer; Triumph Twinseat; twist grip with adjustable tension; shell blue sheen and black finish.

For Technical Details see Back Cover.

ABOVE: With swingarm rear suspension and more power, the TR5 Trophy was considerably updated for 1955.

RIGHT: Bud Ekins rode a TR5 to victory in the 1955 Catalina Grand Prix. He would become one of the greatest off-road riders for Triumph.

Continued from page 45

1955 TR5 Trophy 500

After more success in the 1954 ISDT, with Jim Alves, Peter Hammond, and Johnny Giles winning gold medals and Triumph the manufacturer's team award, it was no surprise to see a completely updated TR5 for 1955. It not only received a swinging arm frame, but the engine now featured the T110 E3325 sport camshafts and higher-compression pistons. The Amal 276 carburetor received a large, TT-type float chamber, and the primary crankcases shortened as on the other swingarm models. Another update was the replacement of the rear 19-inch wheel with a wider WM3x18-inch. This was also completely new, with a much larger wheel spindle. The TR5 was still available in Trials trim by special order, and it soon excelled in cross-country racing. After Bill Johnson provided Bud Ekins a TR5 Trophy to win the Catalina Grand Prix, the Trophy's reputation was established.

1955 T15 Terrier 150 and T20 Tiger Cub 200

As delivery of both the Terrier and Tiger Cub was delayed, there was no difference in general specification for 1955. The T20 Tiger Cub now had a standard low-level exhaust, and both models benefited from many small improvements during the year, including a reinforced gas tank and heavier-duty oil pump. Although most Terriers and Tiger Cubs were sold in the United Kingdom, Turner still placed significant emphasis on the US market, sending around 16 percent of the first- and second-year production stateside. Performance kits soon appeared, increasing the power up to 80 percent, and Terriers and Tiger Cubs were quite successful in racing events. In the 1955 Catalina Grand Prix, victory went to Hazen Blair on a 165cc Terrier and Don Hawley on a Tiger Cub.

1955 TR5 Trophy (Differing from 1954)

Compression Ratio	8:1
Horsepower	33 at 6,500 rpm
Rear Wheel	WM3x18-inch
Rear Tire	4.00x18-inch
Wheelbase	55.75 inches
Weight	365 pounds
Engine and Frame Numbers	TR5 56153 to 69171

1955 T15 Terrier (T20 Tiger Cub (Differing from 1954)

Color
Black (US T15)

Engine and Frame Numbers
T15 8518 to 17388; T20 8518 to 17388

The World's Fastest Motorcycle

In the 1950s a worldwide battle erupted over setting the outright motorcycle land speed record. Wilhelm Herz, riding an NSU streamliner, had set a record of 180 miles per hour in 1951, and on July 2, 1955, on a narrow country road outside Christchurch, New Zealand, Russell Wright took his Vincent Black Lightning to a new record of 185 miles per hour. But in Fort Worth, Texas, a group of enthusiastic mechanics was busy preparing a Triumph-powered streamliner. During 1954, American Airlines pilot J. H. "Stormy" Mangham and mechanic Jack Wilson built a cigar-shaped streamliner powered by an iron 6T Triumph engine. After running to 155 miles per hour on gasoline in 1954, the following year they went back, this time with Texas Short Track Champion Johnny Allen in the cockpit. Allen had had run 124 miles per hour on a Tiger 100 to take the AMA National Class C stock machine record, and this had persuaded Bill Johnson to get on board the project. Wilbur Cedar, Johnson's JoMo partner, organized Meriden to arrange some special high-speed Dunlop tires, and in September 1955 Allen ran a two-way flying kilometer of 193.7 miles per hour. This was clearly a new record but was only recognized by the AMA, not the Fédération Internationale de Motocyclisme (FIM), as the AMA was not affiliated with the FIM at that time.

The NSU and Herz traveled to Bonneville in 1956 and set a new record of 211 miles per hour on the supercharged *Delphin III*, but this was quashed only days later by Allen. Despite the offer of a factory splayed-port aluminum cylinder head, Wilson retained the iron 6T unit, modified to run with a pair of Amal 1⅜-inch Amal GP carbs. Running on nitro, the naturally aspirated 650 produced around 100 horsepower, enough for Allen to post a best two-way run of 214.17 miles per hour. Again the FIM refused to accept this as a world record, but Edward Turner was unperturbed. He decided every new Triumph would have a decal claiming "World Motorcycle Speed Record Holder," which resulted in a prolonged legal rift with the FIM and ultimately resulted in 1960 in a two-year suspension of Triumph's FIM license. But none of this mattered. In American eyes the record stood, and Triumph benefited enormously.

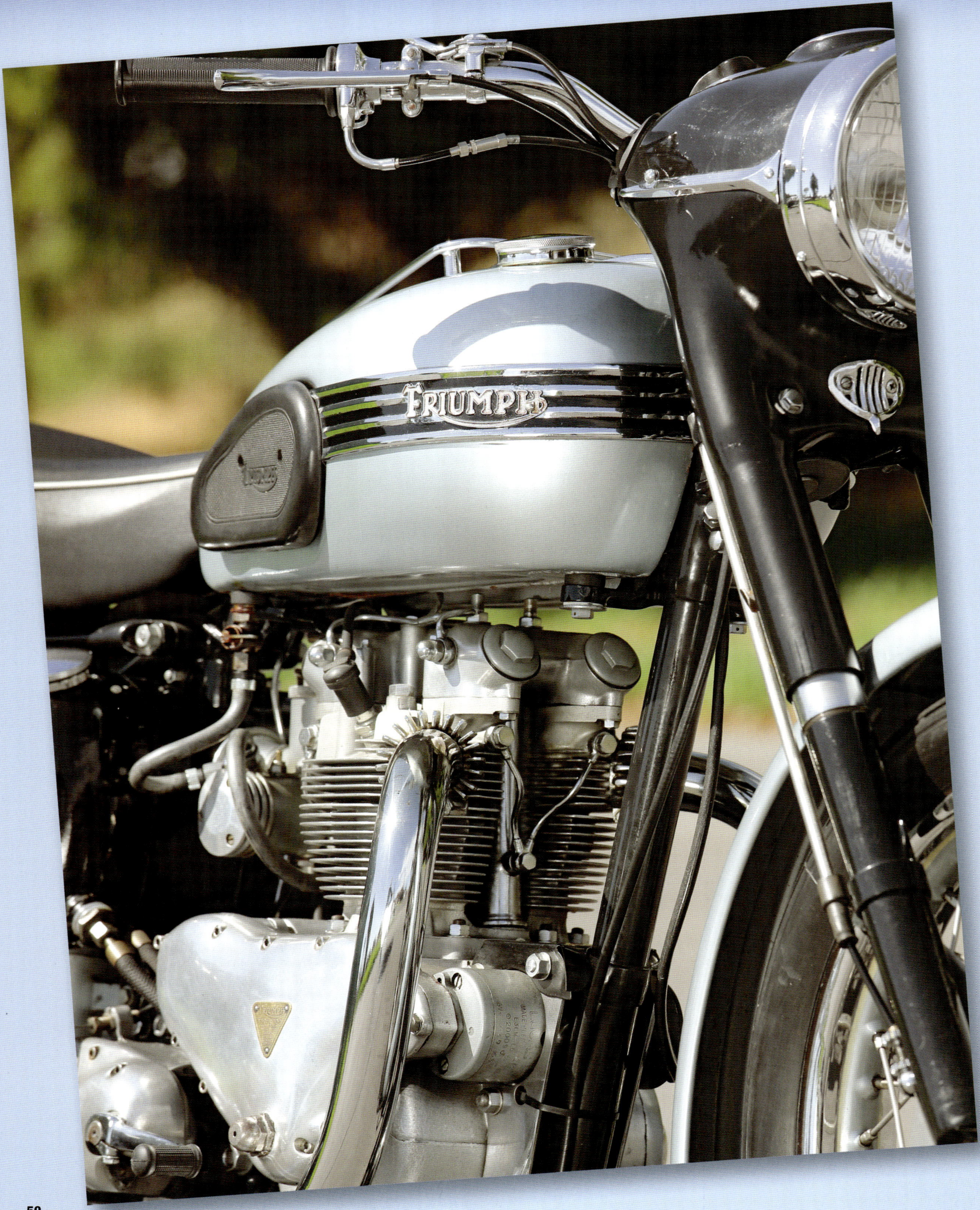
TRIUMPH

1956–1962

Birth of Iconic 650s

The Trophy and Bonneville

A government-created economic recession, followed by the Suez Crisis and more gasoline rationing, made 1956 a particularly grim year in Britain. The automotive industry as a whole was particularly hard-hit, and political machinations within the BSA Group saw Edward Turner's power increasing to include the entire automotive division. Now with Triumph, BSA, and Ariel motorcycles, along with Daimler and Carbodies (makers of London taxis) under his control, and facing a weak domestic market, Turner bowed to American requests for more performance. The result was the 650cc TR6 "street scrambler" (the Trophybird), and a higher-performance T110 Tiger. Johnny Allen's 1955 and 1956 speed records provided considerable publicity, and with Bud Ekins' off-road racing success, Triumph was on a roll in America.

1956

Although Edward Turner was now in charge of BSA and Triumph, he remained a staunch Meriden man, showing particular disdain for many existing BSA designs and, wittingly or unwittingly, hastening their demise. One of these was the sporty Gold Star single, seen by Turner as an annoyance that was beating his own Tiger 110. In the important Thruxton Nine Hour production race, Percy Tait may have won the 750cc class, but half a dozen 500cc Gold Stars finished ahead of him. It was a similar scenario in the US Catalina Grand Prix, where T110s finished fourth, fifth, and sixth behind a trio of Gold Stars. As Turner seemed determined to put a Triumph stamp on BSA, leading BSA engineer Bert Hopwood departed for Norton. Turner's influence soon showed in a new range of BSA singles that bore more than a passing resemblance to the Triumph Tiger Cub. Turner's decisions were generally business-based, and while he had many detractors, he remained unperturbed. As BSA struggled, Triumph went from strength to strength, each year seeing more improvements to the existing range as well as new models. Triumph boasted they were the world's leading motorcycle, and it wasn't an idle claim.

ABOVE: Edward Turner with his T110 twin. This received an aluminum cylinder head and Amal Monobloc carburetor for 1956.

OPPOSITE: The alloy cylinder head of the 1956 T100 Tiger had splayed exhaust ports. Note the grill underneath the headlamp instead of the pilot light this year.

CHAPTER THREE

TOP: In Shell Blue Sheen and black, the 1956 T100 Tiger was an extremely attractive motorcycle.

ABOVE: Tigers had an 8-inch front brake with air scoop.

RIGHT: The one-piece black central unit housed the air cleaner, oil tank, battery, and toolkit.

1956 T110 Tiger 650 and T100 Tiger 500

To cure the overheating problems afflicting the iron head T110, a new diecast aluminum cylinder with cast-in austenitic iron valve seats was introduced. With its splayed exhaust ports, this head was named the "Delta" due to its shape as seen from above. Unlike in the T100 aluminum head, oil drained directly into the pushrod tubes, eliminating the external oil pipes, and the cylinder barrel remained cast-iron, albeit painted silver to maintain aesthetic symmetry and the appearance of the 500. The aluminum head and barrel T100 continued as before, and on both models new Vandervell VP3 shell big-end bearings replaced the previous white-metal type. Both versions received the new Amal Monobloc carburetor, and the T100 was still available with a twin carburetor option. For the US market Meriden supplied the T100R, each with a carefully built, bench-tested "Red Seal" engine. Shared with the TR5R, it was fitted with twin carburetors, high-compression pistons, and an open megaphone exhaust system. The chassis for the T110 and T100 was identical; other detail updates this year included a wider dual seat and a modified headlight nacelle, with the pilot light now incorporated inside the headlight.

While the T110 still exhibited some fragility at racing speeds, as a fast sport roadster it continued to set the class standard. *The Motor Cycle*, in March 1956, achieved a one-way top speed of 109 miles per hour, stating, "Possessing a performance which suggests the analogy of an iron hand in a velvet glove, it is one of the most impressive Triumphs yet produced and is justly popular among sporting riders."

1956 6T Thunderbird 650 and 5T Speed Twin 500

Updates were minor to both these models this year. Shared with all other twins were the Vandervell shell-type big-end bearings, and the 5T now featured 6T crankcases with a new matching cylinder barrel. The 6T continued with an SU carburetor, but renumbered, while US 6Ts were still fitted with the Amal Monobloc as they were in 1955. Thunderbird colors were new this year, a very attractive Polychromatic Gray, the last

1956 T110 and T100 Tiger *(Differing from 1955)*

Carburetor
Amal 376/40 1 1/16-inch (T110); Amal 376/35 15/16-inch (T100)

Engine and Frame Numbers
T110 70930 to 82799; T100 70930 to 82799; from July 5, 1956, numbers went back to 0101 to 0944 (to avoid numbers going into six digits)

TOP LEFT: The 1956 Tiger headlight nacelle. The lever on the right is for advance/retard.

ABOVE: The barrel-shaped silencers were a distinctive Triumph feature. The Girling shock absorbers were unchanged.

LEFT: The Speed Twin was little changed for 1956, the final year for the classic four-bar gas tank styling.

CHAPTER THREE

1956 6T Thunderbird and 5T Speed Twin
(Differing from 1955)

Carburetor
SU 590 (6T)
Color
Polychromatic Gray (6T);
Black or Aztec Red (US 6T)
Engine and Frame Numbers
6T 70970 to 82222 and 0654 to 0943;
5T 71642 to 82443 and 0602 to 0932

year of a single-color finish with four-band gas tank styling. Suspension updates included modified bump stops in the front fork and softer, 100-pound, springs. By now the Thunderbird and Speed Twin had evolved into extremely competent and civilized motorcycles, suitable for sidecars or everyday use.

1956 TR6 Trophy 650 and TR5 Trophy 500

Aimed squarely at the US market, the new TR6 Trophy was cleverly created from mostly existing components, in this case the TR5 chassis and the T110 motor. With the 42-horsepower T110 engine in the relatively lightweight TR5 chassis, the performance was particularly sprightly, and the TR6 was extremely well suited to American cross-country events at that time. Triumph launched the TR6 by taking three bikes straight out of the crates and entering them in the Californian Big Bear Motorcycle Run. Bill Postel, Bud Ekins, and Arvin Cox led the entire race, finishing first, second, and third, respectively. This victory set the stage for the Trophy to dominate off-road racing in America for the next decade.

Updates to the TR5 this year were minor. The 500cc aluminum engine now featured the shell big-end bearings and Amal Monobloc carburetor as on the T100 Tiger. TR5 and TR6s this year featured a high-level siamesed exhaust system exiting on the left. As they came from the factory there were no heat shields. The front fork on both the TR5 and TR6 was now fitted with rubber gaiters, as on Ekins' 1955 Catalina TR5, and the TR5 headlamp increased to 7 inches. The gas tank held 3¼ gallons, as opposed to 4¼ gallons on other models, with 20- and 18-inch wheels and Dunlop Trials Universal as before. Later in the model year some US TR6s received a 19-inch front wheel with the T110 8-inch front brake.

1956 TR6 Trophy

Bore	71mm
Stroke	82mm
Capacity	649cc
Horsepower	42 at 6,500 rpm
Compression Ratio	8.5:1
Carburetor	Amal 376/40 1 1/16-inch
Ignition	Lucas K2FC "wader" magneto
Gearbox	Four-speed
Front Fork	Telescopic
Rear Suspension	Swingarm
Brakes	7x1 1/8-inch front and 7x1 1/8-inch rear
Wheels	WM1x20-inch front and WM3x18-inch rear
Tires	3.00x20-inch front, 4.00x18-inch rear
Wheelbase	55.75 inches
Dry Weight	370 pounds
Color	Blue Sheen and black
Engine and Frame Numbers	TR6 70199 to 82797 and 0598 to 0909

1956 TR5 Trophy and TR5R Trophy
(Differing from 1955)

Compression Ratio	9:1 (TR5R)
Carburetor	Amal 376/35 15/16-inch (TR5); Amal 276 (TR5R)
Horsepower	40 at 7,000 rpm (TR5R)
Engine and Frame Numbers	TR5 70199 to 82797 and 0598 to 0909; TR5 76113 to 76224 (TR5R)

Designed for competition, this is a rare 1956 TR6 with factory-fitted aluminum cylinders. Normal TR6s had cast-iron cylinders painted silver to replicate aluminum.

1956–1962

The TR6, ostensibly a TR5 chassis with T110 motor, was immediately successful.

Handsome and effective, with its off-road tires and high-level exhausts, the TR6 was particularly popular in California.

1956 T15 Terrier (T29 Tiger Cub)
(Differing from 1955)

Wheels
WM2x16-inch front and rear
Tires
3.25x16-inch front and rear
Dry Weight
205 pounds (T20)
Engine and Frame Numbers
T15 17389 to 26275; T20 17389 to 26275

Also available in the US was the TR5R, a ready-made racer with quickly detachable lighting. Based on a TR5 Trophy, this featured a factory Red Seal motor with higher-compression pistons, twin remote float-bowl Amal carburetors, racing camshafts, and downswept exhausts with open megaphones.

1956 T15 Terrier 150 and T20 Tiger Cub 200

The Terrier's sales proved disappointing: 6,240 were built in 1954 and only 2,074 in 1955, and 1956 was the final year for the Terrier. Updates were minimal, and it proved particularly unpopular in America. JoMo sold none, and TriCor imported only 50. The T20 Tiger Cub was more successful and received some

CHAPTER THREE

The T20 Tiger Cub retained plunger rear suspension for 1956 and received smaller, 16-inch wheels.

ABOVE: Two-tone colors were introduced for 1957, the ivory and blue an option on the T110.

BELOW: The Tiger T100 was little changed for 1957.

updates, most notably 16-inch wheels to lower the overall height, and seat height dropped 1½ inches. Other improvements included a stronger crankshaft and con-rod, a larger-capacity oil tank, and a new gas tank. US versions also had the option of a high-level exhaust system, but the flawed plunger rear suspension continued for one more year. Although nearly 60 percent of Tiger Cub production was destined for the United Kingdom, sales grew considerably in the United States this year, to nearly 1,000.

The Tiger Cub was also developed as an off-road racer, in the United States and by Meriden. Ken Heanes won a gold medal in the 175cc class in the 1956 ISDT in Bavaria, his factory Tiger Cub reduced in capacity. In the United States the Tiger Cub continued to dominate the 200cc class of the Catalina Grand Prix.

1957

The most noticeable update for 1957 was the introduction of flamboyant chrome-grill gas tank badges across the entire range. Along with two-tone paint finishes, this styling feature echoed a Buick grille and further cemented Triumph's position as a market leader in America. With this new addition Triumphs simply looked more modern than the competition. This year also saw the release of the 350cc "Twenty-One" (21 years since the establishment of Triumph Engineering and 21 cubic inches of displacement), although it didn't actually make the 1957 lineup. The other significant introduction was range expansion through variations on existing models. So while the basic lineup consisted of 5 models as before, the total number of variants now available in the United States was 13. With 303 dealers in the east, TriCor saw a 33 percent increase in sales during 1957 for Triumph's best year yet.

The new chrome gas tank badges and two-tone colors for 1957 created a style that would continue until 1965.

1957 T110 and T100 Tiger (Differing from 1956)

Compression Ratio
8:1 (T110); 9:1 (T100 HDA)
Horsepower
40 at 6,500 rpm (T110)
Dry Weight
390 pounds (T110); 385 pounds (T100)
Color
Gray/black, ivory/blue (optional)
Engine and Frame Numbers
T110 0945 to 011115; T100 0945 to 011115

1957 6T Thunderbird and ST Speed Twin
(Differing from 1956)

Carburetor
SU 603 (6T)
Dry Weight
395 pounds (6T and 5T)
Color
Gold/black (6T); Aztec Red (US 6T)
Engine and Frame Numbers
6T 01406 to 011110; 5T 01797 to 010253

1957 T110 Tiger 650 and T100 Tiger 500

The new gas tank badges and two-tone colors dominated the 1957 updates, with the T110 now also having a slightly lower compression ratio and less power. In contrast, the T100 was offered with an optional HDA twin-carburetor Delta head with splayed inlet ports. As on the T110, this no longer featured external oil pipes, as the oil drained directly into the pushrod cover tubes. The twin carb head also came with larger inlet valves, E3134 camshafts with racing R followers, and a pair of Amal 376 Monobloc 1-inch carburetors. With this option the T100 was an extremely strong-performing 500, with *Motor Cycling* managing a top speed of 105 miles per hour in April 1957. Both the T110 and T100 retained the single-sided 8-inch front brake as before, but with a new drum and hub. All models now featured an "Easy-Lift" center stand originally developed for police use.

1957 6T Thunderbird 650 and 5T Speed Twin 500

Cosmetic updates dominated the 6T and 5T, which otherwise continued largely unchanged. The Speed Twin continued in Amaranth Red, while the Thunderbird was now Metallic Gold. Both models received a heavily finned, full-width, 7-inch front brake and larger chainguard this year. While the Thunderbird maintained a loyal following, the Speed Twin seemed to be marking time in preparation for a significant update for 1959.

1957 TR6 Trophy 650 and TR5 Trophy 500

As the TR5 and TR6 Trophy were proving exceptionally popular, updates were minor and spawned a number of derivatives. This included the TR5R in various guises and three variants of the TR6. The TR6 engine was shared with the T110, with a slightly lower compression ratio and horsepower output. The TR5 this year gained the full-width, 7-inch brake of the 6T and 5T, while the TR6 received the 8-inch T110 front brake. The front wheel for both was now a 19-inch with either a ribbed or Trials Universal tire, but trials specification was still available for the TR5.

While Meriden listed all TR5 variants as the TR5R, in the United States they were designated the TR5A, TR5B, and TR5R (or T100RS). These were all available for special order from Meriden, the TR5A with low exhaust pipes and street equipment and the TR5B with a high-rise scrambler exhaust system. All featured the new twin-carb splayed-port aluminum cylinder head, high-compression pistons, and E3134 camshafts. Depending on the specification the carburetors were either Amal Monobloc 376 or the earlier 276 with

The TR5 had a close-finned aluminum engine and siamesed exhaust system.

CHAPTER THREE

With its distinctive high-level exhaust and functional appearance, it was no wonder that the TR6 became one of Triumph's most popular models in America, particularly with the more modern-looking two-tone colors and new badges for the 1957 version shown here.

ABOVE: The 1957 TR6, now with the T110's 8-inch front brake and 19-inch front wheel.

RIGHT: The Johnson Motors crew with a TR6 at the 1957 Big Bear Run.

remote float bowls. A few were also built with a single carburetor. A TR5R (T100RS) production racer and T100RR rigid-framed dirt tracker were also available in the United States in small numbers. The dirt tracker first appeared in 1955 and lasted through until 1959. They had a rigid frame, Amal GP carbs, and reverse cone megaphones.

Three TR6 variants were also available in the States: the TR6A with a special (less restrictive) low exhaust system and Smiths tachometer; the TR6B with high-level siamesed pipes and no tachometer;

1957 TR6 and TR5 Trophy
(Differing from 1956)

Horsepower
40 at 6,500 rpm (TR6)
Compression Ratio
8.0:1 (TR6)
Brakes
8x1⅛-inch front (TR6)
Wheels
WM2x19-inch front
Tires
3.25x19-inch front
Dry Weight
380 pounds (375 pounds TR5)
Color
Gray and black (Aztec Red and ivory US)
Engine and Frame Numbers
TR6 0944 to 011471; TR5 0944 to 011471

58

1956–1962

The T20 Tiger Cub received a new frame with swingarm rear suspension for 1957.

and the TR6C with the T110 low exhaust system and no tachometer. Colors for the United States were mostly two-tone, the orange-like Aztec Red and ivory. Success continued in America for the TR6, with Bud Ekins winning the 1957 Big Bear Run for the second successive year. Of the first 25 placed finishers, 20 were Triumph-mounted. The TR6 also took out the Catalina Grand Prix, with Bob Sandgren winning the open class, and Eddie Day won the Greenhorn Enduro. It was all terrific publicity, and sales of the TR6 boomed, particularly on the West Coast.

1957 T20 and T20C Tiger Cub 200

A competition T20C joined the T20 for 1957. Based on Ken Heanes' 1956 ISDT machine, mechanically the

1957 T20 and T20C Tiger Cub (Differing from 1956)

Rear Suspension
Swingarm
Wheels
WM1x19-inch front and WM2x18-inch rear (T20C)
Tires
3.00x19-inch front and 3.50x18-inch rear (T20C)
Color
Crystal Gray and black
Dry Weight
215 pounds (T20); 205 pounds (T20C)
Engine and Frame Numbers
T20 26276 to 35846; T20C 26276 to 35846

T20C was identical to the T20, but with longer forks and 19- and 18-inch wheels shod with either road or trials tires. The exhaust system was high-rise, the gearing lower, and the gas tank slightly smaller in capacity (changing from 3.5 to 3.1 gallons). Both the T20 and T20C also received a new cylinder head, now with finning on the inlet rocker box, Vandervell VP3 big-end bearings, and new drive-side inner and outer covers. The new frame featured swingarm rear suspension with a pair of Girling shock absorbers, and instead of the earlier grease-filled fork, the front fork was an improved oil-damped type. Not only was the Tiger Cub significantly improved as a street bike this year, but its racing success also continued. In the 1957 Big Bear Run, Kenny Harriman won the 200cc class, and Cubs and Terriers dominated the Catalina Grand Prix, with Ralph Adams winning the 200cc class.

1958

With the effects of the Suez Crisis still evident in Britain, sales of motorcycles continued to decline in favor of scooters. While Triumphs were still selling well in America, Turner's main preoccupation now was the design of a new scooter, the BSA-built Tigress, reworking the Daimler car range by creating two V-8 engines (with cylinder heads based on the Triumph Thunderbird) and updating the current Triumph twins by introducing unit construction. Turner saw unit construction as a means of making the engine lighter and more compact and introducing partial enclosure paneling to protect the rider from oil and road grime. Above all, Turner was a stylist, and his idea was to

59

CHAPTER THREE

create a new style, quieter and colorful, countering the negative image of motorcycles. It was another of Turner's pioneering concepts, and first of these new twins was the Twenty-One. This time Turner wasn't in charge of the project; instead he appointed ex-Rolls Royce engineer Charles Granfield as chief engineer at Triumph Engineering. The new styling was dubbed the "Bathtub" and influenced other British manufacturers to follow the same path. Ultimately it was a disaster: the attempt to merge scooter style with a motorcycle proved unpopular with traditional motorcyclists, particularly in Triumph's main market, the United States. The other dubious new feature for 1958 was the introduction of the "Slickshift" automatic clutch on the 650s and 500s. This allowed gear changing with the gear pedal alone, but it was disconcerting and unpopular, and many owners simply disconnected it. All models this year had an optional Neiman steering head lock and a new oil seal to eliminate annoying oil leaks at the mainshaft and kick-start shaft.

1958 T110 Tiger 650 and T100 Tiger 500

This year the T110 Tiger received a modified aluminum Delta cylinder head, with a smaller combustion chamber and smaller-diameter valves. The inlet was reduced from 1$\frac{19}{32}$ inches to 1½ inches and the exhaust from 1$\frac{11}{32}$ inches to 1$\frac{7}{16}$ inches. While this did alleviate the propensity for cracking between the valve seats and cylinder head studs, it diminished performance. Also for 1958, the T110 was offered with an optional twin-carburetor splayed Delta head, similar to the 1957 T100 Tiger option. This provided the T110 the performance it needed to finally win the important Thruxton 500 production race, where a young Mike Hailwood teamed with Dan Shorey to win the race outright, ahead of Bob McIntyre on a Royal Enfield 700.

Both the T110 and T100 Tiger received a full-width, 8-inch front brake, now without an air scoop and vent, and smaller-diameter (1½-inch instead of 1¾-inch) exhaust pipes with new "straight-through" unbaffled silencers. Along with the Slickshift gearbox, the other feature introduced this year was more deeply valanced fenders, the front fender without front stays.

Mike Hailwood on his way to victory in the 1958 Thruxton 500 production race. This was the T110 Tiger's first victory in this significant race.

The 1958 T100 Tiger shared the more deeply valanced fenders with other models.

1958 T110 and T100 Tiger
(Differing from 1957)

Carburetor
Twin Amal 376 1⅛-inch (optional T110)
Color
Black/ivory (optional UK, standard US)
Engine and Frame Numbers
T110 011115 to 020075;
T100 011115 to 020075

1958 6T Thunderbird 650 and 5T Speed Twin 500

As Britons became more affluent, this was the final year for the 6T's miserly SU carburetor. US versions continued with the Amal Monobloc. This was also the last year of the pre-unit 5T. Like other models, both the 6T and 5T received the valanced fenders and Slickshift gearbox.

1958 TR6 Trophy 650 (Trophybird) and TR5 Trophy 500

Along with the suspect Slickshift, the TR6 received the updated T110 engine, with smaller combustion chamber and valves, and the T110's full-width, 8-inch front brake. Only the roadster "Trophybird" TR6A and Scrambler TR6B were offered in the United States, the TR6A with a low exhaust system, rubber-mounted speedometer and tachometer, and quickly detachable headlamp and DC lighting. The TR5 (TR5B in

1958 6T Thunderbird and 5T Speed Twin
(Differing from 1957)

Color
Aztec Red/black (6T US)
Engine and Frame Numbers
6T 011284 to 019824; 5T 011116 to 020074

The UK 6T retained the SU carburetor for 1958.

1958 TR6 and TR5 Trophy
(Differing from 1957)

Colors
Ivory and Aztec Red (US)
Engine and Frame Numbers
TR6 011861 to 019244;
TR5 011861 to 019244

the United States) was ostensibly unchanged for 1958 (apart from the usual Slickshift and so on) but a new, high-performance TR5AD (*D* for "Delta" head) was also available by special order. This came standard with the twin-carb splayed-port head and was the most expensive model in the range. It was also a harbinger of things to come.

When it came to competition, 1958 brought more of the same for the TR6, with Roger White leading another top five for Triumph in the Big Bear Run. Eight of the top ten finishers were Triumph-mounted. The TR6 again took the honors in the final 1950s Catalina Grand Prix. Bud Ekins led initially on his TR5 but retired with a flat tire, and Bob Sandgren eventually took his second successive victory on a TR6.

1958 T21 Twenty-One 350

An underwhelming reception greeted the delayed Twenty-One when it finally appeared a year later than expected. Although the T21 was Edward Turner's idea of providing reliable and affordable transport to a wider range of riders, his arrogance in assuming what was best for the market dictated a style that was out of touch. It wasn't only in the United States that the Bathtub styling met with resistance; in the United Kingdom, younger riders were becoming the primary market, and they wanted leaner-looking motorcycles.

Styling aside, the T21 heralded a number of significant new features for Triumph twins. The engine featured unit construction but still followed the usual layout. The forged crankshaft included a bolt-on flywheel and H-section con-rods, and as on the 650 and 500, two gear-driven camshafts operated the pushrods. The cylinders were cast iron, painted silver to match the aluminum head, and ignition was by a Lucas distributor located behind the cylinders. The single downtube frame was derived from that of the Tiger Cub, and the quest for a low seat height saw the T21 rolling on a pair of small 17-inch wheels. While the reviled Bathtub styling was striking, it did include some interesting innovations. Raising the hinged seat provided easy access to the toolkit, coil, rectifier, battery, and oil tank cap. While reliability problems were few, the T21 wasn't a great-handling motorcycle. The Tiger Cub–based frame was inadequate, especially when combined with the additional weight of the Bathtub pressed-steel bodywork.

1958 T20, T20C and CA Tiger Cub, T20J Junior Cub 200

The introduction of the T21 and Slickshift transmission reflected Triumph's desire to widen motorcycling's

The 1958 TR6 Trophy, TR6/B in the United States, with Slickshift and a full-width front brake with fluted chrome brake drum cover.

1958 Twenty-One (T21)

Bore	58.25mm
Stroke	65.5mm
Capacity	348cc
Horsepower	18.5 at 6,500 rpm
Compression Ratio	7.5:1
Carburetor	Amal 375/32 $^{25}/_{32}$-inch
Gearbox	Four-speed
Front Fork	Telescopic
Rear Suspension	Swingarm
Brakes	7-inch front and rear
Wheels	WM2x17-inch front and rear
Tires	3.25x17-inch front, 3.50x17-inch rear
Wheelbase	51.75 inches
Dry Weight	345 pounds
Color	Blue Sheen and black (Azure Blue US)
Engine and Frame Numbers	3TA H3 101 to H3 760

CHAPTER THREE

New Design!

Completely new
THE STREAMLINED TWENTY-ONE

Here's the newest addition to the family of Triumphs. Has a wonderful new Triumph engine of 21 cu. in. 350 c.c. Representing over 50 years of design leadership by Triumph, it gives turbine-like smoothness, extreme ease of starting, silent performance.

Enclosed streamlined rear sets new standard of cleanliness for both rider and passenger. Full width hub front brake, swinging arm hydraulic suspension — 100 new features. In Azure Blue with black frame.

Years ahead in design

ABOVE: The Twenty-One heralded a new range of Triumph twins, but the "Bathtub" styling proved unpopular.

RIGHT: New features for the 1958 Tiger Cub included one-piece gas tank badges and a larger rear fender.

appeal, and this continued with the T20J Junior Cub offered in the United States. Some states allowed 14- to 16-year-old riders to operate machines with less than 5 horsepower, and the T20J included an intake restrictor to reduce the power output. It seemed like a good idea, but only five T20Js were produced for 1958.

Updates to the 1958 Tiger Cub (from number 39167) included a French-made Zenith carburetor. This provided improved acceleration and torque, with a marginally lower top speed. The primary drive chain was upgraded to a ⅜ x ⁷⁄₃₂-inch Duplex with a cast-iron clutch housing and inner and outer primary covers. This finally overcame the primary drive and clutch problems that had plagued the Tiger Cub since its inception. Also during 1958, the gas tank badge became a one-piece known as the "Mouth Organ," and the rear fender was more valanced. A high-compression T20CA was also available in 1958, and Meriden offered a selection of high-performance parts. These included a larger inlet valve and "R" camshaft. Tiger Cubs continued their success as Don Hawley won ahead of Ralph Adams and Tiger Cubs took the first seven places in the 200cc class of the Catalina Grand Prix. Other Tiger Cub entries were Hollywood stars Lee Marvin and Keenan Wynn, but both retired with mechanical failure.

1958 T20, T20C and CA Tiger Cub, T20J Junior Cub *(Differing from 1957)*

Compression Ratio
9:1 (T20CA)
Horsepower
4.95 at 5,700 rpm (T20J)
Carburetor
Zenith 17 MXZ-CS5; Amal 332 (T20J)
Color
Aztec Red/black (US)
Engine and Frame Numbers
T20 35847 to 45311; T20C 35847 to 45311; T20CA 35847 to 45311; T20J from 36365

1959

Edward Turner may have been more interested in scooters, Bathtub styling, the new, unitized C range, and Daimler V-8 engines, but despite these predilections, Triumph was still the most popular imported motorcycle in America in 1958. Americans didn't really care for the new styling or scooters: what they really wanted was a twin-carburetor 650, and eventually Turner bowed to pressure from both TriCor and JoMo. One of Turner's concerns was that higher performance would stretch the already fragile T110, and he was aware of the need to strengthen the 650's three-piece crank assembly. In March 1958, Meriden tested a T110 with splayed-port cylinder head, twin carburetors, E3134 cams, 8.5:1 compression ratio, and forged one-piece crankshaft with bolted-on flywheel. Testing was successful, the power climbing to 48.8 horsepower, and although Turner initially believed it would lead Triumph

1956–1962

The biggest news for the 1959 season was the late inclusion of the Bonneville, ostensibly a twin-carb Tiger T110. This is a single-carb 1959 Tiger T110 painted in unusual Tangerine and gray Bonneville colors of that year.

to bankruptcy, by August 1958 he had finally agreed to give the project production clearance. In a meeting with TriCor's Denis McCormack and JoMo's Bill Johnson and Wilbur Cedar, it was agreed to call the new machine the "Bonneville." As it was primarily destined for America, and Johnny Allen's performances at Bonneville were still reasonably fresh, the name was enthusiastically received. Not only was the naming of the Bonneville inspirational, the model would become the most significant motorcycle of all time.

This year Triumph divided the lineup into three types: all 650cc machines were known as the B range, the unit-construction 500 and 350 the C range, and the Tiger Cubs the A range. Also new this year was the ill-fated Tigress scooter, designed to take on the popular Vespa and Lambretta. This expensive, BSA-built, 250cc fan-cooled overhead-valve four-stroke twin suffered from poor serviceability. Triumph dealers were reluctant to stock them, so sales were slow. With the Tigress, Triumph's 1959 US range now included 16 models.

CHAPTER THREE

1959 T120 Bonneville

Bore	71mm
Stroke	82mm
Capacity	649cc
Horsepower	46 at 6,500 rpm
Compression Ratio	8.5:1
Carburetors	Twin Amal 376/204 Monobloc 1 1/16-inch and 14/617 float bowl
Ignition	Lucas Magneto K2F with manual advance (auto-advance during the year)
Gearbox	Four-speed
Front Fork	Telescopic
Rear Suspension	Swingarm
Brakes	8-inch front and 7-inch rear
Wheels	WM2x19-inch front and rear
Tires	3.25x19-inch front and 3.50x19-inch rear
Wheelbase	55.75 inches
Dry Weight	404 pounds
Color	Tangerine and gray (Royal Blue later in the year)
Engine and Frame Numbers	T120 020076 to 029633

1959 T110, T100 Tiger and 6T Thunderbird
(Differing from 1958)

Carburetor	Amal Monobloc 376/210 1 1/16-inch (6T)
Rear Wheel	WM3x18-inch (US T100)
Rear Tire	4.00x18-inch (US T100)
Engine and Frame Numbers	T110 020076 to 029633; T100 020076 to 029633; 6T 020581 to 029362

ABOVE: The Bonneville was too late to be included in Meriden's 1959 catalog, but a single-page sheet appeared later.

Johnny Allen also returned to Bonneville in 1959. With JoMo support he endeavored to break his 214-mile-per-hour record, but, at 220 miles per hour and accelerating, the parachute ropes locked the rear wheel. Allen held it for 500 yards before it rolled and flipped for a further half mile. He walked away with a few broken ribs, and the earlier record stood until 1962.

1959 T120 Bonneville 650

As it was ostensibly a modified T110 Tiger, in some respects the early Bonneville was a disappointment. The engine did include a number of significant updates, noticeably the forged one-piece EN16B crankshaft with bolted flywheel, new pistons, and the new twin-carburetor cylinder head. While the "120" designation might have been illusory, the Bonneville was still one of the fastest motorcycles available. But there was no disguising its T110 roots and unfortunate colors, personally chosen by Turner—another example of his distance from the reality of what buyers really wanted. The early Bonneville also suffered from Turner's obsessional cost cutting: the single downtube brazed steel frame and 8-inch single leading shoe front brake were not really up to the task and stretched to the limit. The spindly front fork and unbraced swingarm also did little to inspire confidence at the speeds of which the T120 was capable. Vibration from the more highly-tuned engine also played havoc with the carburetion and marginal Lucas electrics. Fortunately the Bonneville was spared the dreadful Slickshift clutch operation, but the T110's headlight nacelle, high handlebars, large dual seat, and deeply valanced fenders simply didn't strike a chord with buyers. In America dealers were already selling splayed-port twin-carb head kits for the T110 and TR6 Trophy, most ending up on TR6s. As a result many Tangerine 1959 Bonnevilles in the United States ended up unsold and were listed again for 1960. Despite their unpopularity at the time, today the 1959 Tangerine models are among the most sought after of all Bonnevilles.

It didn't take long for Meriden to realize the disaster of the initial color scheme, and before the model year was out they changed the color scheme on UK and general export machines. Royal Blue replaced Tangerine, but the rest of the machine was unchanged. These colors were much more attractive and would continue for 1960.

1959 T110 Tiger 650, T100 Tiger 500 and 6T Thunderbird 650

This year the T110 Tiger was no longer Triumph's leading sport model, but it did benefit from some of the updates introduced on the T120 Bonneville, notably the forged one-piece crankshaft. Teardrop silencers replaced the straight-through type, and during the

The 1959 Tiger T110's heavy-handed styling and Bonneville Tangerine and gray colors weren't particularly popular.

CHAPTER THREE

RIGHT: The Tiger T110 retained a single Amal Monobloc carburetor.

FAR RIGHT: Inside the 1959 engine was a stronger, one-piece, forged crankshaft.

1959 was the final year for the pre-unit 500cc T100 Tiger.

The UK 6T Thunderbird no longer had an SU carburetor.

The Aztec Red and ivory tank colors were reversed for the 1959 TR6 Trophy.

The US 1959 TR6A was much as before, and when fitted with the splayed port head and twin carbs it was more popular than the Bonneville.

66

1956–1962

Powering the uninspiring "Bathtub" 5TA Speed Twin was a short-stroke engine that many believed to one of Triumph's finest. The 5TA was also much lighter and more compact than the earlier Speed Twin.

1959 TR6 and TR5 Trophy
(Differing from 1958)

Colors
Aztec Red and Ivory

Engine and Frame Numbers
TR6 020883 to 029688; TR5 022269 to 022363 and 025724 to 025729

year the Lucas magneto's manual advance retard was changed to an automatic type. In most respects the new Bonneville overshadowed the T110, which now struggled for identity in the lineup. This year also saw the end of the pre-unit T100 Tiger, this also adopting the one-piece crankshaft and baffled silencers. On US export models the fenders were a narrower sports type, and the rear wheel changed from 19 to 18 inches. The 6T Thunderbird also gained the one-piece crank this year, and all models now had an Amal carburetor.

1959 TR6 Trophy 650 (Trophybird) and TR5 Trophy 500

As with the T100 Tiger, this was to be the final season for the TR5, with only two small batches totaling 111 produced. Some of these were fitted with the twin-carb splayed head, racing E3134 camshafts, 9:1 pistons, and a Lucas K2FR racing magneto and were particularly desirable. All Trophys shared the new one-piece crankshaft this year, and most were TR6s. In the United States the TR6A and TR6B Trophybirds were again offered; this year the TR6B Scrambler model came with two individual 1⅜-inch upswept exhaust pipes and mufflers, one on each side. Off-road success continued with Bud Ekins again winning the 153-mile off-road Big Bear Run. This was an amazing victory; Ekins finished over half an hour ahead of second place, despite suffering a puncture and breaking a wheel. Other victories included Buck Smith's win in the Greenhorn Enduro.

1959 5TA Speed Twin
(3TA Twenty-One differing from 1953)

Bore	69mm
Stroke	65.5mm
Capacity	490cc
Horsepower	27 at 6,500 rpm
Compression Ratio	7.0:1
Carburetor	Amal 375/75 ⅞-inch
Gearbox	Four-speed
Front Fork	Telescopic
Rear Suspension	Swingarm
Brakes	7-inch front and rear
Wheels	WM2x17-inch front and rear
Tires	3.25x17-inch front, 3.50x17-inch rear
Wheelbase	51.75 inches
Dry Weight	350 pounds
Color	Amaranth Red
Engine and Frame Numbers	5TA H5 785 to H11032; 3TA H3 761 to H11511

67

CHAPTER THREE

1959 T20, T20C, CA and T20S Tiger Cub, T20J Junior Club (Differing from 1958)

Compression Ratio	9:1 (T20S)
Carburetor	Amal 332 7/8-inch (T20S); Zenith 9.5mm MXZ (T20J)
Wheels	WM1x19-inch front and WM2x18-inch rear (T20S)
Tires	3.00x19-inch front and 3.50x18-inch rear (T20S)
Dry Weight	210 pounds (T20S)
Color	Ivory and Azure Blue (T20S)
Engine and Frame Numbers	T20 45312 to 56359; T20C 45312 to 56359; T20CA 45312 to 56359; T20S 45312 to 56359; T20J 45312 to 56359

The T20 Tiger Cub received the unloved Bathtub rear section for 1959.

Celebrating 50 years of the Scottish Six Day Trials, Tiger Cubs finished first, fourth, and seventh. This is Arthur Ratcliffe, who finished seventh, while ultimate winner Roy Peplow, on the right, watches.

1959 5TA Speed Twin 500 and 3TA Twenty-One 350

The T21 became the 3TA, and replacing the pre-unit 5T Speed Twin this year was the Twenty-One–based 5TA. These two models comprised Triumph's C range. As it was based on the 350, the new 5TA was significantly smaller and lighter than its pre-unit predecessor, and apart from the usual Amaranth Red color, the styling and Bathtub bodywork of both the 5TA and 3TA were identical. The short-stroke 500 engine soon proved extremely reliable and very appealing to racers. In May 1959 dealers received a bulletin describing the process involved in converting the 5TA to Class C road or dirt track specification, and the first win for a unit 500 was in the 1959 Ascot National half mile. Sammy Tanner won on a JoMo-prepared 5TA. TriCor also prepared the 5TA, and Everett Brashear was the second-fastest qualifier for the 1959 Daytona 200. His speed was over 120 miles per hour, but he failed to finish the race after the 350-based gearbox broke. This prompted the factory to improve the gearbox for 1960.

1959 T20, T20C, CA and T20S Tiger Cub, T20J Junior Cub 200

The basic T20 received a significant styling update this year, with the rear now partially enclosed like that of the TA and 5TA. The "Oval" cylinder barrel included more substantial finning, and this year the restricted T20J received a Zenith carburetor. A few more T20Js were built (166), but it wasn't really successful. The T20C and CA were ostensibly unchanged, and new this year was a high-performance T20S with a larger inlet valve, sports "R" camshaft, an Amal Monobloc carburetor, and close ratio gearbox. An optional wider ratio gearbox was also available for trials work. Other updates included the stronger front fork of the C-range twins and a temperamental Energy Transfer ignition system without a battery. While most Tiger Cubs produced this year were T20s, they were not as popular in the United States as the sporty T20S. Despite its modest capacity, the Tiger Cub still had a following. At Bonneville in August 1959, Californian dealer Bill Martin set an AMA 200cc Class C one-mile record of 139.82 miles per hour on a gas-fueled streamliner.

1960

After a boom year, Britain experienced an unexpected motorcycle sales slump in 1960, forcing Triumph to place even more emphasis on the American market. Still intense rivals, the two distributors, TriCor and JoMo, began to request machines built to individual specification, including different seats, exhaust systems, engine tuning, gearing, fenders, and color schemes. Unfortunately this stretched Meriden's production control, the quality began to wane, and 1960 was a terrible year for reliability and warranty claims.

The T120 Bonneville was significantly redesigned for 1960, but problems remained. This UK model doesn't have the optional tachometer that was standard in the United States.

A new threat also loomed for the British industry, that of Japan. Edward Turner was well aware of the growth of the Japanese motorcycle industry—Honda's exploits at the Isle of Man were widely publicized—and during 1960 he made a trip to Japan to observe their industry firsthand. Turner's plan was to gather information so he could plan countermeasures to preserve his share of motorcycle world markets, but he returned to England stunned and shocked. He had found extremely modern factories producing high-quality products, and Turner was particularly impressed by the speed at which Japanese technicians could produce, test, and develop new designs. He subsequently wrote in a report, "I have been giving considerable thought to what we might do to combat this situation, and I must confess these answers are going to be hard to find." At this stage Japan's production was concentrated on motorcycles below 300cc, and Turner went on to say "the machines themselves are more comprehensive than our own in regard to equipment, such as electric starting, and are probably better made, but will not appeal to the sporting rider to anything like the same extent as our own." Turner must have known the Japanese would eventually encroach on Triumph's sport territory but chose to ignore it. Profits were strong, and he was still managing director, but at 60 years of age, with diabetes, Turner appeared to have lost the drive to take Triumph into a new, more competitive, era. In the meantime it was business as usual, with annual updates to existing models seen as enough to quell any foreign threat.

1960 T120 Bonneville 650, TR7A, and TR7B

The Bonneville was significantly updated for 1960, not only mechanically, but also stylistically. Gone was the unloved Tangerine color, replaced by the blue of late 1959, but also eliminated were the headlight nacelle and deep fenders. With its separate, quickly detachable headlight and instruments, plus TR6-style alloy blade fenders, the new Bonneville was lighter and looked far more sporty. A new Duplex frame provided a noticeably shallower steering head angle (27 degrees from 24.5 degrees) to improve agility and reduced the overall wheelbase by more than an inch. Unfortunately, the brazed-lug frame with bolted rear subframe wasn't strong enough and was prone to fracturing around the headstock. Vibration also caused fuel tank straps to break. When Turner saw a rider thrown and killed in the December 1959 Big Bear Run after the frame cracked and broke, he instigated an on-track development

CHAPTER THREE

The US-only TR7B was a high-exhaust scrambler version of the T120 Bonneville.

session. This resulted in an additional frame crossbar added midseason (from D1563), curing the cracking and contributing to an overall improved machine.

The Bonneville might have looked and handled better than its predecessor, but a dubious feature was the adoption of alternator electrics. Always built to a price, the crankshaft-mounted Lucas RM12 alternator lacked proper voltage regulation, and batteries sometimes boiled over. Ignition remained by Lucas magneto, and as fuel surging was still a problem, examples after D5975 were fitted with a pair of Amal Monobloc carburetors with individual float chambers.

Other updates included a new front fork with external rubber gaiters, similar to the C-range twins. This year UK- and US-specification machines diverged considerably, British examples with a larger, 4-gallon gas tank and low handlebars. To distinguish it from leftover 1959 models, in the United States this year the T120 was sold as the TR7A and TR7B. Both had a smaller, 3-gallon tank, higher handlebars, a standard tachometer, and an 18-inch rear wheel as on the Trophy. The TR7A came with low exhaust pipes and the TR7B with Trophy high exhausts and Trials Universal tires. The T120 more than lived up to its name, with Gary Richards clocking 149.51 miles per hour at Bonneville to set a new record for an unfaired motorcycle. He followed this in 1961 with a run of 159.54 miles per hour.

1960 T110 Tiger 650 and 6T Thunderbird 650

Both the T110 Tiger and 6T Thunderbird shared the new Duplex frame and front fork of the T120 Bonneville, but the style deviated considerably from the stripped-look T120. JoMo now described the T110 as a "Road Cruiser Streamliner." Both the T110 and 6T now featured the C-range Bathtub rear enclosure, a large, deeply flared and valanced front fender, and the usual headlight nacelle. Both also received smaller, 18-inch wheels this year, the Thunderbird retaining the 7-inch front brake and featuring a new color, Charcoal Gray. Engine-wise, the T110 Tiger now included the T120's Lucas alternator and also retained magneto ignition. The 6T engine was largely unchanged, and this would be the final year for the venerable cast-iron cylinder head.

1960 TR6A and TR6B Trophy 650 (Trophybird)

As the TR6 was still based on the T110 Tiger, the frame was now the Duplex type, the front fork was new, and the engine included the crankshaft-mounted alternator. The TR6 retained magneto ignition and continued as two variants, the road-going TR6A, ostensibly a single-carburetor TR7A Bonneville, and Scrambler TR6B (single-carb TR7B). The TR6A also received a standard tachometer this year.

1960 T120 Bonneville, TR7A, TR7B
(Differing from 1959)

Carburetors	Twin Amal 376/233 Monobloc 1 1/16-inch (from D5975)
Rear Wheel	WM3x18-inch rear (TR7A, TR7B)
Tire	4.00x18-inch rear (TR7A, TR7B)
Wheelbase	54.5 inches
Dry Weight	393 pounds
Color	Royal Blue and gray
Engine and Frame Numbers	T120 029634 to 030424; then D101–D7726 (D for Duplex)

1960 T110 Tiger and 6T Thunderbird
(Differing from 1959)

Carburetor
Amal 376/244 1 1/16-inch (T110);
Amal 376/246 1 1/16 inch (6T)
Wheels
WM2x18-inch front and rear
Tires
3.25x18-inch front and 3.50x18-inch rear
Wheelbase
54.5 inches
Dry Weight
392 pounds (6T)
Color
Charcoal Gray (6T)
Engine and Frame Numbers
T110 029634 to 030424, then D101–D7726;
6T D507 to D7711

1956–1962

A brand-new 1960 TR6A Trophy, still in the crate nearly 55 years later. This stripped-down model was ostensibly a single-carb Bonneville and unquestionably a superb motorcycle.

1960 TR6A and TR6B Trophy (Differing from 1959)

Wheelbase
54.5 inches
Dry Weight
393 pounds
Engine and Frame Numbers
TR6 029364 to 029688, then D904 to D6500

1960 T100A Tiger 100, 5TA Speed Twin, and 3TA Twenty-One (Differing from 1959)

Horsepower
32 at 7,000 rpm (T100A)
Compression Ratio
9.0:1 (T100A)
Carburetor
Amal 375/35 ⅞-inch (T100A and T5A)
Dry Weight
363 pounds (T100A)
Color
Black and ivory (T100A); Ruby Red (5T)
Engine and Frame Numbers
T100A H11512 to H18611; 5TA H11962 to H18626, 3TA H11512 to H18611

1960 T100A Tiger 100, 5TA Speed Twin and 3TA Twenty-One

The demise of the pre-unit 500 saw the T100 resurrected as the unit T100A. This higher-performance, short-stroke 500 had not only higher compression pistons, but also E3325 camshafts. While retaining the Lucas distributor and alternator, the T100A also

The 1960 TR6C had a left-side siamesed high-rise exhaust system.

71

CHAPTER THREE

New for 1960 was the first performance unit twin, the 5TA-based Tiger 100. A confused mixture of touring styling with a performance engine, it lasted only a year in this form.

offered an unusual Energy Transfer ignition that allowed running without a battery or lights. Considering that the styling was the full Bathtub arrangement of the 5TA, this was a strange feature. The temperamental ET ignition and contentious styling resulted in a flawed machine, and the T100A not really a true sport model in the manner of the previous T100. This year the T100A and 5TA also boasted a primary chain tensioner, another addition that proved unsatisfactory as it required frequent adjustment, which necessitated draining and renewing the chaincase oil each time. After 22 years in Amaranth Red, the Speed Twin was now a bright Ruby Red.

1960 T20 Tiger Cub 200, T20S Scrambler 200, T20W Woods 200, T20J Junior Cub 200

This year there was no longer a competition T20C and T20CA: the T20S Scrambler was now available in three different guises, and the range expanded to include the T20W Woods. Intended for trials and cross-country, this had a low-compression motor and high exhaust system. A number of updates were incorporated over the entire range this year, notably a new cylinder head with a larger inlet valve, a two-piece crankcase, and, on the T20 and T20W, a larger Zenith carburetor. All models received the longer T20S shock absorbers, and the T20S was available as a standard road specification,

The T20S was available in three versions for 1960. This is the Scrambler/Racing specification model with a remote float-bowl Amal carburetor.

1960 T20, T20S, T20W Tiger Club, T20J Junior Club (Differing from 1959)

Carburetor
Zenith 18mm MXZ (T20, T20W)
Engine and Frame Numbers
T20 56360 to 69516; T20S 56360 to 69516; T20J 56360 to 69516

low-compression Trials spec, and Scrambler Racing. These had a variety of different carburetors, from Amal Monobloc and Zenith to an Amal with remote float bowl. All T20Ss had the Energy Transfer ignition system, "R" camshaft and high-level exhaust system. As it could be specified with a range of options for different duties, the T20S was extremely popular in the United States, comprehensively outselling the standard T20. The year 1960 would be the peak for the Tiger Cub, with just over 13,000 sold. But the writing was on the wall. In June 1960, a 305cc Honda CB77 with electric start cost $595, less than a T20S Sports Tiger Cub.

1961

No other year was more pivotal for Triumph Engineering's future than 1961. Jack Sangster stepped down as chairman of the BSA Group, appointing Eric Turner (no relation to Edward) as his successor. An accountant, Eric Turner sought to streamline the BSA and Triumph operations, including distribution and product lines. Charles Granfield retired as chief designer, and Edward Turner persuaded Bert Hopwood to leave Norton and return to Meriden as director and general manager. Both appointments would shape Triumph's next decade, with Eric Turner making some dubious managerial decisions and Bert Hopwood being instrumental in the design of the three-cylinder Trident. The economic climate in Britain was also especially bleak, and motorcycle sales plummeted in the wake of increased taxes, as well as antipathy and negative connotations caused by Rockers and their propensity for street racing. With the home market dying, Triumph became even more US-oriented, with several models now built only for America.

1961 T120R and T120C Bonneville 650

With new, attractive blue and silver colors, the T120R ("Road") this year was arguably the quintessential pre-unit Triumph twin. The T120C ("Competition") was a US-market version with high exhaust pipes. Engine updates were minor, the cylinder head casting on all 650s including bracing to reduce cooling fin ringing, while inside the gearbox needle roller bearings replaced the bronze bushes on the layshaft. The frame was still the modified Duplex type of late 1960, but

1956–1962

the steering head angle was reduced to 25 degrees to provide a better compromise between road and off-road steering. To alleviate the fuel tank strap breakage problems the tank was strengthened at the nose and now mounted on three rubber blocks. The smaller gas tank was also standardized for all markets. Another improvement was to the brakes, both front and rear now with fully floating shoes.

The T120R also had the performance to match its stunning looks. In June 1961, *Motor Cycling* managed a mean top speed of 116.9 miles per hour. This was the fastest road machine they had tested. The T120R's performance was also vindicated in the Thruxton 500 race for production motorcycles. This year Tony Godfrey and John Holder won at 67.29 miles per hour, with Bonnevilles also finishing second, fourth, fifth, and sixth.

ABOVE: By 1961 the Bonneville had evolved into a beautiful and highly efficient motorcycle.

LEFT: 1961 Bonnevilles had a pair of Amal Monobloc carburetors without air filters.

73

CHAPTER THREE

1961 T110 Tiger 650 and 6T Thunderbird 650

In the wake of the decline in the British market, this was the final year for the once-mighty T110 Tiger, and antipathy toward the nacelle and Bathtub styling saw it no longer represented in the US lineup. The T110 engine was largely unchanged and the chassis still shared with the 6T. As the 6T Thunderbird received an aluminum cylinder head this year, there was now little separating the two apart from new colors: attractive Kingfisher Blue and silver for the T110.

In addition to the aluminum cylinder head, the 6T Thunderbird received a slightly higher compression ratio and, initially, sporty E3325 camshafts. These were later changed to a milder "Dowson" cam, but in neither form did the claimed horsepower rise. This was probably to protect sales of the more expensive T110, and some at the factory believed the 6T's power closer to 37 horsepower this year. Both the T110 and 6T received the new, strengthened Duplex frame with steeper steering head angle, and the 6T also shared the other 650 twins' full-width, 8-inch single leading shoe front brake.

1961 T110 Tiger and 6T Thunderbird (Differing from 1960)

Compression Ratio
7.5:1 (6T)
Carburetor
Amal 376/255 (6T)
Wheelbase
54.75 inches
Color
Kingfisher Blue and silver (T110); black and silver (6T)
Engine and Frame Numbers
T110 D7727 to D15788; 6T D9259 to D15754

1961 TR6A and TR6B Trophy (Differing from 1960)

Compression Ratio
8.5:1
Horsepower
45 at 6,500 rpm (with a straight-through exhaust system)
Wheelbase
55.25 inches
Dry Weight
383 pounds
Color
Ruby Red and silver
Engine and Frame Numbers
TR6, TR6R, TR6C, TR6SR, TR6SC; D8432 to D15756

The 1961 T120R looked good from any angle.

1961 T120R Bonneville (Differing from 1960)

Carburetors
Twin Amal 376/257 Monobloc 1 1/16-inch
Horsepower
50 at 6,500 rpm (with a straight-through exhaust system)
Wheelbase
55.25 inches
Color
Sky Blue and Silver Sheen
Engine and Frame Numbers
T120R D7727 to D15788

1961 TR6R (SR) and TR6C (SC) Trophy 650

The TR6 wasn't listed in the United Kingdom this season but continued in the United States in two versions, the TR6R and high-pipe TR6C. Both shared the engine updates and frame of the T110 and T120R, and the compression ratio was increased to the Bonneville's 8.5:1. At some stage during the year an additional S appeared in the designation, but there was no change to the specification.

1961 "C" Range: T100A Tiger, 5TA Speed Twin, TR5AR, TR5AC, 3TA Twenty-One 350

The 5TA Speed Twin and 3TA Twenty-One continued ostensibly unchanged, and the T100A received a larger carburetor, hotter cams, and more power, and it no longer had the Energy Transfer ignition. The frame on all C models also had a slightly steeper steering head (to 27 degrees), and the front and rear brakes included the floating-shoe arrangement.

T20, T20S T20SL T20T Tiger Cub, T20J Junior Cub (Differing from 1960)

Horsepower
14.5 at 6,500 rpm (T20SL)
Compression Ratio
9:1 (T20SL)
Carburetor
Amal 375/317 $^{15}/_{16}$-inch (T20SL)
Colors
Silver Sheen and black (T20); Ruby Red and silver (T20SL and T20T)
Engine and Frame Numbers
T20 69517 to 81889; T20SL 64592 to 81889; T20T 69570 to 81889; T20J 69570 to 81889

ABOVE: While the sport models eschewed the traditional headlight nacelle, it continued on the Thunderbird and Tiger 110.

TOP: The Thunderbird received an aluminum cylinder head for 1961 and higher-compression pistons.

CHAPTER THREE

1961 T100A Tiger 100, 5TA Speed Twin, TR5AR, TR5AC, 3TA Twenty-One
(Differing from 1960)

Horsepower	34 at 7,000 rpm (T100A, TR5AR, TR5AC)
Carburetor	Amal 376/273 1-inch (T100A, TR5AR, TR5AC)
Wheels	WM2x19-inch front and WM3x18-inch rear (TR5AR, TR5AC)
Tires	3.25x19-inch front and 4.00x18-inch rear (TR5AR, TR5AC)
Wheelbase	53.5 inches
Dry Weight	323 pounds
Color	Kingfisher Blue and silver (T100A, TR5AR, TR5AC)
Engine and Frame Numbers	T100A H18612 to H25251; TR5AC H18635 to H23103 5TA H19215 to H24755; 3TA H18612 to H25251

Although it wasn't popular, particularly in the States, Triumph persevered with the Bathtub styling in 1961.

New for the States this year were two Tiger 500cc unit-construction models, the TR5AR and TR5AC. Both shared the higher-horsepower T100A engine, the TR5AR with twin low exhaust pipes and the TR5AC with a siamesed low-level system. The TR5AR gas tank was a stressed frame member, and the TR5AC, with a tiny 2¼-gallon tank, included a bolted-on tubular strengthening strut. In other respects they were smaller Trophys, with 19- and 18-inch wheels and similar sport styling. As the TR5AC was significantly lighter than the earlier (1957) pre-unit Trophy, it was very successful in competition, winning gold medals in European International Six Day Trials events and victories in several enduros in the United States. Both the TR5AR and TR5AC were only offered for 1961 before evolving into more specific versions.

1961 T20 Tiger Cub, T20S, T20SL Scrambler, T20T Trials, T20J Junior Cub 200

This year saw a small explosion in the 200cc A-range. The T20T Trials replaced the T20W Woods, and the T20SL Scrambler Light joined the T20S Scrambler. The standard T20 Tiger Cub and the few T20Js continued much as before, with the T20SL similar to the T20S. With an Amal Monobloc carburetor, 9:1 compression, "R" camshaft, and larger inlet valve, the power increased 45 percent over the T20. The front fork was a stronger 3TA type, and it came with distributor points, Energy Transfer ignition, and battery-less 6-volt DC lighting. When tested by Motor Cycling in April 1961, the T20SL managed a top speed of 77.8 miles per hour and was the fastest 200 they had tested. They summed it up as "truly difficult to fault." The T20SL came in two versions, a road version with a low exhaust and the scrambler with a high exhaust, and like the TR5AR and TR5AC it was only offered for 1961. Over the next few years the T20S would spawn several new models, often in Bonneville and Trophy colors, and in three subgroups: road, off-road, and all-around use. Despite the Japanese challenge, Tiger Cub sales remained strong this year at 12,339 sold, but it was a downhill slide from there.

1962

Eric Turner engaged international management consultants McKinsey and Company to study the streamlining of BSA and Triumph, and McKinsey suggested an amalgamation of the two companies. This led to increased tension between Meriden's union and management, something that would define the next decade and beyond. The first evidence of amalgamation was the transfer of Triumph's publicity department to BSA's Small Heath. Also during 1962, Edward Turner's ingenious Tina scooter was released and a new wing built at the factory to produce it. With a 100cc, single-cylinder, two-stroke engine and automatic transmission, the Tina was groundbreaking, but like many of Turner's designs it wasn't properly tested before entering production. Over the next few years Meriden spent a lot of time and effort sorting out the Tina's problems, and by the time they were fixed, with the T10 of 1965, the scooter boom was virtually over.

Also in 1962, JoMo founder Bill Johnson died of a heart attack. His partner Wilbur Cedar assumed control in the west, while TriCor's Denis McCormack retired, with Earl Miller taking over as general manager.

1962 T120, T120R, and T120C Bonneville 650

This was the final version of the pre-unit Bonneville, and there were only a few updates. To quell engine vibration and increase pulling power the engine received a heavier flywheel and 71 percent balance factor. During the year a new crankshaft with an even wider flywheel and 85 percent balance factor was introduced. Vibration was reduced, but at the expense of crisp throttle response. As on the 1961 T120R, Triumph finally had the frame and carburetors in order, but it was all to no avail, as an all-new unit T120 would appear for 1963. But the pre-unit T120 maintained a loyal following, particularly among the Triton and café-racer crowd, which preferred the earlier, charismatic 650.

The crowning glory for the pre-unit T120 was the 224.57-mile-per-hour motorcycle world speed record

The road version of the T20SL. This was another model made only for one season.

1962 T120, T120R, and T120C Bonneville
(Differing from 1961)

Color
Flame and Silver Sheen (US)
Engine and Frame Numbers
T120 D15789 to D20308

With its very small gas tank and siamesed right-side exhaust, the TR5AC only lasted one year but evolved into a new range of very successful models.

CHAPTER THREE

With a standard tachometer and higher handlebars, the 1962 American T120R differed from British versions. The front fender was without the "pedestrian slicer" number plate, and a two-tone seat distinguished 1962 versions.

set by Bill Johnson (no relation to JoMo's founder) at Bonneville on September 5, 1962. Owned and built by Joe Dudek, Johnson's streamliner was powered by a Pete Colman–built 667cc nitro methane-powered T120 engine. This time the record was fully recognized by the FIM.

1962 6T Thunderbird 650, TR6SS, TR6SR, TR6SC Trophy 650

The final pre-unit Thunderbird was produced in two versions, the UK model still with the earlier Bathtub rear enclosure and the US model without. Both retained the headlight nacelle and deeply valanced front fender. All 6Ts received the heavier flywheel and altered crank balance factor of the T120, and for this year only the 6T exhaust system was siamesed, exiting on the right. The unpopular Slickshift automatic clutch was finally dropped; the 6T was the last model to feature it.

The TR6 Trophy was reintroduced for the UK market this year and designated the TR6SS. This was very much a single-carb Bonneville, incorporating the Bonneville's engine updates and sporty street profile.

1962 6T Thunderbird; TR6SS, TR6SR, TR6SC Trophy (Differing from 1961)

Compression Ratio
7.5:1 (6T)
Carburetor
Amal 376/285 (6T)
Dry Weight
371 pounds (6T)
Engine and Frame Numbers
6T D16389 to D20088; TR6SS, TR6SR, TR6SC D16189 to D20308

For the United States two TR6s continued, the road TR6SR (very similar to the TR6SS) and the high-pipe TR6SC. After picking up a TR6 from Meriden, Bud Ekins won a gold medal in the 1962 ISDT at Garmisch-Partenkirchen in West Germany.

1962 C Range: T100SS Tiger, 5TA Speed Twin, T100SR, T100SC, 3TA Twenty-One 350

This year a new, sporty 500cc roadster, the T100SS, evolved from the TR5AR. Replacing the confused T100A Tiger, it no longer included the headlight nacelle and Bathtub rear enclosure and now featured abbreviated rear side panels, a smaller front fender, and front fork gaiters like the T120. The 34-horsepower engine was unchanged, but rolling on the TR5A's 19- and 18-inch wheels, and with a siamesed exhaust, the

In September 1962 Bill Johnson set a new motorcycle world speed record of 224 miles per hour on Joe Dudek's streamliner. The AMA also recognized a 230-mile-per-hour average over a measured mile.

Steve McQueen and *The Great Escape*

Actor Steve McQueen was an accomplished motorcycle rider, purchasing his first new bike, a TR5 Trophy, from Don Brown of Johnson Motors. Brown told him to get it serviced at Bob Ekins' dealership in North Hollywood, and Ekins and McQueen soon became great friends. McQueen successfully competed in a number of off-road events, including Hare and Hound Scrambles, and the Mint 400 and Baja 1000 desert races. For the World War II movie *The Great Escape*, McQueen requested a motorcycle chase scene, and Ekins supplied two 1961 TR6s to the film studio, Mirisch, during 1962. These were modified to look more German by painting them army green and adding a large front fender and pannier racks. The motorcycle chase scenes culminated in the jumping of the barbed wire fence and were shot outside Füssen in Bavaria. As McQueen wasn't allowed to perform the jump, Ekins undertook it. With a shortage of qualified motorcyclists available as extras, however, McQueen did ride a motorcycle in the chase sequence.

The Great Escape was released in July 1963 and was one of the highest-grossing films that year. It made Steve McQueen a superstar, and the motorcycle jump became legendary. When McQueen needed another stuntman for the 1968 film *Bullitt*, he turned to his mate Bud Ekins to drive the Mustang in the famous car chase through the streets of San Francisco.

Steve McQueen with the TR6 used for the jump in *The Great Escape*. The bike was a 1961 TR6 made to loosely look like a military BMW R12 or Zündapp KS750. As McQueen wasn't allowed to perform the jump in the movie, Bud Ekins was the stunt rider. *Triumph Motorcycles*

CHAPTER THREE

The US 6T Thunderbird for 1962 no longer included the rear Bathtub enclosure, but retained the headlight nacelle and deep front fender. The handlebars were much wider than UK-spec versions.

T100SS looked much sportier and more functional than the earlier T100A. The light T100SS was also sprightly: *Motor Cycling* managed a best top speed of 98 miles per hour late in 1961. The T100SS was a UK and general export model, with US versions being the T100SC Trophy and similar T100SR. These were ostensibly identical to the earlier TR5AC and TR5AR, while the 5TA Speed Twin and 3TA Twenty-One continued unchanged for the year.

After Don Burnett finished second in the 1961 Daytona 200 on a TriCor Cliff Guild–tuned Tiger, both US distributors offered complete 500cc racing engines for 1962. These were built at TriCor and JoMo and sold to dealers and privateers. Available as either "Ready to Race" or "Exchange," the engines were cataloged as the T100S/RR. Burnett returned to Daytona in 1962, again on the Guild-built TriCor Tiger, this time narrowly beating Dick Mann's Matchless G50 to take Triumph's first victory in this important event. This heralded the beginning of a highly successful racing career for the short-stroke 500.

1962 T20 Tiger Cub, T20SS Street Scrambler, T20SH Sports Home, T20SC, T20SR, TR20 Trials, TS20 Scrambler

Honda was now making serious inroads into the American market and in 1962 sold more motorcycles in the United States than the rest of the market combined. The Tiger Cub was particularly vulnerable, and this year Triumph tried to stem the tide by expanding the Tiger Cub range considerably. They also incorporated several updates, gradually introduced during the year, including new crankcases, with a ball bearing supporting the crankshaft on both the timing and drive sides, and a larger journal two-piece crankpin. The T20J, T20SL, and T20T were discontinued, and the T20SL evolved into the US T20SS (for JoMo only in 1962). The T20SS had an optional high-level exhaust and was an all-purpose road/sport model with the T20SL high-compression engine and 19- and 18-inch wheels. The home market T20SH was a pure road

But for a tachometer, higher handlebars, and a seat strap, the US TR6SR was similar to the TR6SS. Both featured a right-side siamesed exhaust system.

80

1956–1962

"An ideal fast light sports 500"
says JOHN GILES

This sleek new "Tiger 100" offers high performance with light weight and great ease of handling. Ultra modern unit construction engine/gearbox with alloy head, special camshafts and high compression pistons. A model designed for the sportsman.

—famous Triumph trials and scrambles star, winner of the 1961 Experts Grand National and countless events in U.K. and on the Continent. Gold Medal winner in 1961 International Six Days Trial riding a Tiger 100SS. Member of British Trophy Team.

The sporty 1962 T100SS evolved out of the TR5SR and was a much more successful effort than the earlier T100A.

TIGER 100 500 c.c. T100S/S
Patent Nos. 475860, 723073, 684685

1962 T100SS Tiger 100, 5TA Speed Twin, T100SR, T100SC, 3TA Twenty-One
(Differing from 1961)

Dry Weight
336 pounds (T100SS)

Engine and Frame Numbers
T100SS H25252 to H27932; T100SC H26772 to H28552; T100SR H25252 to H27932; 5TA H25904 to H29727; 3TA H25252 to H27932

model, also with the high-performance engine and 19- and 18-inch wheels.

Four other versions were also built in 1962, two almost exclusively for the East Coast. The T20SC ("Scrambler Competition") was the East Coast equivalent of the West Coast T20SS, and the T20SR ("Road") had the complete high-performance package of high-compression piston, larger inlet valve, larger carburetor, low exhaust, and road tires. Two specialist competition models were also created out of the T20SS. Described as "Works Replica" models, the TR20 was a trials machine, and the TS20 was for scrambles. But it was all a little late to save the aging Tiger Cub. Arguably outdated, its sales in 1962 were half the previous year's (6,412), and Triumph simply couldn't sell them at a competitive price and make a profit. The future for the Tiger Cub lay in specialization, the route Triumph would eventually take in the future.

1962 T20, T20SS, T20SC, T20SR, TR20, TS20 *(Differing from 1961)*

Carburetor	Amal 32/1 $^{11}/_{16}$-inch (T20); 376/272 $^{15}/_{16}$-inch (T20SS, T20SH, T20SC, T20SR)
Colors	Burgundy and Silver Sheen (T20SS, T20SH, T20SC, T20SR, TR20, TS20)
Engine and Frame Numbers	T20 81890 to 88346; T20SS 82276 to 88346; T20SH 82756 to 88346; T20SC 84912 to 88346; T20SR 84967 to 88346; TR20 85108 to 88346; TS20 85323 to 88346

This is the brilliant new 650 c.c. twin cylinder Triumph engine used with varying specifications in the Thunderbird, Trophy and Bonneville 120.
Whilst retaining all the major and well tried engine and gearbox components from the design it supersedes, it encloses them in a completely new unit which is stiffer, cleaner in design and more efficient. At the same time a considerable saving in overall weight has been made. New features incorporated include a duplex primary chain, a more robust clutch with improved shock absorber and a new ignition system in which twin contact breakers are fitted in the timing cover.

H.W. PERKINS

TRIUMPH

1963–1970
Unit Construction 650s

Bonnevilles, Trophys, and Tigers

As the C range of twins had been unit construction since 1957, and BSA had their unit-construction A65 ready to go, it was inevitable that Triumph would unitize Edward Turner's original parallel twin. Unit construction made sense on several fronts, notably in reducing manufacturing costs, and considering Turner's predilection for economic rationalization, it was surprising a unit 650 took so long to appear. Turner and Bert Hopwood created the evolutionary and revolutionary new engine over an eight-month period, and much of engine layout followed that of the Twenty-One, with the internals similar to the pre-unit 650. The unit 650 engine retained the attractive earlier shape of the gearbox and timing chest, but received significant updates including stiffer crankcases, a ninth cylinder barrel-fixing stud (hence the nickname "nine-stud"), and a new aluminum cylinder head with increased fin area. The crankshaft received a lighter flywheel and 85 percent balance factory, while the primary drive chain was increased to a ⅜-inch Duplex. To the chagrin of traditionalists, the magneto went, and all 650s now featured coil ignition with dual points driven off the end of the exhaust camshaft. Brian Jones joined Triumph from Norton and was responsible for providing a new single downtube frame, and while the first unit 650s were not initially rated as highly as their pre-unit predecessors, they soon caught on.

Late in 1962 Doug Hele joined Triumph as development engineer. Hele had left Norton after they closed their Birmingham factory, and one of his first projects at Triumph was to plan the 750cc Trident with Bert Hopwood. Hele would also have a major impact on the development of the Bonneville, making gradual improvements to handling that contributed to Triumph's success during the 1960s and established it as the most iconic British sport motorcycle of the era. Such was the Bonnie's allure that in the 1960s it sold three times as well as comparable Norton and BSA twins.

OPPOSITE: The new 650cc unit-construction engine followed the lines of the earlier Twenty-One. Internally it was similar to the pre-unit version.

CHAPTER FOUR

1963 T120, T120R, T120C, T120C TT Bonneville

Bore	71mm
Stroke	82mm
Capacity	649cc
Horsepower	46 at 6,500 rpm
Compression Ratio	8.5:1
Carburetors	Twin Amal 376/286-287 Monobloc 1 1/16-inch
Ignition	Lucas MA6 coils, 4CA contact breakers
Gearbox	Four-speed
Front Fork	Telescopic
Rear Suspension	Swingarm
Brakes	8-inch front and 7-inch rear
Wheels	WM2x18-inch front and WM3x18-inch rear (UK)
Tires	3.25x18-inch front and 3.50x18-inch rear (UK)
Wheelbase	55 inches
Dry Weight	363 pounds
Color	Alaskan White
Engine and Frame Numbers	T120 DU101 to DU5824; T120R DU101 to DU5824; T120C DU101 to DU5824

1963

Three B-range 650s shared the new unit-construction engine this year: the twin-carburetor Bonneville and the single-carb Trophy and Thunderbird. The smaller C range expanded to include a sport 350, the T90 Tiger, and, in a fruitless endeavor to counter the new Hondas and Yamahas, the struggling Tiger Cub range received still more updates. Triumph also pioneered the offering of official accessories, with windscreens, mirrors, and fairings available. As usual, Triumph concentrated on the US market, and 6,300 Triumphs were sold in America during 1963. This total would quadruple over the next four years.

1963 T120, T120R, T120C, and T120C TT Bonneville 650

With its new single downtube frame, the unit-construction Bonneville harkened back to the first T120 of 1959. Retaining the 1962 Duplex frame's 25-degree steering head angle, the new frame had a much thicker 1 5/8-inch downtube, with the previously unsupported swingarm lug now integrated with stiffer rear engine

After several years of two-tone color schemes, the first unit-construction Bonneville was painted only one color, Alaskan White. This UK example has an 18-inch front wheel.

84

1963–1970

LEFT: The unit 650 engine followed the style of the earlier pre-unit version, with extremely attractive aluminum outer engine casings. UK versions such as this often weren't fitted with a tachometer.

ABOVE: The Amal Monobloc carburetors still weren't fitted with air filters on 1963 UK Bonnevilles.

plates. This provided an immediate boost in handling, but there was still room for improvement. Like the first Bonneville of 1959, the 1963 unit-construction T120 included a number of carryover parts, notably the weak 1962-style front fork, resulting in some high-speed instability. The engine specifications for the unit T120 were largely unchanged from the previous year, with the same compression ratio, valve sizes, and carburetors. The revised longer silencers incorporated larger baffles, but the claimed power was the same.

As in 1962, the T120 was produced in a number of different variants. UK versions received an 18-inch front wheel, a smaller rear tire, heavier spring (145-pound) rear shocks, and a larger, 4¼-gallon gas tank, while US examples retained the 19-inch front wheel, 100-pound shock springs, and a 3½-gallon tank. The T120C off-road model continued with an engine skid plate and higher exhaust pipes, while JoMo specifically requested the T120TT Special to cater for the demand for desert races on the West Coast. This had a high-performance engine based on that of Bill Johnson's World Record streamliner and was only built in small numbers. As competition requirements varied across the country, so did specification for the T120Cs. West Coast examples were usually without lights and mufflers, while East Coast versions were more of a Street Scrambler, the smaller 500 generally preferred for Enduro competition. On a JoMo–prepared T120C, Eddie Mulder won the TT national championship.

The unit T120 may have been an improvement, but because the entire engine structure was more rigid, vibration increased. And although significantly lighter and better handling, it still had some reliability issues, primarily electrical and centering on the Lucas contact breakers and voltage regulator. Despite these trifles, the Bonneville was still the bike to have in 1963, and over the next few years it came to define the 1960s sport motorcycle.

1963 6T Thunderbird, TR6SS, TR6SR, TR6SC Trophy 650

Both the single-carburetor 6T Thunderbird and the TR6 Trophy featured the Bonneville's single downtube frame, and their engines included the nine-stud cylinders, Duplex chain primary drive, and coil and points ignition. The Trophy was still ostensibly a single-carb Bonneville, while the Thunderbird continued

85

CHAPTER FOUR

An entirely new sports 350 c.c. twin with a specification similar to the well-known 500 c.c. model Tiger 100. With a brilliant performance, smooth, fast and easy to handle, this interesting addition to the Triumph range will undoubtedly prove a popular model with the rider who prefers a 350 and can appreciate the "plus" performance of this new model.

TIGER 90
350 CC

ABOVE: The 1963 UK and general-export specification TR6SS Trophy had a low exhaust system.

RIGHT: The T20 engine had new castings for 1963, with finned rocker covers. This is a Meriden publicity photo.

much as before, albeit with some styling updates. The Thunderbird was nearing the end of its life, and the styling was now less bulky, with the side panels enclosing only part of the rear. The headlight nacelle continued, as did the deep front fender, and this year the US version also included rear side panels. The Thunderbird evolved into a single-seat, police-specification 650, the Saint, in 1963. When tested by *The Motor Cycle* in August that year, they achieved a top speed of 102 miles per hour, and the low-compression 650 twin provided power "that was so smooth as to be nigh-on-liquid." As vibration was reduced, in many respects the lower-compression 6T unit-construction engine was more pleasant than the TR6 and T120 versions.

The unit-construction TR6SS Trophy for the United Kingdom and general export was a street bike with a siamesed right-side low exhaust system, and as usual the United States received a TR6SR (with individual, Bonneville-style silencers) and a TR6SC (with

1963 6T Thunderbird; TR6SS, TR6SR, TR6SC Trophy (Differing from 1962)

Wheelbase
55 inches (6T); 55.5 inches (TR6)
Dry Weight
369 pounds (6T); 363 pounds (TR6)
Colors
Regal Purple and silver (TR6)
Engine and Frame Numbers
6T DU764 to DU5824; TR6SS, TR6SR, TR6SC DU102 to DU5790

1963 T100SS Tiger 100, 5TA Speed Twin, T100SR, T100SC, T90 Tiger, 3TA Twenty-One (Differing from 1962)

Horsepower 27 at 7,500 rpm (T90)
Compression Ratio 9:1 (T90)
Carburetor Amal 376/300 Monobloc $^{15}/_{16}$-inch (T90)
Ignition Lucas MA6 coils, 4CA contact breakers (T100SS, T100SR, T100SC, T90)
Wheels WM2x18-inch front (T100SS and T90)
Tires 3.25x18-inch front (T100SS and T90)
Wheelbase 53.5 inches (T90)
Dry Weight 333 pounds (T90)
Colors Alaskan White (T90); Regal Purple and silver (T100SR, T100SC); Silver Bronze (3TA)
Engine and Frame Numbers T100SS H27933 to H32464; T100SC H30214 to H30284; T100SR H27933 to H32464; 5TA H30289 to H32361; 3TA H27933 to H32464

individual high pipes like the T120C). Success continued for the TR6 in the ISDT, held in Czechoslovakia this year, with Bud Ekins and Ken Heanes both winning gold medals.

1963 T100SS Tiger, 5TA Speed Twin, T100SR, T100SC 500, T90 Tiger 90, 3TA Twenty-One 350

The smaller twin range expanded with the sporty Tiger 90 350 joining the lineup for 1963. This followed the general style of the T100SS, the 3TA-based engine with a new cylinder head with larger valves, a larger carburetor, and higher-compression pistons. The ignition was also updated, with a Bonneville-style twin contact breaker system as the 3TA's distributor. This year the T100SS, T100SR, and T100SC also adopted the twin contact breaker ignition system, this requiring a new timing cover. T100SC production was very limited, with possibly only around 70 produced this year. The T90 chassis was identical to the T100SS, both now with an 18-inch front wheel, but neither model was sold in America. All 1963 350 and T100 models received a new three-vane shock absorber unit, providing improved shock absorption.

The 5TA Speed Twin and 3TA Twenty-One retained the distributor ignition for one more year and also reverted to two individual exhaust pipes. They were also the last models to retain the unpopular Bathtub rear enclosure, and this would be the Bathtubs' final year. The 3TA was no longer offered in America this year, and the Speed Twin disappeared from the 1964 US lineup. In competition the T100SC continued where it left off in 1962. Bill Baird won his second National Enduro Championship, and John Giles won a gold medal at the ISDT. In the hands of Brian Davis and Bill Scott, a T100SS Tiger won the 500cc class in the Thruxton 500 production race.

1963 T20 Tiger Cub, T20SS Street Scrambler, T20SH Sports Home, T20SC, T20SR, TR20 Trials, TS20 Scrambler 200

Several engine updates were incorporated in the Tiger Cub range for 1963. A Lucas 4CA points assembly located at the end of the camshaft replaced the earlier distributor type on top of the crankcase. Known as the "side points" engine, it also featured new crankcases and closed rocker boxes with finned aluminum covers. The T20SS was now built in the same specification for JoMo and TricCor, with a new high or low exhaust system and adjustable rear suspension, while in the home market the T20SH, T20SC, T20SR, TR20, and TS20 continued much as before. It was enough to maintain sales nearly at the 1962 level.

1963 T20, T20SS, T20SC, T20SR, TR20, TS20 (Differing from 1962)

Colors
Flame and Silver Gray (T20, T20SS)
Engine and Frame Numbers
T20 88347 to 94599; T20SS 88347 to 94599; T20SH 88347 to 94599; T20SC 88347 to 94599; T20SR 88347 to 94599; TR20 88347 to 94599; TS20 88347 to 94599

1964

This year marked the end of the Edward Turner era, but although he retired as chief executive of the BSA Automotive Division, he retained a directorship and continued to work as a freelance designer. Although he had his detractors, Turner was a keen motorcyclist and understood the industry. Bert Hopwood had expected to take Turner's place, but Eric Turner appointed Harry Sturgeon, a former de Havilland Aircraft executive who was then serving as director of a BSA grinding machine subsidiary. Sturgeon was

CHAPTER FOUR

The "Desert Sled" was a Trophy, successfully modified by dealers and privateers for Californian desert racing. For 1964 the factory produced their own version, the TR6SC Desert Sled shown here, with battery-less ignition and open exhaust pipes.

unusual in that he never moved to the Midlands, but preferred to live on a Hertfordshire farm, staying in a hotel when he visited Birmingham, and he immediately implemented changes that would have a profound effect. He wanted production to increase dramatically, eventually doubling Edward Turner's optimum of around 2,000 a month. This resulted in deterioration in build quality, and parts suppliers were unable to meet deadlines, which caused problems with spare parts supply. Another problem associated with the dramatic increase in production was an expansion of Meriden's workforce and thus the power of the unions representing them; in the quest for increased production, Sturgeon weakly agreed to wage demands. Sturgeon's policy of sales first was ultimately shortsighted, as was his implementation of McKinsey's report recommending the amalgamation of BSA and Triumph. And with the domestic market still slow, even more of Triumph's production would be for export to the United States.

But despite these faults, Sturgeon brought some positive input. He differed significantly from Edward Turner in that he promoted racing, particularly that of road-based machines, and he was quoted as saying, "Win on Sunday, sell on Monday."

1964 T120, T120R, T120C, and T120C TT Bonneville 650

Doug Hele's influence became more apparent this year, with the Bonneville receiving a stronger external spring front fork to improve high-speed stability. The engine was also updated, including new crankcases with an improved crankcase breather, and modifications to counter the perennial problem of oil leakage. Oil leaks plagued Triumph's reputation throughout the 1960s, and they never were fully solved. Other engine updates this year included larger inlet and exhaust valves, larger Amal Monobloc carburetors, and an improved clutch operating system. When tested by *Motor Cycle* in May 1964, the new Bonneville impressed with a best one-way top speed of 112 miles per hour (on a damp track). They summed up the Bonneville with "the overall picture is of a machine which must not be far short of the ultimate in super-sports luxury."

Several T120 variants continued for the United States, with the East and West Coast T120Cs differing in minor details. All US T120s retained a 19-inch front wheel. As in 1963, a T120C TT Special was also available, with high-rise open exhaust pipes, a higher compression ratio (12:1), and a pair of 1 3/16-inch Amal Monoblocs. The claimed power was 54 horsepower at 6,500 rpm.

1964 6T Thunderbird, TR6SS, TR6R, TR6SR, TR6C, TR6SC Desert Sled Trophy 650

The 6T Thunderbird's styling was further refined, with a smaller front fender for UK versions and US examples now with a sporty Trophy-style front fender. The front fork was now an external spring type, and while the nacelle continued, it was altered to accommodate the new fork. Engine updates were minimal, and the 6T pioneered 12-volt electrics for Triumph. This early arrangement featured two 6-volt batteries in series, with Zener diode charge control. *Cycle World* tested the Thunderbird in August 1964, commenting, "We simply cannot find fault with the ride, and everything operates with great precision. In all, the Thunderbird impressed us as being very much the gentleman's motorcycle. It has been as pleasant a motorcycle as ever offered to us for test."

For 1964 the TR6 also featured the front fork with external springs and a new Smiths magnetic speedometer with anti-vibration mountings. The Amal Monobloc carburetor was increased to 1 1/8 inch, and the TR6C and SC were fitted with the new

1964 T120, T120R, T120C, T120C TT Bonneville
(Differing from 1963)

Carburetors
Twin Amal 389/203 Monobloc 1 1/8-inch (T120R); Amal 389/95 (T120C TT)
Color
Gold and Alaskan White
Engine and Frame Numbers
T120 DU5825 to DU13374; T120R DU5825 to DU13374; T120C DU5825 to DU13374

1964 6T Thunderbird, TR6SS TR6R TR6SR TR6C TR6SC Trophy
(Differing from 1963)

Carburetor
Amal 373/303 (6T); 389/97 Monobloc 1 1/8-inch (TR6)
Colors
Hi Fi Scarlet and silver (TR6)
Engine and Frame Numbers
6T DU6329 to DU13210; TR6SS, TR6R, TR6SR, TR6C, TR6SC DU6127 to DU13287

Steve McQueen at the ISDT

The 1964 ISDT was held in Erfurt, East Germany, at the height of the Cold War, and actor Steve McQueen joined four of America's best off-road riders to represent the United States in the Vase A team. Along with McQueen's friend Bud Ekins were Bud's brother Dave, John Steen, and Cliff Coleman. In the wake of *The Great Escape*, McQueen was at the height of his popularity, yet he found the time to travel to England prior to the event to help prepare the two TR6SCs and three T100SCs with Meriden factory technicians. These were 1964 spec bikes with 1965 front forks. Each was fitted with an extra coil and condenser, and 7:1 pistons to cope with the low-octane fuel. At the official opening McQueen proudly carried the Stars and Stripes for the US team. McQueen rode a TR6SC in the event but retired after a crash on the third day. Bud Ekins crashed his T100SC, breaking a leg, but Dave Ekins (T100SC) and Coleman (TR6SC) both won gold medals and John Steen won a silver medal. Triumph completed a highly successful event, with Johnny Giles and Ken Heanes (TR6), as well as Roy Peplow and Ray Sayer (T100s) also winning gold medals. Despite his fame and superstar status, Steve McQueen always seemed relaxed and at ease in the company of motorcyclists.

ABOVE RIGHT: Steve McQueen was a Triumph fan and an extremely competent off-road rider. *Triumph Motorcycles*

RIGHT: McQueen about to set off on an ISDT stage on the Triumph Trophy.

FAR RIGHT: Hot and sweaty and a long way from Hollywood. McQueen during the 1964 ISDT. *Triumph Motorcycles*

CHAPTER FOUR

Bob Dylan with his 1964 T100SR. He crashed this in 1966, an event that affected his career for eight years. *Triumph Motorcycles*

For 1964 the T100SS lost its skirts and looked similar to the TR6.

1964 T100SS Tiger, 5TA Speed Twin, T100SR, T100SC 500, T90 Tiger 90, 3TA Twenty-One 350

After the confusing mixture of ignition systems on the 1963 smaller twin range, all now featured the camshaft-driven contact breaker ignition arrangement. All models also included an improved clutch operating mechanism, redesigned pushrod tubes with superior oil sealing, and a new front fork as on the 650cc twins. The 5TA Speed Twin and 3TA Twenty-One now featured abbreviated rear enclosures (similar to this year's Thunderbird), while the Tiger 100 and Tiger 90 were stripped completely, now in the style of the TR6 Trophy. As the Speed Twin neared the end of its life the color also changed, from red to silver and black. The 3TA was silver and beige this year. As in 1963, the only C-range twins offered in the United States for 1964 were the T100SR and T100SC. Racing success continued for the T100 Tiger, with Gary Nixon winning the 20-mile National Championship Dirt Track Race at Sacramento, California, in September. Nixon would go on to become one of Triumph's most successful racers.

One high-profile 1964 T100SR owner was Bob Dylan. On July 29, 1966, Dylan was riding near Woodstock and crashed his Tiger T100 after being blinded by the sun. He suffered facial lacerations, and after the crash Dylan said, "When I had that motorcycle

Energy Transfer battery-less ignition and lighting system. There were a number of variants, with different East and West Coast versions. One new version this year was a stripped-down factory TR6SC Trophy Special "Desert Sled" with polished alloy fenders, straight-through high exhausts, and Energy Transfer battery-less ignition. TR6s were totally invincible in Californian off-road racing this year, winning every major race, and Bud Smith took the 500-mile AMA Enduro National Championship.

accident, I woke up and caught my senses. I realized that I was just working for all these leeches and I didn't want to do that. Plus I had a family and I just wanted to see my kids." Dylan then retreated from his frenetic schedule and, for the next eight years, did not play much in public.

1964 T20 Tiger Cub, T20SS Street Scrambler, T20SH Sports Home, T20SC, T20SR, TR20 Trials, TS20 Scrambler, T20SM Mountain Cub 200

Tiger Cub sales were now pretty stagnant in America, and with less than 1,300 of all variants sold stateside in 1963, Edward Turner was considering discontinuing the Cub in the United States. However, JoMo Sales Manager Don Brown persuaded him there was a market for a lightweight trail bike that could climb steeper hills and go faster than the underpowered Honda CA105T Trail 55 that was currently in vogue with the Northern California hunting and fishing crowd. Turner agreed to an initial order of 400 of a new model based on the East Coast T20SC, incorporating the wide-ratio TR20 Trials Cub gearbox, Trials tires, and a high exhaust. The engine was low-compression (7:1) but with a sport camshaft and larger Amal carburetor. Titled the "Mountain Cub," it soon proved immensely popular and staved off the Tiger Cub's premature extinction.

1965

In January 1965 the BSA Group purchased a controlling interest in Johnson Motors. With both the West and East Coast operations owned by BSA, Harry Sturgeon had the impetus to further implement the amalgamation of BSA and Triumph as recommended in the earlier McKinsey report. This occurred first in England, with spare parts and production materials consolidated at BSA's Small Heath plant during 1964. Sturgeon also wanted Meriden to concentrate solely on twin-cylinder models and in July 1964 decided to move Tiger Cub production to Small Heath. From February 1965 the Tiger Cub was produced at Small Heath, assembled initially from components supplied by Meriden. This was first of several contentious decisions resulting from the McKinsey report, with the workforce at both factories not exactly endorsing the move. In the meantime, Sturgeon's production boom and concentration on the US market earned BSA and Triumph the Queen's Award for Export. And with production booming and sales strong, there was only limited time and resources to develop the twins further.

With the Japanese threat unabated—Honda's double overhead camshaft twin-cylinder CB450 appeared in March—Hopwood and Hele continued developing the 750cc Trident. This followed Hopwood's "modular" theme, where one cylinder of the existing Speed Twin 500 could become three, and a prototype was built by late 1965.

1964 T100SS Tiger 100, 5TA Speed Twin, T100SR, T100SC, T90 Tiger, 3TA Twenty-One *(Differing from 1963)*

Colors	Hi Fi Scarlet and silver (T100SS, T100SR, T100SC); silver and black (5TA); gold and Alaskan White (T90); silver and beige (3TA)
Engine and Frame Numbers	T100SS H32465 to H35986; T100SC H32793 to H35489; T100SR H32465 to H35986; 5TA H32918 to H35986; 3TA H32465 to H35986

1964 T20, T20SS, T20SC, T20SR, TR20, TS20, T20SM *(Differing from 1963)*

Carburetor	Amal 376/314 Monobloc $15/16$-inch (T20SM)
Dry Weight	223 pounds (T20SM)
Colors	Hi Fi Scarlet and silver (T20, T20SH, TR20, TS20); Crystal Blue and silver (T20SS, T20SM); Kingfisher Blue or scarlet (T20SC, T20SR)
Engine and Frame Numbers	T20 94600 to 99719; T20SS 94600 to 99732; T20SH 94600 to 99732; T20SC 94600 to 99732; T20SR 94600 to 99732; TR20 94600 to 99732; TS20 94600 to 99732; T20SM 94600 to 99732

After the dour 1964 color scheme, the Bonneville received attractive new colors for 1965. This was the final year for the "Mouth Organ" 1950s grille-type tank badge.

CHAPTER FOUR

ABOVE: A beautiful US spec 1965 T120R. US versions had the classic short muffler and air filters for the Amal carburetors.

OPPOSITE ABOVE LEFT: The classic Bonneville instrument layout. UK examples also received a tachometer this year, and 1965 was the final year for the Bonneville's parcel rack on US versions.

OPPOSITE ABOVE RIGHT: This unusual larger taillight was fitted to US models for 1965 and only lasted one year.

OPPOSITE BELOW: Another view of the 1965 T120R. Most had a two-tone seat like this one; with its blue and silver colors, this year's Bonneville is very desirable.

1965 T120, T120R, T120C, and T120C TT Bonneville 650

Along with a classic new color scheme, the Bonneville for 1965 included a number of small updates. In order to reduce primary drive chain wear and drive-side main bearing failure, the fixed end of the crankshaft was moved to the left, with the drive sprocket butting directly against the main bearing. With finding top dead center critical in setting the correct ignition timing, the right-side crankcase now included a plugged, threaded hole through which a metal plunger could engage a slot in the flywheel at TDC. A redesigned front fork also appeared this year, with redesigned internals and more travel, and all 650s featured a redesigned rear brake linkage with the rod inside the engine plate, improving brake action. US versions received a larger taillight and UK models a standard tachometer.

1965 T120, T120R, T120C, T120C TT Bonneville
(Differing from 1964)

Compression Ratio
11:1 (T120C TT)

Horsepower
47 at 6,700 rpm, 52 at 6,500 rpm (US)

Dry Weight
360 pounds (T120C TT)

Color
Pacific Blue and silver

Engine and Frame Numbers
T120 DU13375 to DU24874; T120R DU13375 to DU24874; T120C DU13375 to DU24874

1965 6T Thunderbird, TR6, TR6SS, TR6R, TR6SR, TR6C, TR6SC Trophy
(Differing from 1964)

Horsepower
37 at 6,700 rpm (6T)
Colors
Burnished Gold and white (TR6)
Engine and Frame Numbers
6T DU14635 to DU24874; TR6, TR6SS, TR6R, TR6SR, TR6C, TR6SC DU14226 to DU23732

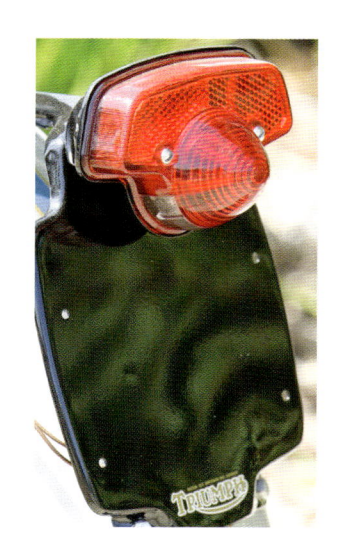

The long resonator silencer also continued for the UK, while the US T120R received a shorter, more attractive teardrop-styled muffler.

As in the previous year in the United States, the street Bonneville carried the T120R designation, with a scrambler T120C for the East Coast and T120C TT for the West Coast. The T120C TT Special featured larger valves, racing camshafts and tappets, and downswept, tucked-in, 1¾-inch open exhaust pipes. Only a few T120C and T120C TTs were built this year.

1965 6T Thunderbird, TR6, TR6SS, TR6R, TR6SR, TR6SC Desert Sled Trophy 650

The 6T Thunderbird incorporated the other updates shared with the B range this year (crank locator, TDC plug, modified front fork), and apart from a larger taillight for the US version was otherwise unchanged. Somehow the claimed power was increased. The TR6 continued as a single-carburetor Bonneville in the United Kingdom, with the TR6SR and TR6SC in the United States. It was still up against the more popular

CHAPTER FOUR

Thruxton Bonneville

With production class racing extremely important in England and Doug Hele's belief that racing improved the breed, Triumph displayed a Thruxton Bonneville at the Earls Court Show at end of 1964. Named after the Hampshire racing circuit where the Bonneville tasted success in 1961, the Thruxton Bonneville commenced production in May 1965. Only around 50 were produced, and the specification was impressive. Each featured a factory blueprinted engine with special camshafts and followers, chopped Amal Monobloc carburetors with a central remote float chamber (as on the 1960 T120), and a special free-flowing exhaust system. The claimed power was 54 horsepower. The frame's steering head angle was steepened, and the front fork included shuttle-valve damping. Both these features would later make their way to the production Bonneville. Other changes included 19-inch wheels front and rear and a ventilated front brake. A few examples were also produced in 1966 with aluminum wheel rims and an Avon fairing.

For 1965 the 500-mile production race was moved from Thruxton to the Castle Combe Circuit in Wiltshire. After losing three years in a row to Norton, Dave Degens, with Barry Lawton, won on a Thruxton Bonneville. Degens, with Rex Butcher, repeated this the following year (at Brands Hatch), and today the Thruxton Bonneville is one of the rarest and most sought after of all production Triumphs.

The Thruxton Bonneville was first displayed at the end of 1964 with an Avonaire fairing.

twin-carb Bonneville, but *Cycle World* found buyers "are making a mistake with the T120R; The TR6SR is easier to start, smoother running, and has the sort of low speed torque that makes it a pleasure to ride. This is one of the most pleasant motorcycles to ride all day, every day."

The TR6SC Desert Sled, or Trophy Special, continued as before. *Cycle World* also tested the TR6SC Trophy Special this year, stating, "The Triumph may not be the best-handling scrambler in the whole wide world, but it has more hair on its chest than King Kong and has proved time and again that it will simply out-muscle its better-handling opposition if its rider is similarly hairy-chested."

1965 T100SS Tiger, 5TA Speed Twin, T100SR, T100SC 500, T90 Tiger 90, 3TA Twenty-One 350

Updates to the 350 and 500 twins centered on the chassis, the frame including a stiffening strut (as on the 1961 TR5AC) to prevent gas tank fractures and a softer front fork with more progressive damping. The gas tank was now rubber-mounted on the frame in four points. The front fender of the 5TA and 3TA was now a smaller, sportier type, and the two versions for the United States were as before, the street T100SR and the off-road T100SC, known as the "Jack Pine"

1965 T100SS Tiger 100, 5TA Speed Twin, T100SR, T100SC, T90 Tiger, 3TA Twenty-One (Differing from 1964)

Tires
3.25x19-inch front and 4.00x18-inch rear (T100SR); 3.50x19-inch front (T100SC)

Color
Burnished Gold and white (T100SS, T100SR, T100SC); Pacific Blue and silver (T90)

Engine and Frame Numbers
T100SS H35987 to H40527; T100SC H36434 to H40098; T100SR H35987 to H40527; 5TA H36615 to H39838; 3TA H35987 to H40527

model. As was usual for American examples, both had a 19-inch front wheel. *Cycle* magazine tested the T100SC in September 1965, finding that "there are few, if any, improvements that could be practically applied to this bike to make it more ideally suited to its dual purposes."

94

1965 T20, T20SS, T20SC, T20SR, TR20, TS20, T20SM *(Differing from 1964)*

Colors
Pacific Blue and silver (T20SS);
Hunting Yellow and black (T20SC, T20SM)

Engine and Frame Numbers
T20 101 to 2001 (BSA numbers); T20SS 99733 to 100013 and 101 on (BSA numbers); T20SH 99733 to 100013 and 101 on (BSA numbers); T20SC 99733 to 100013 and 101 on (BSA numbers); T20SR 99733 to 100013 and 101 on (BSA numbers); TR20 99733 to 100013 and 101 on (BSA numbers); TS20 99733 to 100013 and 101 on (BSA numbers); T20SM 99733 to 100013 and 101 on (BSA numbers)

1965 T20 Tiger Cub, T20SS Street Scrambler, T20SH Sports Home, T20SC, T20SR, TR20 Trials, TS20 Scrambler, T20SM Mountain Cub 200

The Tiger Cub was now assembled at BSA's Small Heath factory in Birmingham, and while the T20 and T20SH continued largely unchanged, the sporty T20SS, T20SC, T20SR, TR20, TS20, and T20SM received an externally sprung front fork. Sport Cubs also received a new frame with strengthened steering head. The kick-start lever was cranked and pivoted for this year only, and the cylinder head and barrel shaped square rather than oval (after production moved to Small Heath in February 1965). Early in 1965 the T20SS, TR20, and TTS20 were discontinued, and the T20SR and T20SC finished in July. This year the yellow T20SM Mountain Cub was by far the most popular of all Tiger Cubs, with 3,219 produced.

1966

Early in 1966 Sturgeon further implemented his plans for the amalgamation of BSA and Triumph, moving Johnson Motors from their Pasadena headquarters to a new group headquarters at Duarte, California. He also announced that Triumph dealers would have to sell BSA and vice versa. This was unpopular in the United Kingdom, where BSA dealers outnumbered Triumph more than two to one, but it caused a riot among Triumph dealers in the United States, where there was actual antipathy between the two brands. Eventually Sturgeon backed down in the United States, with Pete Colman and Meriden's Service Manager John Nelson placating Triumph dealers, promising they wouldn't have to sell BSA. After Wilbur Cedar died in May 1966, Sturgeon succeeded him as the president of Johnson Motors. The BSA/Triumph merger was now complete, and with even more emphasis on the US market, Triumph again won the Queen's Award for Export.

This year also saw other dubious managerial decisions. Eric Turner vetoed Bert Hopwood's promising modular single overhead camshaft 250cc triple in favor of Edward Turner's P30 overhead camshaft 350. While the second version of the 750cc triple, the P2, was developed, with three prototypes on the road by mid-1966, at this stage Eric Turner still favored boring the existing 650 Triumph and BSA twins and fitting overhead camshafts.

1966 T120, T120R, T120C, and T120 TT Bonneville 650

The policy of the Bonneville's continual development reached its zenith in 1966, and this year the T120 incorporated many updates. Inside the engine was a new crankshaft and flywheel assembly, 2½ pounds trimmer and providing more throttle response at the expense of teeth-shattering vibration at higher speeds. Other updates included higher-performance Thruxton-style sports camshafts and followers,

TOP: The 1965 T100SC, this year with aluminum fenders and the ugly taillight.

ABOVE: Specifically built for the US market, the T20SM Tiger Cub was incredibly popular. For 1965 it was yellow, with polished aluminum fenders and a new kick-start.

CHAPTER FOUR

The Gyronaut X-1

Triumph's success with streamliners at Bonneville continued in 1965 and 1966 with the radical twin-engine Gyronaut X-1. With its custom fiberglass aerodynamic shell and big-budget sponsorship, the Gyronaut X-1 ushered in a new era of professional speed record specialization. The seeds were sown in 1959 when Detroit Triumph dealer Bob Leppan built a dual-650 engine Triumph dragster, the successful *Cannibal Mark II*. But by 1962 this was outclassed by the new wave of car dragsters, and Leppan turned toward Bonneville speed records. He took a single-engine 650 streamliner to Bonneville in 1963, and there he met the former head of Ford's advanced styling, Alex Tremulis. With Jim Brufoldt they set about creating a dual-engine streamliner capable of beating the then-current Johnson and Dudek record of 230 miles per hour. Tremulis came with impressive credentials, having been a stylist with Auburn Cord Duesenberg in the 1930s, and with Tucker after the war. He envisaged the X-1 as a stepping-stone to larger car-engine machines, but this wasn't to be.

Powering the Gyronaut X-1 were two pre-unit TR6 twins coupled together with a single-row Renolds chain. These engines had proved extremely reliable over the years and featured 10.5:1 Hepolite pistons, high-lift Harman & Collins cams, and four Amal 289 1⅛-inch carburetors with 510 GP remote float bowls. With Lucas racing magnetos, and running on methanol, the engines produced around 60 horsepower each in 1965. The engines were mounted in a specialized tubular chassis with center-hub steering and originally an Airheart rear disc brake. Borrani WM3x19-inch alloy wheels were shod with special Goodyear Land Speed tires. At Bonneville in 1965 it set a new AMA gas-powered record of 217.624 miles per hour before crashing.

Bob Leppan with the Gyronaut X-1 at Bonneville in 1965. This year he set a new gasoline record of 217.62 miles per hour. *Triumph Motorcycles*

Leppan returned to Bonneville in August 1966 with the engines now producing around 65 horsepower each. This time he set a new world motorcycle speed record of 245.667 miles per hour, a record that would unofficially stand for four years. As the Gyronaut X-1 displaced more than 1,000cc, the record wasn't recognized by the FIM. For the next four model years, Triumph T120R Bonnevilles received a new decal proclaiming "World's Fastest Motorcycle." After setting seven records with the new Trident in 1969, Leppan took the Gyronaut X-1 to Bonneville in 1970. This year, with the engines displacing 820cc and producing around 180 horsepower, Leppan crashed at around 270 miles per hour when his front suspension collapsed. Leppan sustained serious injuries, nearly losing his left arm, and the earlier record was lost, first to Don Vesco's Yamaha and then to Cal Rayborn's Harley-Davidson. Although Leppan recovered, it was the end of the road for the Gyronaut X-1.

1966 T120, T120R, T120C, T120TT Bonneville
(Differing from 1965)

Compression Ratio
9:1
Carburetors
Twin Amal 389/95 Monobloc 1³⁄₁₆-inch (from DU29738)
Dry Weight
365 pounds
Color
White and Grenadier Red
Engine and Frame Numbers
T120 DU24875 to DU44393; T120R DU24875 to DU44393; T120TT DU24875 to DU44393

higher-compression pistons, and larger Amal carburetors (during the year).

Doug Hele's gradual frame improvement through race development was also evident this year, with a steering head angle of 28 degrees (as on the Norton Featherbed) to cure the dangerous weaves that could occur with the previous 25-degree rake. Other updates included the 6T's 12-volt electrical system, a larger (3-quart) oil tank, and a new full-width, 8-inch single leading shoe front brake claim to be more rigid and delivering a 44 percent increase in braking area. Stainless steel fenders replaced the painted steel type, and a much more attractive, cast and polished aluminum taillight housing was introduced this year. US versions received a beautiful but impractical 2½-gallon gas tank. There was no parcel rack, although one did feature on the larger-tank UK models, and new "Eyebrow" tank badges and strange, light gray

1963–1970

ABOVE: Even if the range was pitiful, the 1966 Bonneville looked extremely purposeful with its small, "slim-line" gas tank.

LEFT: Flamboyant colors of the US T120R. Unlike the predominantly white US version, UK examples had red tanks with white underneath.

CHAPTER FOUR

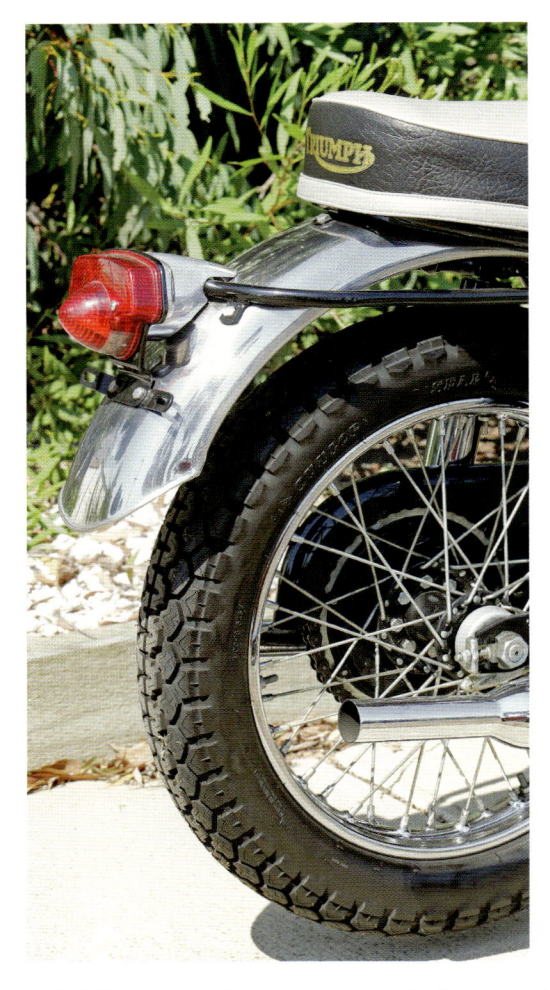

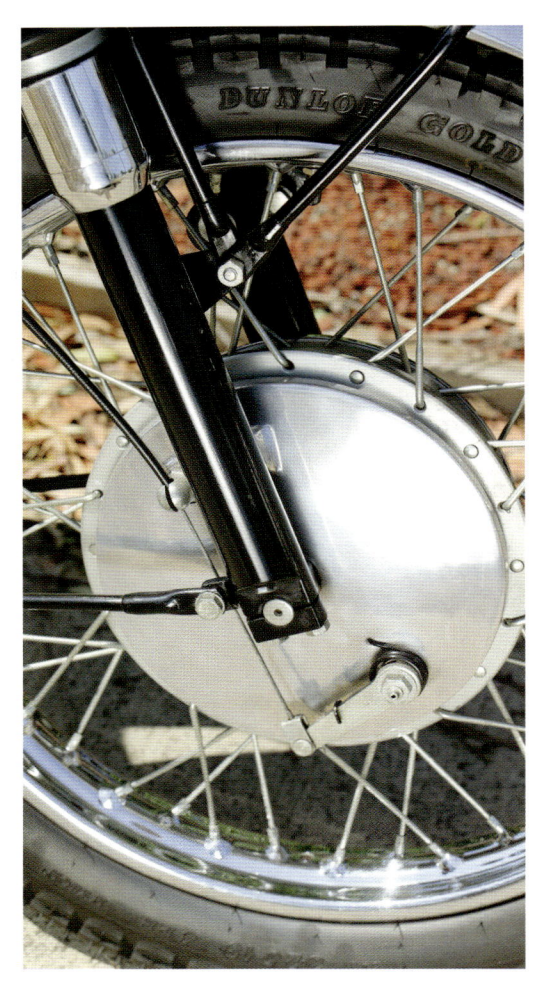

ABOVE LEFT: The cast-aluminum taillight bracket was a welcome improvement over 1965. This was also the final year for the two-tone seat.

ABOVE MIDDLE: The T120R received larger carburetors during 1966, plus light gray handgrips for this year only.

ABOVE RIGHT: The front brake was also improved for 1966, but remained a single leading shoe.

handgrips heralded a new era. This year the TT Special was coded the T120TT, and it still proved extremely popular. The 1966 Bonneville was greeted with enthusiasm: *Cycle* magazine stated, "The 1966 Bonneville is proof that Triumph is going all out to back up its claim of 'world's best and fastest motorcycle.'"

1966 6T Thunderbird, TR6, TR6SS, TR6R, TR6SR, TR6C, TR6SC Trophy 650

The three single-carburetor 650s shared the Bonneville's new frame, improved front brake, larger oil tank, and lighter crankshaft. This was the final

1966 6T Thunderbird; TR6, TR6SS, TR6R, TR6SR, TR6C, TR6SC Trophy
(Differing from 1965)

Dry Weight
365 pounds (TR6)
Color
Pacific Blue and white (TR6)
Engine and Frame Numbers
6T DU25877 to DU44393; TR6, TR6SS, TR6R, TR6SR, TR6C, TR6SC DU24876 to DU43161

The rear view of US 1966 TR6SR. This would originally have had light gray handgrips, as all models did this year.

The 1966 Daytona 200

With a real threat of Honda's new 450 eating into Triumph's US market, Bert Hopwood and Doug Hele persuaded Harry Sturgeon that a major roadracing success was needed in the States. They then concentrated their efforts on winning the 1966 Daytona 200, at that stage for 500cc overhead-valve twins and 750cc side-valves. A single overhead camshaft 500 was built, but with not enough time for development, they settled on creating racers out of the existing T100. The cylinder was adapted for twin carbs with splayed ports and included larger valves and a narrower 78-degree included valve angle. The compression ratio was 9.75:1, and carburetion was by a pair of Amal GPs. After much development the engines produced 45.6 horsepower at 8,500 rpm. This engine was installed in a new frame, built of 531 Reynolds tubing and lower and stronger than the production version, with 19-inch wheels and a BSA aluminum gas tank.

Meriden prepared four special T100Rs for the 1966 Daytona 200. These had modified frames and new cylinder heads. The Amal GP carburetors were mounted on long rubber intakes. Underneath the blister on the timing cover was a special Lucas racing contact breaker unit connected by an Oldham coupling to the exhaust camshaft.

Four T100 racers were sent from Meriden for Daytona, with TriCor and JoMo each receiving two. TriCor's Cliff Guild also built another two bikes, these and the JoMo versions having different frames. In the lead-up to the race, it was soon evident the Hele's Meriden-built machines were fast but fragile, all suffering connecting rod and aluminum tappet black failure. This prompted Gary Nixon to ride one of Guild's slower but more reliable machines in the race, while Buddy Elmore and JoMo's Dick Hammer stayed with the Meriden versions. As they had proved fragile and didn't qualify well, with Elmore only 46th at the start, the Triumphs were not expected to win. But they were fast. Early on in the race Hammer's engine broke, and after the leading Harleys retired, Elmore and Nixon swapped the lead until Nixon retired with a puncture. Elmore went on to win at a record speed of 96.582 miles per hour.

The rest of the 1966 road racing season was also good for Triumph in the United States, where Elmore won at Loudon and Nixon at Greenwood, Iowa. Nixon also excelled in dirt track nationals, winning at Springfield and finishing second in the national championship. After winning the Grand National Enduro Championship four years in a row, Bill Baird took out the 175 Mile Enduro at Schererville, Indiana, and Eddie Mulder won three TT Nationals on a Bonneville TT Special. Triumph was now set to challenge Harley-Davidson for the number-one plate.

The US-spec 1966 TR6SR in new colors of Pacific Blue and white.

CHAPTER FOUR

This 1966 TR6SR still featured the TR6SR engine designation.

The attractive green and white 1966 T100R, still with a single Amal Monobloc carburetor.
Lorin Smith

1966 T100 Tiger 100, 5TA Speed Twin, T100R, T100C, T90 Tiger, 3TA Twenty-One
(Differing from 1965)

Wheels
WM2x18-inch front and rear (5TA and 3TA)
Tires
3.25x18-inch front, 3.50x18-inch rear (5TA and 3TA)
Wheelbase
53.5 inches (3TA)
Dry Weight
337 pounds (T100SS, T90); 341 pounds (3TA)
Colors
Sherbourne Green and white (T100, T100R, T100C); Grenadier Red and white (T90); Pacific Blue and white (3TA)
Engine and Frame Numbers
T100 H40528 to H49832; T100C H40528 to H48215; T100R H40528 to H49832; 5TA H42227 to H46431; 3TA H40528 to H49832

year for the 6T Thunderbird, and while it retained the headlight nacelle to the end, the rear panels were now deleted. As more production was destined for America, where stripped-down performance models were more popular, the Thunderbird simply faded away with a whimper.

The TR6 Trophy also received some engine updates, including new valve springs and cam followers, along with the larger T120 valves. Ostensibly the TR6C and TR6SC were identical, as were the TR6R and TR6SR, and later in the model year the *S* was dropped from the engine designations. The TR6C and TR6SC now shared the Bonneville's 2½-gallon tank, while the TR6, TR6R and TR6SR received slimmer 4½-(TR6) and 3½-gallon (TR6R and TR6SR) gas tanks. All had the new Eyebrow badge, and the larger versions retained the parcel rack. The TR6C and SC aside, all TR6s now featured 12-volt electrics with Zener diode voltage control. The East Coast TR6C and SC continued with dual high-level exhausts, while the West Coast desert racing TR6C and SC had ignition-only electrics, twin siamesed open exhausts on the left, aluminum fenders, and folding footpegs (to comply with AMA regulations). This rather confusing TR6 lineup would be simplified for 1967.

1963–1970

Not only made by BSA, but now half BSA itself: the 1966 Tiger Cub with a BSA Bantam chassis.

1966 T100 Tiger, 5TA Speed Twin, T100R, T100C 500, T90 Tiger 90, 3TA Twenty-One 350

Updates to all of the smaller twins included a 12-volt charging system, larger oil tank, new tank badges, light gray handgrips, and a cast-alloy taillight bracket. A new frame included a welded additional top tube instead of the previous bolted-on type. The Tiger was now known as the T100, with US versions the T100R and T100C, and this year the controversial rear panels on the 5TA and 3TA were finally discarded. Both received 18-inch wheels front and rear (replacing the earlier 17-inch), and along with the Thunderbird, 1966 would be their final year. As the 5TA and 3TA still retained the headlight nacelle, their styling was an unsuccessful and confused mixture of old and new. By 1966 the shrouded fork and headlight nacelle looked decidedly old-fashioned, and the reality for these rather mundane low-performance models was that the market had changed. Most Triumphs were now exported to America, where the 5TA and 3TA were no longer sold, and everywhere the Japanese were starting to dominate the small and mid-capacity commuter market. After nearly 30 years the illustrious line of Speed Twins came to an end in 1966, and while the Speed Twin started out in a blaze of glory, it died with a whimper. However, the T100R and T100C were cruising along nicely in America, and the C range would get a huge shot in the arm next year as Triumph made the most of their Daytona success.

1966 T20B Tiger Cub, T20SH Sports Home, T20SM Mountain Cub 200

With Tiger Cub production now in full swing at Small Heath, the powers at BSA decided to discontinue the basic Triumph-framed Tiger Cub, replacing it with a hybrid BSA Bantam D7 with a Cub engine and creating the T20B. While this might have made good economic sense, the T20B was always doomed as both Triumph and BSA enthusiasts treated it with suspicion. The Sport Cubs retained their Meriden frames, and the lineup was considerably streamlined this year, with the T20SH finishing in February. As the only Sports Cub left, by the middle of the year the T20SM was renamed the T20M. And with the serial numbers now with a BSA sequence, numbers were often duplicated.

Despite the move to Small Heath, some updates were incorporated on the Tiger Cub engine, the primary improvement being an increased-capacity oil pump. With the basic Cub now in a BSA Bantam chassis, the wheels were 18 inches front and rear, with wider brakes. But the whole machine was dimensionally larger, and it was unpopular. Already Tiger Cub sales were sliding in the wake of new Japanese competition, and it was only the US T20SM Mountain Cub that sustained it this year.

1966 T20B, T20SH, T20SM
(Differing from 1965)

Carburetor
Amal 375/61 Monobloc $25/32$-inch (T20B)
Wheels
WM1x18-inch front and rear (T20B)
Tires
3.25x18-inch front and rear (T20B)
Wheelbase
51.1 inches (T20B)
Dry Weight
220 pounds (T20B)
Colors
Nutley Blue and white (T20B); Metallic Blue and white (T20SH); Alaskan White (T20SM)
Engine and Frame Numbers
T20B 101 on (BSA numbers); T20SH 101 on (BSA numbers); T20M 2001 on (BSA numbers)

CHAPTER FOUR

New pistons and improved electrics made 1967 another great year for the Bonneville. Colors were initially aubergine and gold, with white later replacing the gold.

1967

Despite increasing quality-control issues caused by stretching Meriden's production facilities to the limit, Triumph was riding the crest of a wave in America, and it wasn't difficult to see why. Triumphs were successful in racing, and they were seen as the most desirable and versatile performance bikes on the market. Of the 33,406 British motorcycles imported into America this year, 24,700 were Triumphs, and over half of those T120Rs. Triumph twins, particularly the sport versions, were always beautiful, bikes you could sit and admire when it was too cold to ride outside. But continual refinement had also seen them evolve into truly useable, all-around performance motorcycles. The Japanese superbikes had yet to hit town, BSA was struggling with chronic quality control, Norton was still developing their Commando, and Triumph was ready to make the most of it. Triumph was without peer in 1967.

But it wasn't all good news, and after Harry Sturgeon left due to ill health in February 1967 (he died of a brain tumor in April), the new managing director of BSA Group was another appointment from the aircraft-oriented industry, Lionel Jofeh. Even more than Sturgeon, Jofeh was determined to pursue the complete amalgamation of BSA and Triumph. He also wanted to further increase Triumph's production and, according to Bert Hopwood, "literally hated Triumph's racing success." One of Jofeh's first projects was the establishment of a research and development center at Umberslade Hall. Although not operational until late 1968, this employed 300 staff, most without a motorcycle background, and had an annual budget of around £1,500,000. Although profits were still high, so were expenses, and many at Meriden felt they were bearing the brunt of it. The powerful unions were also increasingly mobilizing the workforce. In the United States, TriCor's Denis McCormack retired in July at the age of 65; Earl Miller succeeded him, but it wasn't the end for McCormack, who would later return to take over in more troubled times. After a short stint with Suzuki, Don Brown returned to the group as vice president and general manager of BSA, Inc., in New Jersey.

Work was also progressing on the new Trident, and while early prototypes followed the lines of the Bonneville, the BSA board decided to outsource the styling to Ogle Design in Hertfordshire. As an independent industrial design house Ogle had no experience with motorcycles, and their resulting Trident and BSA Rocket 3 styling wasn't greeted enthusiastically. It was originally planned to release the new triples at the

1963–1970

end of 1967, but indecision over the styling saw this delayed a year. In the meantime, four preproduction triples were sent to the United States at the end of 1967 for evaluation.

1967 T120, T120R, T120TT Bonneville 650

Despite the political uncertainty regarding Triumph's status within the BSA Group, the Bonneville continued to go from strength to strength. Each year saw successive improvements, and during 1967 there was a new E3134 profile inlet camshaft, a redesigned oil pump to reduce wet sumping, and Hepolite pistons (replacing Meriden's own). Also during the year, new Amal Concentric carburetors replaced the Monobloc and a 160-degree auto-advance dwell cam finally solved the ignition gremlins. UK versions now received the US 19-inch front wheel, and this was to be the final year for the T120TT Special Competition Bonneville.

When it came to performance, the 1967 Bonneville didn't disappoint. *Motor Cycle* managed a top speed

1967 T120, T120R, T120TT Bonneville
(Differing from 1966)

Carburetors
Twin Amal R/L930 Concentric 30mm (from DU59320)

Wheels
WM2x19-inch front and WM3x18-inch rear (T120 UK)

Tires
3.00x19-inch front and 3.50x18-inch rear (T120 UK)

Color
Aubergine and gold (white from DU48157)

Engine and Frame Numbers
T120 DU44394 to DU66245; T120R DU44394 to DU66245; T120TT DU44394 to DU66245

The US T120R with small gas tank and stainless steel fenders. UK versions were larger and painted.

103

CHAPTER FOUR

Early 1967 Bonnevilles such as this still had Amal Monobloc carburetors.

1967 TR6 TR6R TR6C Trophy (Differing from 1966)

Horsepower
43 at 6,300 rpm, 45 at 6,500 rpm (US)
Compression Ratio
9:1
Carburetor
Amal 389/239 Monobloc 1 3/16-in, Amal R30/9-B Concentric 30mm (after DU63043)
Colors
Mist Green and white
Engine and Frame Numbers
TR6, TR6R, TR6C, DU46201 to DU66246

of 110 miles per hour, and ex-factory MV Agusta rider John Hartle piloted a T120 to victory in the Hutchinson 100 at 83.87 miles per hour and the Isle of Man Production TT at 97.1 miles per hour. The Bonneville also won the 500-mile production race at Brands Hatch for the third successive year, factory tester Percy Tait teaming with Rod Gould. They won at 79.15 miles per hour, their Bonneville sporting a new twin leading shoe front brake that would be seen on the production version for 1968. And it wasn't just performance that kept the Bonneville on top. The Bonneville epitomized style and in the 1968 film *Coogan's Bluff,* in which Clint Eastwood rode an aubergine and white 1967 model T120R through New York's Central Park.

One of Evel Knievel's most widely publicized jumps was on a T120TT. On New Year's Eve 1967 Knievel attempted to jump the Caesars Palace fountains in Las Vegas. Almost unknown at the time, Knievel arrived in Las Vegas as a cowboy from Montana, but he left as a superstar—and it was on a Triumph. Knievel later said his T120TT slowed unexpectedly after hitting the takeoff ramp, resulting in a leap somewhat short of the projected 141 feet. He landed on the safety ramp supported by a van, sustaining a crushed pelvis and femur and multiple fractures. Knievel was in a coma for a month, but was this failure launched his career. Full of crazy ideas, Knievel also built an experimental nitro-powered Triumph Bonneville rigged with wings and twin jet engines that he hoped could jump the Grand Canyon. Although Knievel used Harleys for his stunts after 1970, from 1966 to 1968 he rode Triumphs and later said the Triumph was the best bike he ever jumped with.

1967 TR6, TR6R, TR6C Trophy 650

This year the single-carburetor Trophy also featured the T120's Hepolite pistons and updated oil pump, as well as its larger valves and initially a larger single Amal Monobloc carburetor. During the model year this became a new Amal Concentric. Maintaining the lower tune and broad spread of power, the exhaust camshaft was still the earlier, softer type. The TR6C exhaust pipes were now a dual staggered type on the left, and while the TR6R was ostensibly a single-carb Bonneville, the gas tank was a larger 3½-gallon type, still with a parcel rack. The Trophy was designed for riders considering their first high-performance heavyweight or those trading up from a lightweight motorcycle, and it proved extremely popular. Of Triumph's US sales this year, 30 percent were Trophys.

1967 T100S Tiger 100, T100T Tiger Daytona, T100R Daytona Super Sport, T100C 500 T90 Tiger 90 350

With the demise of the mundane Speed Twin and Twenty-One, the T100 embarked on a three-year period of refinement resulting directly from the racing program. The most significant update this year was the introduction of Doug Hele's twin-carb cylinder head for the

1963–1970

ABOVE: Designed as an entry into the big-bore motorcycle market, the TR6C was the most popular Triumph in America during 1967.

FAR LEFT: The pair of high-rise staggered exhausts on the left undoubtedly accentuated the TR6C's appeal.

LEFT: Early 1967 TR6Cs retained the older-style Amal Monobloc carburetor.

CHAPTER FOUR

Gary Nixon and the #1 Plate

As they weren't really expected to win, Triumph's 1966 Daytona victory surprised many, and it spurred Doug Hele and his team to increase their efforts for 1967. They built seven bikes for the 1967 season, six for the United States and one for Percy Tait in the United Kingdom. Updates this year included higher-compression (11:1) pistons, a squish-type cylinder head, and new polished cranks. The power didn't improve much, to 48.5 horsepower at 8,700 rpm, but the spread was wider. Other improvements saw a new low-level exhaust system, lower frame, shortened forks, a 210mm Fontana four leading shoe front brake, and a slimmer fiberglass fairing and gas tank. They were shipped from Meriden with 18-inch wheels, but at Daytona these were swapped for a 19-inch set (with Goodyear tires). Painted in blue and white, colors that would typify Triumph road racers for the next six years, the 301-pound racers handled superbly and topped 130 miles per hour in testing. Three machines were provided to TriCor, for Gary Nixon, Larry Palmgren, and Buddy Elmore, and three went to JoMo for Eddie Mulder, Dick Hammer, and Gene Romero. The race was a Triumph show from start to finish, with Nixon and Hammer trading the lead until Hammer crashed. Elmore then settled in to take second place, and Nixon won at a record 98.227 miles per hour.

After Daytona Nixon continued to dominate the AMA Championship, winning at a rain-soaked Loudon in addition to surprise flat-track wins at Portland, Oregon, a short-track win at Santa Fe (on a Tiger Cub), and a final road race victory at Carlsbad. A second place finish at the season half-mile finale at Oklahoma sealed the AMA Grand National Championship, Triumph's first.

Twenty-six-year-old Gary Nixon with TriCor's legendary tuner Cliff Guild after winning a wet Loudon-Lacona Classic in June 1967. *Triumph Motorcycles*

T100T Tiger Daytona and the US T100R Daytona Super Sport. This was the first twin-carb 500 since the pre-unit TR5AD of 1959. A single-carb Tiger T100S was also available in the United Kingdom, with a single-carburetor high-pipe T100C for the United States. This retained the battery-less ignition and was still a competitive Enduro machine, as Bill Baird proved by clinching his sixth straight National Enduro Championship.

The Daytona's new cylinder head featured the narrower included valve angle of the racer and larger inlet valves, as well as parallel, rather than splayed, inlet ports. The stubs were splayed to allow a pair of "pancake" air filters. The carburetors were Amal Monoblocs, slightly larger for the US T100R, with a pair of E3134 camshafts and racing tappets. Accompanying the engine updates was a new frame, also incorporating features from the Daytona racing bikes, that was shared across the range. The top and front tube diameter was increased, with the top tube now leading straight to the steering head, and the steering head angle increased to 28 degrees. With a stronger swingarm, a braced pivot, and an improved front fork, the T100 promised vastly improved handling. *Cycle* magazine tested a T100R in February 1967, saying, "It is a great motorcycle in almost every respect. Steady as a rock even on fast, bumpy bends that make some bikes act as though their frames had broken." Their

1967 T100S Tiger, T100T, T100R, T100C, T90 Tiger (Differing from 1966)

Horsepower	39 at 7,400 rpm (T100T); 41 at 7,200 rpm (T100R); 38 at 7,000 rpm (T100C)
Carburetors	Twin Amal 376/325/5 Monobloc 1 1/16-inch (T100T); 1 1/8-inch (T100R)
Dry Weight	341 pounds (T100T and T100R)
Color	Pacific Blue and Alaskan White (T100); Hi Fi Scarlet (T90)
Engine and Frame Numbers	T100T H49833 to H57082; T100R H49833 to H57082; T100C H49834 to H57082; T100S H49833 to H57082

The 1967 T100C shared the chassis improvements with the Daytona but still featured a single Amal Monobloc carburetor. The seat was all black this year. *Lorin Smith*

main criticism was of the 7-inch front brake: "such a strong and sporting motorcycle deserves a maximum effort in the braking department." Triumph paid heed, and this feature would appear for 1968.

1967 T20B Tiger Cub, T20B Super Cub, T20M Mountain Cub 200

As the T20B Bantam Cub was proving unpopular, a deluxe version, the T20B Super Cub was offered for 1967. This was mechanically identical to the T20B but ostensibly a BSA D10 Bantam with a Cub engine. The basic T20B continued alongside, and although production ended in June 1967, unsold stock was available for more than a year. In the States, the T20M Mountain Cub was the only version offered. After years of modifications to the plain big-end bearing, it had finally became a trouble-free single-row caged needle roller big-end bearing. The engine appearance was also changed by the adoption of a square cross-section barrel and cylinder head. But as far as BSA management was concerned, there was no future for the Tiger Cub, and they already planned to replace it with new BSA offerings. The Tiger Cub was doomed in America as well, with 650 twins outselling it by nearly 12 to 1.

1967 T20B, T20B SC, T20M
(Differing from 1966)

Colors
Bushfire Red (T20B SC); Grenadier Red and Alaskan White (T20SM)

Engine and Frame Numbers
T20B 101 on (BSA numbers); T20SC 101 on (BSA numbers); T20M 3001 on (BSA numbers)

CHAPTER FOUR

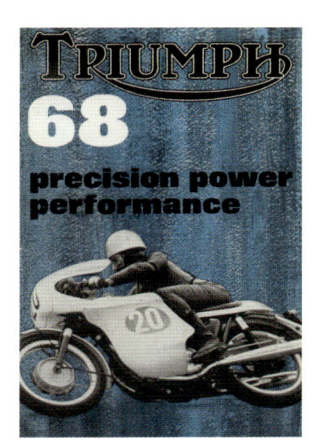

Triumph was on top in 1968 and proudly proclaimed it in their advertising literature.

1968

JoMo and TriCor optimistically thought they could sell 35,000 Triumphs in the United States during 1968, but Meriden still suffered workforce and quality-control troubles. However, Meriden's difficulties were nothing compared to BSA's at Small Heath. As the Bonneville entered its golden period, BSA's future looked grim. Not only was Umberslade Hall draining resources, but 4,000 US-bound water-damaged BSAs had to be returned to Small Heath for repair. Along with US warranty claims that were now billed to the parent company, profits slumped. The one shining light was that finally the new 750cc triple was being readied for production, although the decision to produce separate BSA and Triumph versions would prove costly and inefficient. And, after considerable deliberation, much of the controversial Ogle styling would be incorporated, ultimately to the detriment of sales and profits. In the wake of the failure of the T20B Cub, further evidence of dubious badge engineering occurred with the introduction of the TR25W single. This was virtually a BSA B25 and was targeted at the US market, where BSA and Triumph dealers were still independent.

This year was the calm before the storm and the final year of the Bonneville's superbike preeminence. In April Norton released their innovative 750cc Commando, and in October Honda announced the CB750 Four. By the next season, Triumph's own T150 Trident would replace the Bonneville as the flagship model. While the Bonneville now epitomized the finest attributes of race-bred development, its days of supremacy were over.

1968 T120, T120R Bonneville 650

As with fine wines, the best motorcycles also have standout vintage years, and for the T120 Bonneville these were between 1968 and 1970. The Bonneville came of age in 1968, now with a race-developed 8-inch twin leading shoe front brake and a front fork

The new 8-inch twin leading shoe front brake provided vastly superior braking performance. The 1968 version had straight pull cable actuation that was sometimes problematic, as the cable could jump off the anchor back plate. A split pin was later fitted to the cable abutment. *Triumph Motorcycles America*

Beautiful from any angle, the 1968 Bonneville was one of the greats; some consider this the finest year of all. This is beautiful restoration by Graham Cousens of Restoration Cycle Works, Michigan. *Triumph Motorcycles America*

with floating shuttle valves providing two-way damping. The twin cam actuated brake had grilled air scoops and outlets and provided vastly improved braking performance over its single leading shoe predecessor. Later in the model year a longer and stronger swingarm further improved the handling. Several engine updates were also incorporated, gradually solving small problems that had appeared over the years. These included new Hepolite pistons, improved cylinder barrel location, improved oil feed to the camshafts to alleviate the wear problem, and modified outer valve springs. Another significant update was the introduction of Lucas 6CA contact breakers that allowed for independent setting. Along with numerous gearbox

1968 T120, T120R Bonneville *(Differing from 1967)*

Front Brake
8-inch twin leading shoe
Color
Hi Fi Scarlet
Engine and Frame Numbers
T120 DU66246 to DU85903; T120R DU66246 to DU85903

108

1963–1970

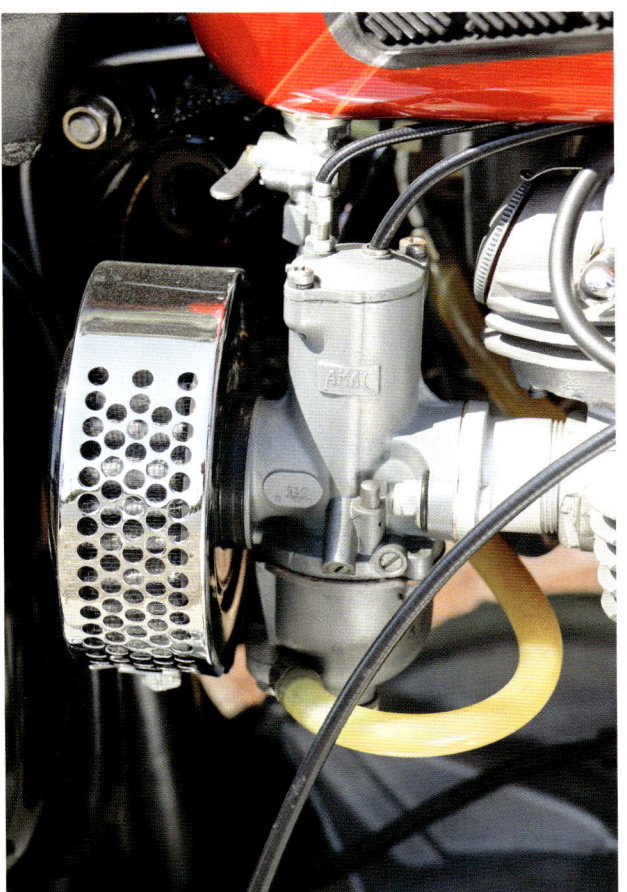

improvements, the Bonneville finally had the substance to match the style. The 2½-gallon gas tank on US versions may have limited the range to 100 miles, but this didn't detract from its popularity. As usual, UK examples differed in detail, still with the longer mufflers, larger gas tank with a parcel rack for the final year, and optional air filters.

1968 TR6, TR6R, TR6C Trophy 650

Harmonizing the Trophy with the Bonneville this year saw a racing E3134 camshaft to match the inlet, along with the Bonneville's valve springs, new cylinder barrel nuts, rocker oil feed, and Lucas 6CA contact breakers. The Trophy also received the Bonneville's new front fork and 8-inch twin leading shoe front brake, and,

ABOVE: The 1968 dual seat had a chrome trim band around the base. US models such as this still featured an alloy taillight housing, but now with red reflectors to comply with federal regulations. UK versions retained the earlier taillight bracket. *Triumph Motorcycles America*

LEFT: Although Amal Concentric carburetors were fitted to some 1967 Bonnevilles, 1968 was the first full year of their use. Initial teething problems, such as sticking slides, were solved this year.

1968 TR6, TR6R, TR6C Trophy *(Differing from 1967)*

Front Brake
8-inch twin leading shoe
Colors
Riviera Blue and silver
Engine and Frame Numbers
TR6, TR6R, TR6C, DU68364 to DU85903

109

CHAPTER FOUR

The 1968 TR6R in blue and silver. US versions had side reflectors this year. The front brake was fitted with garish chrome hub plate that was purely cosmetic.

later, the longer, strengthened swingarm. The TR6R continued with painted fenders and a 3½-gallon gas tank with parcel rack, while the TR6C retained stainless steel fenders and the smaller gas tank this year.

When testing a TR6R in February 1968, *Cycle* magazine found the new Amal Concentric carburetor flawed: surging, dying while idling, and overheating in traffic. They wrote, "Alas, the Amal (latest series '900') just doesn't work satisfactorily." Covering a standing quarter mile in 14.2 seconds at 92.1 miles per hour indicated that the single-carb touring model gave away little to the more frenetic Bonneville. *Cycle* summed the Trophy by saying, "The plain truth is we like the Triumph Trophy because it does well what it was made to do."

1968 T100S Tiger 100, T100T Tiger Daytona, T100R Daytona Super Sport, T100C 500 T90 Tiger 90 350

Looking for their third win in three years, Triumph returned to Daytona in 1968 with four bikes. This year they were a development of the 1967 version, the engines including through-bolts fixing the cylinder head and barrels, and to improve weight distribution the oil tanks were moved in front of the crankcases. The new frame was all welded, without the usual bolted-on rear subframe, but from the outset the Triumphs were outclassed. Harley-Davidson reportedly spent $500,000 developing new bikes, and they ran 150 miles per hour in qualifying compared to Nixon's 136 miles per hour. In the race all the Triumphs suffered misfiring caused by faults in the Lucas Energy Transfer ignition system. The best-placed Triumph was Elmore's seventh. Although Triumph was not happy with the Daytona result and threatened to quit all forms of competition, Nixon did go on to provide Triumph their second number-one AMA Grand National Championship. This time he only won at Houston (on a TR25W) and at Columbus, but consistency earned him the title.

The T100 also received many of the improvements incorporated on the 650cc twins this year, notably the two-way damped front fork and Lucas 6CA

110

1968 T100S Tiger, T100T, T100R Daytona, T100C, T90 Tiger
(Differing from 1967)

Carburetors
Twin Amal R626/10 L626/9 Concentric 26mm (T100T, T100R); single R626/8 Concentric 26mm (T100S and T100C); Amal Concentric R624/2 24mm (T90)

Front Brake
8-inch (T100T, T100R)

Color
Aquamarine Green and silver (T100); Riviera Blue and silver (T90)

Engine and Frame Numbers
T100T H57083 to H65572; T100R H57083 to H65572; T100C H58017 to H65572; T100S H57083 to H65572

contact breakers. The carburetors were now Amal Concentrics, and the single-carburetor T100S and T100C adopted the Daytona's cylinder head with the narrower included valve angle and larger inlet valve. The T100T and T100R Daytona received the 8-inch single leading shoe brake from the previous T120 and TR6. Another improvement on the T100C this year was 12-volt coil ignition replacing the troublesome Energy Transfer system. The 350cc T90 also received an Amal Concentric carburetor; this was to be the T90's final year, with the styling now mimicking the T100S Tiger.

1968 TR25W Trophy 250, T20B Super Cub 200

With only the T20B Super Cub remaining, and sales gradually diminishing, BSA decided to fill the gap for a single-cylinder Triumph with the introduction of the TR25W Trophy. No one was really fooled by the TR25W's origins, and while it carried the familiar Triumph Eyebrow tank badges and T90 front fork, there was no mistaking that the TR25W was actually a BSA B25 dressed to look like a trail bike. The B25 wasn't a new design: its origins went back to Edward Turner's BSA C15 250 of 1958. Inspired by the Terrier and Tiger Cub, the C15 evolved into the BSA B25 in 1967. New crankcases housed a one-piece forged crankshaft, the power was increased 50 percent, and with 12-volt electrics, the TR25W looked well equipped, on paper at least. Unfortunately, the quest for higher engine performance resulted in serious unreliability. Con-rod bearings self-destructed, and oil leaks were endemic. On top of this, the vibration was hammering at higher speeds. Although BSA in Birmingham built the TR25W, warranty and the associated costs were borne by Meriden, and all this did was fuel resentment between the two organizations.

Gary Nixon again provided Triumph the AMA #1 plate in 1968, but Harley-Davidson was much stronger competition this year and Nixon only won through consistency.

Ostensibly a BSA, the TR25W Trophy was unreliable and unloved.

CHAPTER FOUR

1968 TR25W, T20B SC (Differing from 1967)

Bore	67mm (TR25W)
Stroke	70mm (TR25W)
Capacity	247cc (TR25W)
Horsepower	22 at 8,250 rpm (TR25W)
Compression Ratio	8.5:1 (TR25W)
Carburetors	Amal R626 26mm (TR25W)
Gearbox	Four-speed (TR25W)
Front Fork	Telescopic (TR25W)
Rear Suspension	Swingarm (TR25W)
Brakes	7-inch front and rear (TR25W)
Tires	3.25x19-inch front and 4.00x18-inch rear (TR25W)
Wheelbase	52 inches (TR25W)
Dry Weight	285 pounds (TR25W)
Colors	Hi Fi Scarlet (TR25W); Firecracker Red (T20B SC)

1969

1969 was the Year of the Superbike, with nearly all manufacturers producing big, fast motorcycles, and Triumph and BSA were at the forefront with their new triple. But despite this, the triples weren't initially received with particular enthusiasm in America, and demand for Triumph twins peaked. Meriden was now building 900 bikes a week, selling 24,407 motorcycles in the United States this year (compared to 2,143 in the United Kingdom). BSA's woes continued, and after some poor business decisions, their group shares plummeted and profits fell dramatically. In July 1969 Jofeh announced the redundancy of 1,200 workers at Small Heath, and one month later he released plans to merge the four separate US distributors into one organization, BSACI, with a new president, Peter Thornton. Thornton came from an advertising background and

The new Trident triple finally arrived for 1969, but the boxy styling proved unpopular. Although a Dunlop K81 "TT100" tire was fitted on the rear, a matching front didn't appear until 1970. The fiberglass side covers enclosed a single air filter serving all three carburetors.

RIGHT: Although the Trident and Rocket 3 engines were similar, the Trident cylinders were upright while the BSA's were canted forward. The oil cooler was a standard fitting.

RIGHT BELOW: This exploded drawing of the T150 engine displays the Triumph origins and overall complexity of the design.

wreaked havoc by alienating many of Triumph's loyal and long-serving personnel.

1969 T150 Trident 750

Triumph Trident and Rocket 3 production commenced in August 1968 for the 1969 model year. The new triple was badly needed, particularly by BSA, which hadn't had a new model for seven years, but also for Triumph, as the Bonneville was now close to its limit for racing. Originally considered a five-year stop-gap model, the Trident was similar to the Rocket 3, but the two differed in appearance and details. This requirement came from the United States, with dealers requesting the triple be produced in two distinct versions. The three-cylinder engine was designed at Meriden by Bert Hopwood, Doug Hele, and Jack Wickes and was ostensibly a "Tiger and a half," with a third cylinder added to the existing twin. The valve gear followed Edward Turner's twin camshaft pattern, but with a smaller bore and longer stroke, identical to the TR25W, to minimize engine width. At the heart of the engine was a forged crankshaft, reheated and twisted to provide 120-degree crank throws, and unlike the twins the triples featured aluminum cylinders. As these were upright on the Trident and slanted forward on the BSA, many engine castings weren't shared. The engine lower end was still vertically split like a twin, and as it consisted of seven different castings, oil leaks were still a problem.

The Trident's engine/gearbox unit was mounted in a single downtube, bolted-up, Bonneville-based frame (the Rocket 3's all-welded frame was Duplex), with a front fork and 8-inch twin leading shoe front brake also similar to the Bonneville's. And although the Ogle styling was generally disliked, the boxy gas tank and "ray-gun" silencers remained.

While the three-cylinder Trident was a Meriden conception based on existing Triumph designs, most of its manufacture was at Small Heath. Many components were extremely difficult and complex to manufacture, and as it was considered a stopgap, retooling was kept to a minimum. Manufacture was very labor intensive: engine assembly required 56 stages, and as a consequence the finished Trident was expensive, at $1,750 compared to $1,439 for a Norton Commando and $1,495 for Honda's CB750. Additionally, early examples suffered from a series of manufacturing issues resulting in a number of expensive warranty claims.

The Trident may have been expensive, poorly manufactured, and afflicted with unusual styling, but

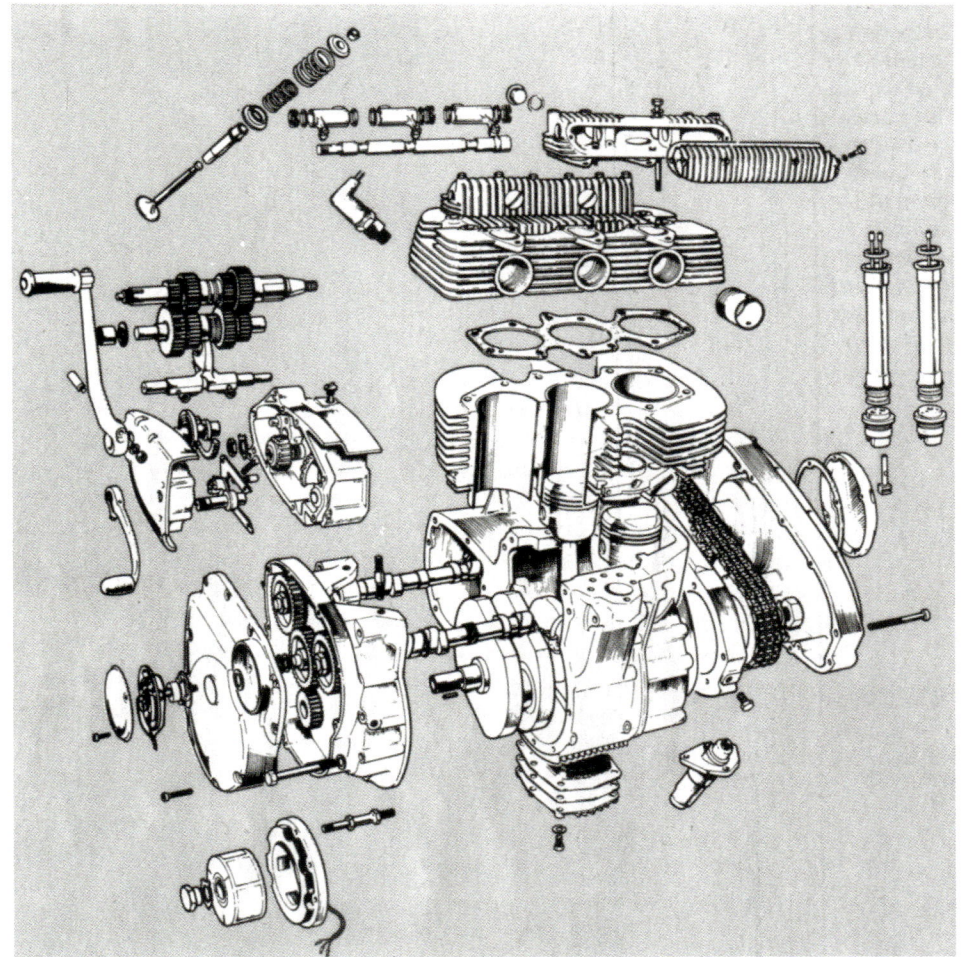

CHAPTER FOUR

The 750cc Trident was the first superbike, and it ushered in a new era. But with its right-side gearshift, lack of electric start, and drum front brake, it didn't seem as modern as Honda's new 750 Four.

1969 T150 Trident

Bore	67mm
Stroke	70mm
Capacity	740cc
Horsepower	58 at 7,250 rpm
Compression Ratio	9.5:1
Carburetors	Triple Amal 626 Concentric 27mm
Ignition	Lucas 7CA contact breakers
Gearbox	Four-speed
Front Fork	Telescopic
Rear Suspension	Swingarm
Brakes	8-inch TLS front and 7-inch rear
Wheels	WM2x19-inch front and WM3x19-inch rear
Tires	3.25x19-inch front and 4.10x19-inch rear
Wheelbase	56.25 inches
Dry Weight	470 pounds
Color	Aquamarine

it was undeniably fast. In *Cycle* magazine's seminal March 1970 Superbike comparison test of seven 1969 models, the T150 posted the second-fastest quarter mile (behind the lighter Norton Commando) of 12.78 seconds at 103.92 miles per hour. After four stock BSA Rocket 3s set several new speed records at Daytona, including an astonishing 131.790 miles per hour over 5 miles, Triumph's West Coast subsidiary offered $25,000 in prize money for any record by a Trident-powered machine during Bonneville Speed Week. The new records included Rusty Bradley's 168.891 miles per hour for a stock-engine, partially faired, gasoline-powered 750, but it still didn't result in increased sales. In America this year BSA/Triumph sold 7,000 triples, compared to Honda's 30,000 Fours. And in a reader's poll in *Cycle* magazine, only 4.6 percent rated the Trident as the best superbike, against 36 percent for the Honda Four. And they were considered so ugly dealers couldn't give Tridents away. Something had to be done—and something would happen for 1970.

1969 T120, T120R Bonneville 650

This year saw the Bonneville further refined into arguably the finest of all. By now nearly all the threads were unified to UNF, the Unified Fine Thread standard, a process that had been gradually introduced since 1967, and work with Amal had improved the unreliable carburetion. The carbs were rubber-mounted this year and the camshafts hardened to cure the cam wear problem; the crankshaft included a heavier flywheel, and to reduce noise, the exhaust header pipes included a balance tube. This allowed UK versions to finally run with the shorter US mufflers. An electric oil pressure switch was now located on the front crankcase, operating a warning light in the headlamp shell. US versions were also fitted with unreliable German-made SIBA ignition coils. Chassis updates saw the front brake cable rerouted, solving the earlier cable problems, and the Eyebrow gas tank emblem made way for a new, smaller and simplified type in aluminum.

Not only was 1969 a high point for the Bonneville in specification, it was also a high point for racing success. Malcolm Uphill won the Isle of Man Production TT on a Bonneville, at an average speed of 99.99 miles per hour miles per hour, with the first-ever

One of the Ogle design features that continued on the production version was the distinctive "ray-gun" muffler. These may have looked unusual, but they were surprisingly effective and emitted an exhilarating sound.

Serial Numbers from 1969

With the Trident and TR25W assembled at Small Heath, early in 1969 a new BSA-style serial number system was introduced. This featured a two-letter prefix indicating the month and year of build followed by a number series, irrespective of the model. In 1981 a third letter (A) was added to avoid repetition.

Month		Year (model year from previous September to August)	
A	January	C	1969B
	February	D	1970
C	March	E	1971
D	April	G	1972
E	May	H	1973
G	June	J	1974
H	July	K	1975
J	August	N	1976
K	September	P	1977
N	October	X	1978
P	November	A	1979
X	December	B	1980
		DA	1981–Jan 1982
		EA	Feb 1982–1983

115

CHAPTER FOUR

ABOVE: The front brake cable routing was improved for 1969.

RIGHT: The fenders were painted for 1969, and the rear Girling shock absorbers had a more modern look with exposed chrome-plate springs.

BELOW: The 1969 Bonneville was painted in new colors of an orangey Olympic Flame and silver. It was still a beautiful motorcycle.

1963–1970

For 1969 the TR6R became the Tiger and was finished in Trophy Red. It was separated from the Bonneville only by the single Amal concentric carburetor.

1969 T120, T120R Bonneville (Differing from 1968)

Color
Olympic Flame and silver
Engine and Frame Numbers
T120, T120R DU85904 to DU90282, then NC00100 to HC24346

100-mile-per-hour lap on a production motorcycle. Because of this Dunlop renamed their K81 tire the TT100. Uphill also teamed with Steve Jolly for a class victory in the Barcelona 24-hour endurance race and with Percy Tait to take the Thruxton 500 race at a record speed. This was a great event for the Triumph Bonneville, with T120s following in second, fifth, sixth, and seventh.

The Trident may have arrived as Triumph's new leader of the pack, but the Bonneville was still a serious contender. As *Cycle* magazine summed up in their March 1969 test, "Just like it has been for a long time now, the Bonneville is one of the two or three most *desirable* motorcycles being made in the world today."

1969 TR6, TR6R Tiger, TR6C Trophy 650

This year the Tiger replaced the Trophy as the name for the TR6 and TR6R, although Trophy continued for the US TR6C. Engine and chassis updates were shared with the Bonneville, but the TR6, with its larger gas tank, lost the parcel rack after Triumph was sued for damage to a rider's testes following an accident.

The classic Triumph Bonneville and TR6R instrument layout included a gray-faced Smiths speedometer and tachometer, with an ammeter in the headlamp shell.

117

CHAPTER FOUR

The TR6R also shared the Bonneville's 8-inch twin leading shoe front brake with revised cable routing. The TR6's color was arguably more attractive than the Bonneville this year.

1969 TR6, TR6R Tiger, TR6C Trophy
(Differing from 1968)

Color
Trophy Red

Engine and Frame Numbers
TR6, TR6R, TR6C, DU87124 to DU88524; then NC02352 to HC2434

1969 T100S Tiger 100, T100T Tiger Daytona, T100R Daytona, T100C Trophy 500

After the disappointing 1968 Daytona result, Meriden decided to concentrate on production racing, and Gary Nixon's ninth-place finish in the 1969 Daytona 200 was their worst result since 1960. Nixon lost the number-one plate, and Triumph's racing future now lay with the Trident. As far as the production T100 went, updates including a stronger bottom end, hardened camshafts, new timing side ball bearing, and improved sealing of the pushrod tubes resulted in a significantly improved machine. With the T120 and TR6's 8-inch twin leading shoe front brake, the T100T and T100R was undoubtedly the finest T100 yet. The T100S and T100C received a smaller, 7-inch twin leading shoe front brake. As far as performance went, the Daytona was one of the fastest 500s available, with *Motor Cycle* achieving a one-way top speed of 107 miles per hour miles per hour in October 1968.

1969 TR25W Trophy 250

Due to the many reliability issues, the TR25W's 1968 introduction was inauspicious, and numerous minor engine updates were incorporated for 1969. These still didn't solve the oil-leak problem, but the drive-side roller main bearing was an improvement. Also new this year were an exhaust system running outside the frame and a twin leading shoe front brake. But the TR25W Trophy still suffered from very poor build quality, and although it was envisaged as an entry-level model, it was plagued with many problems that had a negative effect on promoting Triumph.

1969 T100S, T100T, T100R, T100C
(Differing from 1968)

Front Brake
8-inch twin leading shoe (T100T, T100R); 7-inch TLS (T100S, T100C)
Dry Weight
352 pounds (T100T, T100S)
Color
Lincoln Green and Silver Sheen (T100)
Engine and Frame Numbers
T100C H66124 to H66976 then XC07583 to EC19426; T100T, T100R, T100S H65573 to H67330 then XC06297 to HC24527

1969 TR25W (Differing from 1968)

Carburetor
Amal R928 28mm
Brakes
7-inch twin leading shoe front
Dry Weight
311 pounds

For its second season, the unloved TR25W received a new front brake and revised exhaust system.

1970

Toward the end of 1969 Lionel Jofeh was so concerned about the efficiency of Umberslade Hall he asked Bert Hopwood to have a look at the situation and rectify any problems. Hopwood was dismayed with what he found, and in January 1970 he resigned in disgust. Umberslade Hall was sucking BSA's resources with dubious projects such as the Ariel 3 Trike, and more effort was wasted on developing Edward Turner's 350cc double overhead camshaft twin. As overall motorcycle sales in America continued to climb, reaching 638,763 for the 1969–1970 period, BSA boss Eric Turner remained optimistic. A new Trident/Rocket 3 assembly section was built at Small Heath, and with Meriden going flat-out to satisfy US demand for 500 and 650cc twins, often working seven days a week, the first half of 1970 saw UK motorcycle export records shattered. But Triumph supply to the United States was still a concern, and in June 1970 Peter Thornton placed a full-page advertisement in *Cycle* magazine claiming, "Since December we have doubled production, airfreighted units as they come off the line, and your dealer will shortly be able to provide your machine from stock."

1970 T150 Trident 750

US dealer feedback forced Triumph to introduce 124 engineering updates to the T150 and Rocket 3 for 1970. These ranged from new castings to reduced porosity and precision lapping machines and new gaskets to reduce oil leakage to the purchase of an expensive new German-made crankshaft-grinding machine. While the UK T150 was visually unchanged, retaining the Ogle-inspired styling, for the United States the most significant update was the introduction of a "beauty kit," or 1970 North American variant, in February 1970. This comprised a traditional Tiger 650 3½-gallon gas tank and mufflers, rounder side covers, and individual air filters, and it was a success. *Cycle Guide* stated in their July 1970 test, "The bike looks lighter, leaner, and much more contemporary."

1970 T120, T120R Bonneville 650

By 1970 the Bonneville had evolved into the quintessential Triumph twin. As the 1969 T120 was already

1970 T150 Trident
(Differing from 1969)

Horsepower
60 at 7,250 rpm (US)
Tire
3.60x19-inch front Dunlop K81
Colors
Aquamarine or Olympic Flame and silver (UK); Spring Gold and black (US)
Engine and Frame Numbers
T150 JD24849 to ND60540

The 1970 model UK T150 was very similar to the 1969 version, still with covered rear shock absorber springs. The fiberglass side covers were black.

119

CHAPTER FOUR

TOP: For many, the 1970 Bonneville represented the end of the line for the pure Meriden twins.

ABOVE: The ugly UK-spec Bonneville with a larger, oval-shaped gas tank and a slightly different color scheme.

finely developed, updates were minor, the engine now including a crankcase breathing arrangement similar to the Trident's. Chassis-wise, the front engine mounts were now bolted on, and the front fork received minor changes. Unfortunately, not all the updates were improvements, and the new gearbox cam plate was inferior and would cause problems until its replacement in 1973. Later in the model year, black-faced Smiths instruments replaced the gray-faced set.

As the AMA sanctioned 750cc overhead-valve engines for dirt track competition, and 200 examples were required for homologation, during 1970 TriCor joined forces with drag racer Sonny Routt to produce 204 750cc T120RTs for homologation. Rod Coates masterminded the collaboration, and Routt had excellent credentials, for at the time his dual-engined

1963–1970

Early Racing Triples

After losing at Daytona in 1968, Meriden wasn't too enthusiastic about further involvement in AMA racing, but a change in the rules allowing overhead-valve 750s brought a change of heart. The push to race the Trident came from Pete Colman in the States, who persuaded BSACI President Peter Thornton to sanction a joint BSA/Triumph effort working out of Duarte, California. Thornton agreed to a budget of $440,000 for the 1970 season, and the BSA/Triumph board in England decided to mount a serious assault on the new Formula 750 class. Toward the end of 1969, Doug Hele began developing the triple at Meriden, with the intention of entering the 1970 Daytona 200. Incorporating many of the lessons learned from racing the 500 twins, Hele raised the compression ratio to 11:1 and bored the cylinders to maximum oversize (as allowed by the regulations), and with a trio of 1 3/16-inch Amal GP carburetors and a special 3-into-1 exhaust system, the engines produced around 84 horsepower at 8,250 rpm. A Quaife five-speed gearbox transmitted the power, and Hele commissioned independent chassis builder Rob North to provide six special frames. On the front was a Fontana 10-inch four leading shoe front brake; on the rear a Lockheed disc brake. As aerodynamics was becoming increasingly important, the fairing was developed at the RAF wind tunnel at Farnborough. In early testing Percy Tait managed a top speed on the 375-pound triple, and it looked good for Daytona.

ABOVE: Fred Swift working on one of the 1970 Triumph racers in the Meriden race shop. Jack Shemans is in the center and Arthur Jakeman in the background.

BELOW LEFT: Blessed with movie-star looks, Gene "Burrito" Romero outlasted his competitors in a grueling 1970 AMA season to become AMA Grand National Champion and provide Triumph their third #1 plate.

BELOW RIGHT: Malcolm Uphill rode a Trident to victory in the 1970 Isle of Man TT Production race.

For Daytona, BSA/Triumph assembled an all-star team that included an out-of-retirement Mike Hailwood, Dave Aldana, and Jim Rice on BSAs, with Gene Romero, Gary Nixon, and Don Castro on Triumphs. Percy Tait was also at Daytona as an advisor, but he persuaded Rob North to build an additional frame, and Triumph quickly provided a seventh bike. Nicknamed "Burrito," Romero qualified fastest at 157.342 miles per hour but lost the race by three seconds to Dick Mann on the Honda Four.

The Trident may have lost at Daytona, but the triples were more successful during the rest of the 1970 AMA roadracing season, with Aldana winning at Talladega and Nixon at Loudon. The triple was also adapted for dirt track racing, but all the riders found them too wide and heavy, with unsuitable power delivery. This season Romero battled with BSA's Jim Rice for the AMA number-one plate, narrowly taking the championship.

Although Triumph was involved in some exploratory production racing outings on the Trident during 1969, they took a more serious approach for 1970. After Percy Tait finished second behind Uphill's Bonneville at the North West 200 production race, two Tridents were prepared for the Isle of Man TT production race in June. This year Uphill rode a Trident, beating Peter Williams' Norton Commando by only 1.6 seconds after 188.65 miles of racing. Factory BSA and Triumph triples were also entered in the 1970 Bol d'Or 24 Hour race at Montlhéry, where Paul Smart and Tom Dickie won at 76.55 miles per hour. But while the 1970 season had proved the Trident was a competitive racer, it would become dominant in 1971.

121

CHAPTER FOUR

Setting the 1970 Bonneville apart from the 1969 version was the new engine breather, a thick tube exiting the rear of the crankcase and running to the top of the rear fender. The black-and-chrome passenger grab rail was integral with the rear fender stay this year.

1970 T120, T120R Bonneville *(Differing from 1969)*

Color
Astral Red and silver
Engine and Frame Numbers
T120, T120R JD24849 to ND60540

Bonneville nitro-methane-fueled 280-horsepower dragster was the quickest motorcycle in the world, running a quarter mile in 9.10 seconds at 165 miles per hour. His 750 kits included new barrels and 76mm pistons and were fitted into standard T120Rs. It was such a clean modification, the T120RT looked virtually indistinguishable from the original.

1970 TR6, TR6R Tiger, TR6C Trophy 650

As usual, TR6 updates mirrored those of the T120, with the TR6 also including the revised engine breather and

1970 TR6, TR6R Tiger, TR6C Trophy *(Differing from 1969)*

Carburetor
Amal R930/45 30mm Concentric
(From AD39329)
Color
Spring Gold
Engine and Frame Numbers
TR6, TR6R, TR6C HD23795 to ND60540

frame engine plates. The US TR6C retained a speedometer only, plus stainless steel fenders.

1970 T100S Tiger 100, T100T Daytona Super Sport, T100R Daytona, T100C Trophy 500

The 500 twin also received the new engine breather arrangement, but ostensibly development had stalled.

LEFT: New for the 1970 Bonneville were the dome-shaped Windtone horns. Unlike the earlier welded front engine mountings, these were now bolted-on plates.

ABOVE: The 1969 and 1970 US Bonnevilles featured gas tank scallops inspired by 1950s custom painter Gurley.

BELOW: Although overshadowed by the flashier Bonneville, the Tiger shared all of the T120's 1970 updates.

CHAPTER FOUR

The year 1970 would also be the final one for the Triumph 8-inch twin leading shoe front brake on the TR6R Tiger and Bonneville.

1963–1970

As work was progressing on the new 350 Bandit/Fury with the obvious intention of replacing the T100, the T100 simply marked time. In February 1970 *Cycle World* noted, "In terms of development the T100 is at the same stage as the Austin-Healey sports car. Both have been refined so conservatively, that they are on the verge of extinction."

1970 T100S, T100T, T100R, T100C
(Differing from 1969)

Dry Weight
354 pounds (T100R); 340 pounds (T100C)
Color
Jacaranda Purple and Silver Sheen
Engine and Frame Numbers
T100C KD28652 to KD60261; T100T, T100R, T100S HD27850 to KD60280

1970 TR25W (Differing from 1969)

Compression Ratio 10:1

1970 TR25W Trophy 250

For its third year, the TR25W received more unsuccessful modifications to solve the terrible oil leaks, as well as a new exhaust system, this time on the left. But by now the damage was done: some Triumph dealers branded the bike "BSA's revenge," and it wouldn't survive the coming 1971 upheaval.

ABOVE: By 1970 the T100 was at the end of its development cycle, but circumstances dictated that it continue for a few more years. *Triumph Motorcycles America*

BELOW: The 1970 TR25W had a new large silencer on the left, with an ugly fiberglass heat shield. There was no center stand.

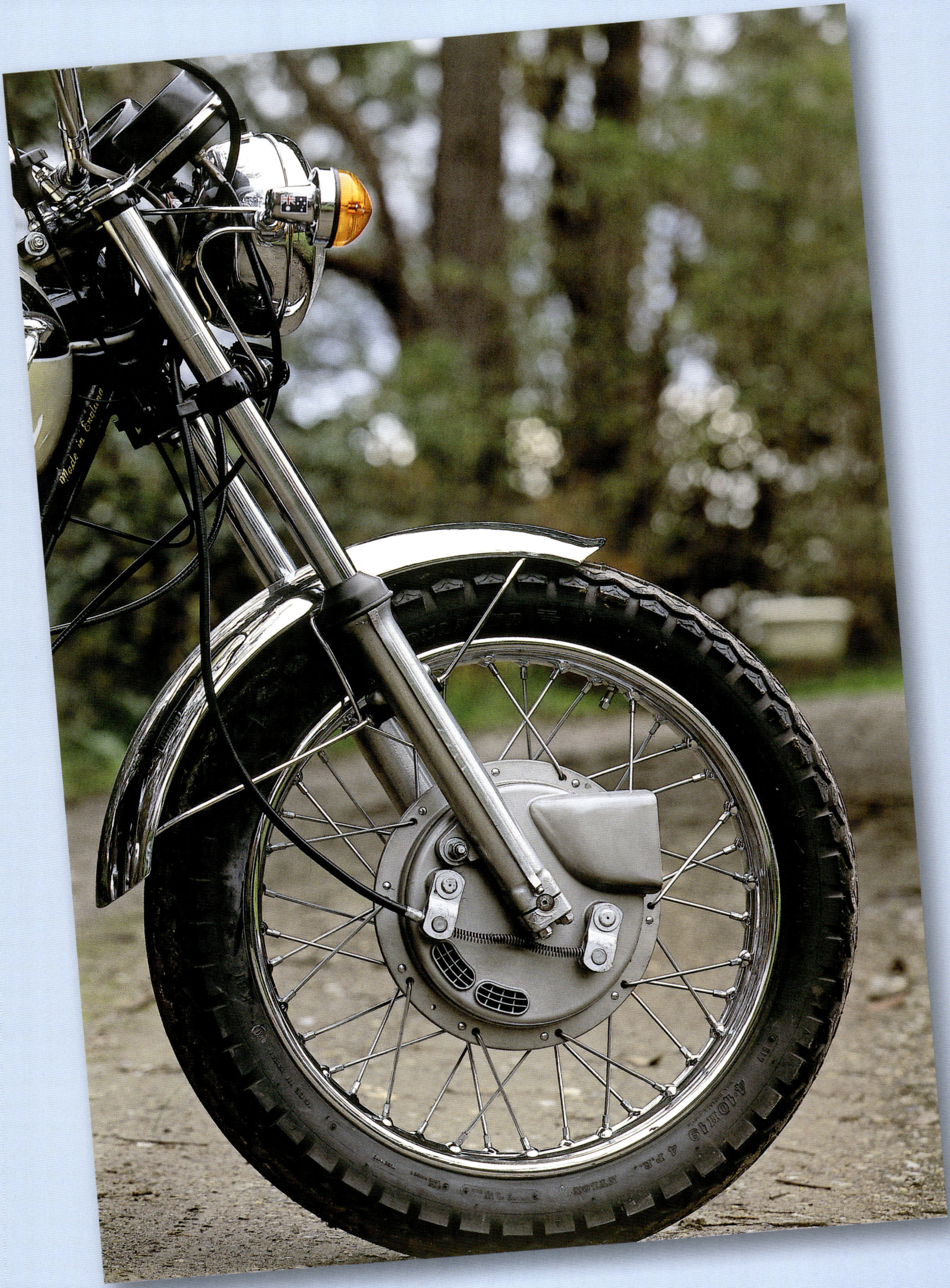

5 1971–1975
Turmoil Times

Norton Villiers Triumph and the Meriden Blockade

1971

In November 1970 BSA heralded the 1971 model range by hosting lavish press launches on both sides of the Atlantic. These signified excess over substance and were a harbinger of the future. Among the new models was Edward Turner's final design, the ill-fated Triumph Bandit/BSA Fury, a double overhead camshaft 350cc twin that would never make it into production, and the new Bonneville that would be plagued by production delays. After experiencing a sell-out year in 1970, BSACI boss Peter Thornton anticipated sales of 50,000 for 1971, but this proved unrealistic. By 1971 BSA/Triumph's joint share of the US market had fallen to just 6.9 percent, fifth in the market behind all the Japanese manufacturers. Back in the United Kingdom, the BSA Group was in serious financial trouble, and a loss of more than £3 million resulted in Barclays Bank demanding that heads roll. The extravagant Thornton was forced to resign in June, Jofeh followed shortly after in July, and respected financier Lord Shawcross took over from Eric Turner at the end of the year. Three thousand employees were sacked at BSA's Small Heath factory, and amid all this executive turmoil long-term Triumph man Denis McCormack came out of retirement to head the US BSACI operation. But it was all a temporary stay of execution. By the end of 1971, BSA Group's debt was £22 million, and in early 1972 Small Heath manufacturing was scaled down and Umberslade Hall closed. BSA was effectively dead, but while industrial problems and petty strikes were rife at Meriden, Triumph survived. And to make it even sweeter for Triumph, in December 1971 Bert Hopwood was asked to rejoin the BSA board as an executive responsible for motorcycle design and engineering.

ABOVE: Edward Turner's 350cc double overhead camshaft Bandit was his final design. Here he is with his right-hand man, Jack Wickes (left). The Bandit was released in a blaze of publicity but never made it into production.

OPPOSITE: Not only was the front fork new, but the front fender was now shorter, lighter, and less cumbersome. The wire stays for the headlight and fender contributed to a small loss in weight, while the new, conical 8-inch front brake looked efficient but was actually less effective than its predecessor.

CHAPTER FIVE

The Trident was restyled for 1971, all versions now with the smaller gas tank. Other updates included a Ceriani-style front fork with exposed stanchions.

1971 T150 Trident 750

Due to the difficulties afflicting the introduction of the 1971 650 twins, which effectively halted production for three months, the Trident received mainly cosmetic updates this year. Later in the year a five-speed gearbox was also listed as an option, but it wasn't available until 1972. The success of the 1970 "beauty kits" saw all Tridents now with the traditional gas tank and simplified side covers, but new was a more modern-looking Ceriani-style front fork without gaiters, minimal fenders, wire fender and headlight stays, dual megaphone-type mufflers, and a fade-prone conical 8-inch front brake. In July 1971 *Cycle* magazine found the new brake "no match for the Trident's speed and mass" and suggested "the technological spinoff from the road racing effort should result in a super effective disc brake for the Trident." They did find the power impressive, as the Trident ran the quickest quarter mile yet for a stock motorcycle, 12.90 seconds at 102.73 miles per hour. And they were impressed with the engine, saying, "The Trident has that magnetic, hypnotic quality that comes from its power to sustain the upward rush."

1971 T150 Trident (Differing from 1970)

Compression Ratio	9:1
Front tire	4.10x19-inch Dunlop K81
Wheelbase	57 inches
Dry Weight	460 pounds
Color	Spring Gold or black
Engine and Frame Numbers	T150 NE01436 to HE30869

The P39 Frame: 1971 T120R Bonneville, TR6R Tiger, TR6C Trophy 650

Back in 1969 BSA had announced their way forward was to incorporate updated chassis with the existing engines, and the first of these new designs was the P39. Designed at Umberslade Hall over a three-and-a-half-year period, the new frame was Triumph's first all-welded production frame and their first Duplex type

Road Racing 1971; The Year of the Triple

BSA/Triumph may have been in a financial crisis, but they still saw racing as an important sales tool, and 1971 would be Triumph's greatest year ever on the track. For the 1971 Daytona 200, Meriden prepared four new bikes, two Triumph and two BSA, to run alongside six American-based updated 1970 models. This year Rob North built new frames and, as the Fontana drum brake would fade during a race, fitted a wider, lower front fork to accommodate twin Lockheed brake calipers and lightweight metal-sprayed discs. As the front end was lower, these new racers were unofficially termed the "low boy." Other updates saw special "squish" pistons and combustion chambers and Amal Concentric carburetors, although most bikes at Daytona ran with the earlier GP carbs. The power with the "squish" engine this year was around 85 horsepower, but the high-compression engine was potentially unreliable.

The BSA/Triumph lineup for Daytona was formidable, with Triumph headed by Paul Smart and Gene Romero, along with Don Castro, Tom Rockwood, and the injured Gary Nixon. BSA had an equally strong team, including 1970 winner Dick Mann, the great Mike Hailwood, Jim Rice, Dave Aldana, and Don Emde. Only Smart, Romero, Mann, and Hailwood received the new "low boy" versions, and Smart put himself on pole with a 105.80-mile-per-hour lap. In the race Smart looked a winner until his high-compression "squish" engine blew with 11 laps to go; Mann took the lead to win at 104.737 miles per hour with a more conservatively tuned, non-squish engine and GP carburetors. Romero finished second again, with a young Don Emde third, providing BSA/Triumph a 1-2-3. Dick Mann went on to win the 1971 Grand National Championship on his BSA triple, with Romero finishing second, marking the end of the era of British bike success in the AMA Championship.

On the other side of the Atlantic, the BSA/Triumph circus rolled on for the Easter Anglo-American Match races. Run over three legs at Brands Hatch, Mallory Park, and Oulton Park, this pitted a team from Britain against a team from America. With all riders mounted on BSA and Triumph 750 triples, this series was very much a BSA/Triumph publicity stunt. As the Americans were unused to British short circuits, the home team easily won the series, with Paul Smart and Ray Pickrell the leading scorers. The Transatlantic Trophy races may have been staged for BSA/Triumph's benefit, but at the Isle of Man this year Tony Jeffries took his Triumph Trident to victory in the F750 race, winning at 102.86 miles per hour, and Ray Pickrell won the 750cc production race on the legendary *Slippery Sam*. Throughout 1971 the BSA/Triumph triples completely dominated British 750cc Championship short circuit racing, and as the North-framed racers were eligible for the Bol d'Or at Le Mans this year, Percy Tait teamed with Ray Pickrell to take the victory. But while the 1971 results were spectacular, the costs were exorbitant and unsustainable, and for 1972 the ax came down hard. For the 1972 AMA season, Triumph and BSA would retain only one full-time rider per team.

Hat trick for BSA/Triumph at Daytona. In the victory lane after the race, from the left, are second-place Gene Romero, winner Dick Mann (on a BSA), and third-place Don Emde, also on a BSA. Emde's bike is a 1970 version, still with a Fontana front brake.

More attractive, tapered, megaphone-style mufflers replaced the "ray-gun" mufflers for 1971.

CHAPTER FIVE

For 1971 only the 650 models received the new P39 frame. Apart from the attractive blue color and single carburetor, the TR6R was identical to the T120R.

since 1962. But the most significant feature was a 4-inch backbone tube that acted as an oil reservoir. The large spine made the frame much stronger than before, but the design met with considerable resistance from Meriden engineers Bert Hopwood and Doug Hele. And it was beset with problems from the outset. It was three months before Meriden received the drawings from Umberslade Hall; they marked time by producing 500cc Daytonas, stockpiling Bonneville engines, and playing chess. When the frames were finally constructed, the first examples badly leaked oil and the engines wouldn't fit. After several modifications, production finally commenced just before Christmas of 1970.

Unfortunately, the delayed introduction was only one problem facing the 1971 Bonneville. The P39 frame raised the seat height by 3 inches, making it the tallest large-capacity road bike, and as the frame was more rigid, engine vibration was also more pronounced. Later in the year a five-speed gearbox, as on the Trident, was listed as an option on the T120RV, but it didn't become available until 1972.

Other chassis components were similar to those of the T150 Trident, including the Ceriani-style two-way

1971 T120R, TR6R, TR6C
(Differing from 1970)

Carburetor
Amal Concentric R930/66, R930/67 30mm (T120R); R930/60 (TR6)

Wheelbase
56 inches

Dry Weight
382 pounds (T120R); 381 pounds (TR6)

Color
Tiger Gold (T120R); Pacific Blue (TR6)

Engine and Frame Numbers
T120R NE01436 to HE30869; TR6R, TR6RV, TR6C PE003157 to HE028817

damped front fork, conical hub brakes, skimpy fenders, and tapered silencers. Turn signals and new cast-aluminum Lucas switches were also standard equipment this year. The center stand design was poor, leading to side-stand reliance, but as this was welded to the thin wall frame tubing, it broke prematurely. According to dealer John McCoy, "The futuristic thin gauge wire fender and headlamp stays broke from

1971–1975

FAR LEFT: With its narrow gas tank, the TR6R style continued a traditional theme.

LEFT: The 8-inch conical front brake didn't provide any improvement over the previous twin leading shoe brake.

the engine vibration, as did the gas tank's underside seams, but the more modern front fork and mufflers, designed to meet increased noise restrictions, were positive updates. So many of the new designs were rushed into production without any testing, that the cumulative effect of all of the 1971 model changes was a disaster that set the tone for the downfall of the entire English motorcycle industry."

Although now virtually universally despised, at the time the oil-in-frame Bonneville met with a positive press response, *Cycle World* commenting in May 1971 on the T120s "precise handling qualities," and stating, "This machine is one of Triumph's best." Also receiving the P39 frame this year were the TR6R Tiger and TR6C Trophy. Only different paint and a different single-carburetor cylinder head distinguished the Tiger from the Bonneville, while the single-carb Trophy featured a distinctive left-side high-rise exhaust system and chrome-plated fenders.

1971 T100R, T100C
(Differing from 1970)

Dry Weight
356 pounds (T100R); 342 pounds (T100C)
Color
Olympic Flame
Engine and Frame Numbers
T100C KE00058 to GE25135; T100R KE00001 to HE25250

1971 T100R Daytona, T100C Trophy 500

While Meriden awaited the arrival of the P39 frame, production concentrated on the T100R 500 Daytona and T100C Trophy. This year the T100R Daytona was standardized for both Britain and the United States, and it and the T100C were basically carryovers from 1970. Engine updates included thicker con-rods, and all models received standard turn signals.

The 1971 T100C was basically a carryover from 1970 but now included turn signals and new switches and taillight. This would be the final year for the T100C.

131

CHAPTER FIVE

The T25T Trail Blazer was a new model for 1971 and only lasted one year. Its aluminum gas tank, 20-inch front wheel, and small front brake differentiated it from the T25SS. The frame also functioned as an oil reservoir, with chain adjustment via a cam plate at the swingarm pivot.

1971 T25SS Blazer, T25T Trail Blazer 250

Among the massive group range launch at the end of 1970 were two new 250s, the T25SS Blazer Street Scrambler and the T25T Trail Blazer. These were identical to equivalent BSA Victor versions, and while the all-aluminum 250cc single-cylinder engine was carried over from the previous TR25W, the oil-bearing frame was a version of the Mark IV BSA works motocross type. The front downtube served as the main oil tank, and the frame included a number of quality features, notably tapered roller bearings for the steering head and swingarm and eccentric chain adjustment from the swingarm pivot. Both featured a new aluminum front fork, and the T25SS boasted an 18-inch front wheel and the same 8-inch conical front brake as the Trident and Bonneville. The trail bike T25T had a more dirt-oriented 20-inch front wheel and small single leading shoe front brake. But while the chassis was an improvement, the troublesome BSA-built engine remained, and realistically the new single was no match for the new generation of more refined Japanese street and trail bikes. After only one year the Blazer was discontinued, and there would be no more lightweight Triumphs.

1972

Production delays caused by the implementation of the P76 frame resulted in BSA and Triumph missing the bulk of the 1971 US sales season, and by May 1971 11,000 1971-model-year BSAs and Triumphs were stockpiled in the States. On top of this, warranty claims on sold 1971 models were alarming. Dealer John McCoy recalls, "The average USA BSA/Triumph dealer was so disenchanted with the bikes, that they spent the minimum amount of effort possible to prepare them for sale, and the customer experience with their new bike was not always what it should have been. George Hall, the owner of a Gilroy, California BSA/Triumph dealership, joked that a particular example had been pushed from inside the store, to the outside display area, so many times that the mileage reading on the odometer, at 30 feet per day, exceeded the limits for the retail warranty, even before the bike even found a buyer."

In April Denis McCormack finally retired and was replaced by another Barclays Bank appointee, Dr. Felix Kalinski. Kalinski immediately instigated a fire sale

1971 T25SS, T25T
(Differing from 1970)

Horsepower
22.5 at 8,250 rpm
Front Brakes
8-inch TLS (T25SS); 6-inch SLS (T25T)
Tires
3.25x18-inch front and 3.50x18-inch rear (T25SS); 3.00x20-inch front and 4.00x18-inch rear (T25T)
Wheelbase
54 inches
Dry Weight
290 pounds (T25SS); 287 pounds (T25T)
Colors
Tangerine

of unsold stock, but this did little to redress the group's serious financial situation. Furthermore, much of the 1972 US selling season was also lost due to continuing industrial disputes at the overstaffed Meriden plant. In April 1972 strikes ensured only 16 days out of 20 were worked, with 65 percent of the production target output achieved. It was estimated the production shortfall from Meriden this year was 5,000 machines.

To raise cash, BSA sold the Armoury Road site and sports fields in Birmingham and then looked to the future by considering Hopwood's new modular range of machines, from a 200cc to 1,000cc five, a license from Audi-NSU to develop a rotary engine, and an 830cc Thunderbird III triple. But it was all in vain, and with losses mounting, Lord Shawcross approached the British government for a bailout. As Britain as a whole was in the midst of an economic crisis, the Conservative government refused to help BSA but offered to provide up to £20 million if BSA merged with their main competitor, Dennis Poore's Norton-Villiers. Talks between BSA and Norton-Villiers began at the end of November 1972, but Poore, the self-appointed savior of British motorcycling with strong Tory-party connections, won. The merger ultimately disadvantaged BSA, with a negative flow-on affect for strongly unionized workforce at Meriden. For Triumph, loyalists Poore and Norton were extremely dubious saviors. Ever since the Edward Turner days, Norton and Triumph had been rivals, much in the mold of Chevrolet and Ford. So it wasn't surprising that the entire deal was treated with considerable suspicion at Meriden.

1972 T150, T150V Trident 750

Although the BSA Rocket 3 finished in January 1972, the Small Heath plant continued to build three-cylinder engines, and production of the Trident was moved to Meriden. Here production was delayed by frequent industrial disputes. New this year was larger, slab-sided gas tank for the UK market, and later in the year a five-speed gearbox for the T150V. Based on a racing Rod Quaife design, early examples were extremely problematic, breaking under only moderate use and causing a warranty headache. But the narrower-ratio five-speed gearbox provided positive benefits of improved engine response and flexibility. *Cycle Guide* found the 1972 T150 a vast improvement over the earlier Rocket 3 they had tested, summing it up with, "If we were in the market for a large displacement touring machine, this is the bike we would be most apt to purchase."

Despite the BSA Group's problems, 1971 racing triples were still made available to UK riders, along with factory mechanics. In the United States, Romero and Mann continued with Duarte support, their 1972 machines with US-made versions of the North frame. But at Daytona the BSA/Triumphs struggled to match their speed of the previous year, and Eddie Mulder was the highest-placed finisher in sixth. In the United States, at least, the days of the triple's dominance were now over as four-strokes declined under the new wave of two-stroke power. In the United Kingdom, Ray Pickrell kept his Arthur Jakeman–prepared F750 Trident in the forefront, winning the Formula 750 TT at the Isle of Man at 104.23 miles per hour and setting a new lap record of 105.68 miles per hour. Tony Jeffries was second on a similar machine.

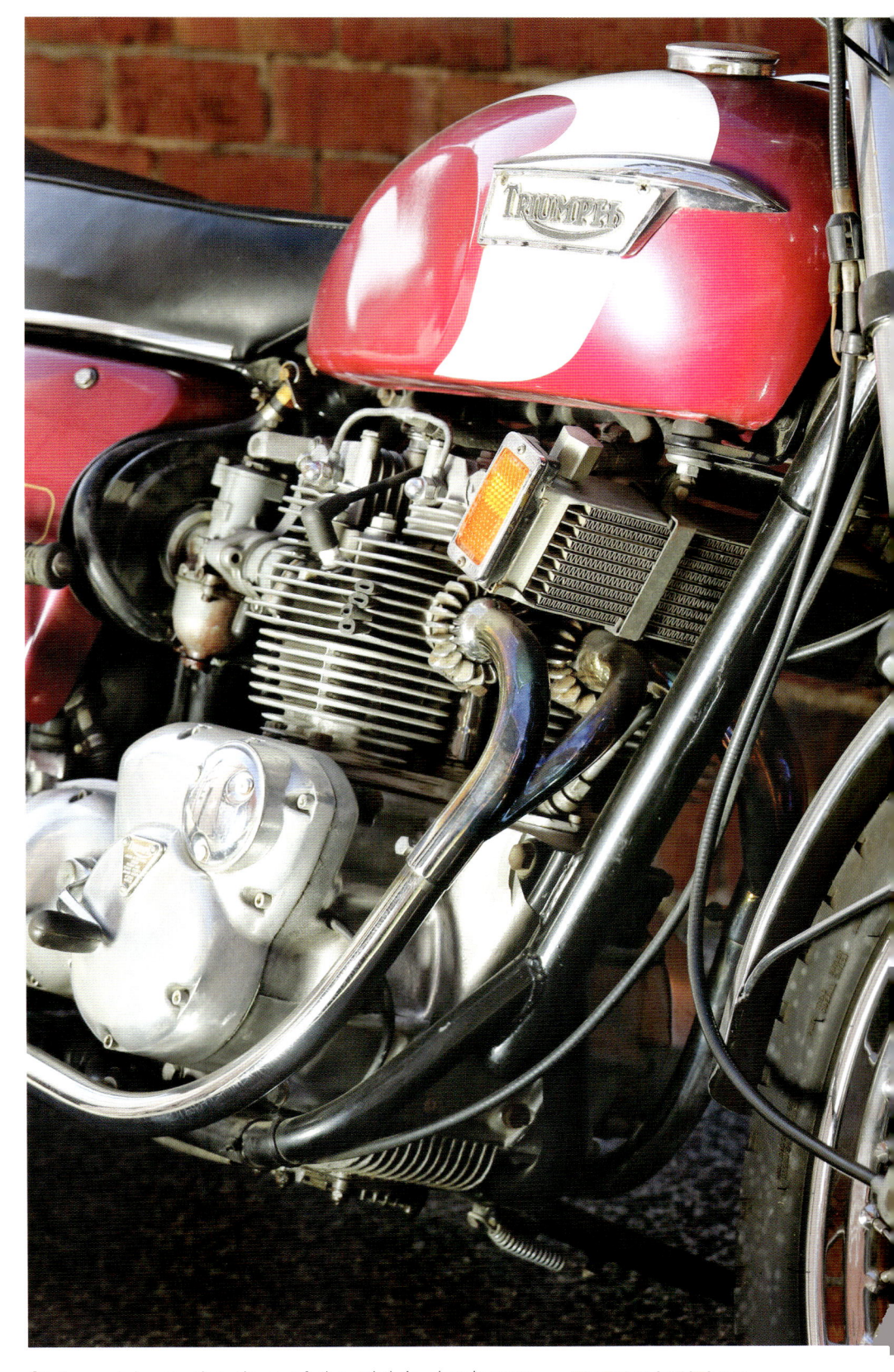

The Trident for 1972 was painted a trendy Regal Purple. US versions such as this retained the smaller gas tank, but UK examples had a larger, slab-sided tank.

CHAPTER FIVE

New side cover decals featured this year only.

The five-speed T150V appeared later in 1972. The five-speed transmission was initially problematic, as many of the gears were poorly heat-treated and brittle.

1972 T150, T150V Trident
(Differing from 1971)

Gearbox
Five-speed (T150V)
Color
Regal Purple

Although the factory wouldn't endorse production racing this year, Doug Hele allowed Les Williams to resurrect the previous year's racer *Slippery Sam* for Ray Pickrell to ride at the TT. Pickrell didn't disappoint, again winning the Production TT, this time at an average speed of 100.00 miles per hour.

1972 T120R, T120RV Bonneville, TR6R, TR6RV Tiger, TR6C Trophy 650, T100R Daytona 500

Due to the stockpile of 1971 models, the 1972 650cc twins were initially little changed. Engine updates later in the year included a new cylinder head and a new rocker box without the separate inspection caps that were always prone to vibrating loose. The problematic five-speed T120RV Bonneville (and similar TR6V Tiger) also became available, but at a price ($200). But the

1971–1975

The Bonneville had a significantly lower seat height for 1972, and a five-speed gearbox was an option.

1972 T120R, T120RV, TR6R, TR6RV, TR6C, T100R
(Differing from 1971)

Gearbox
Five-speed (T150V)
Dry Weight
387 pounds (T120R); 386 pounds (TR6)
Colors
Tiger Gold and white (T120R); Polychromatic Blue and white (TR6); Cherry and white (T100R)
Engine and Frame Numbers
T120R T120RV from HG30870; TR6R, TR6RV, TR6C JG033084 to EG057252; T100R JG32303 to GG59646

most significant update was to the chassis as Meriden finally accepted that the seat height was too high. After two uncataloged interim chassis, the final version provided a 31-inch seat height, achieved through a combination of shorter fork springs and shock absorbers, a thinner seat, and lowering the rear subframe 3 inches. UK examples again featured a large, slab-sided gas tank, and late in the season a small 2½-gallon slimline tank was offered for the United States. This was also the final year for the TR6C Trophy. As production problems continued at Meriden, the only 500 Triumph this year was the T100R Daytona, virtually unchanged from the previous year. This still included

ABOVE: The North-framed Triumph Trident F750 racer for 1972 was ostensibly a 1971 version. Compact and effective, they were arguably the finest 750cc four-stroke racing machines of the early 1970s.

LEFT: Ray Pickrell was one of the triple's leading racing exponents, shown here at Cadwell in 1972. A serious accident at the 1972 Mallory Race of the Year ended Pickrell's illustrious career, but he was so popular he still won the British *Motor Cycle News* Man of the Year award.

135

CHAPTER FIVE

With its five-speed gearbox and front disc brake, the Trident was finally a match for the other superbikes. The UK version received new styling for 1973, with two different types of muffler offered.

all the earlier features such as bolted-up frame and pre-conical twin leading shoe front brake.

Generally the 1972 updates were well received, but *Cycle World* balked at the price of $1,725 for the T120RV. Although they found the Bonneville "unequivocally one of the best handling roadsters in its size category," and the first Bonneville to run a standing-start quarter mile under 14 seconds (13.91 seconds at 94.53 miles per hour), the Bonneville was no longer Edward Turner's affordable twin. *Cycle* magazine also questioned the relevance of the Bonneville in a changing world, commenting, "Sure it shakes, and sure it leaks oil. It's British. What the hell."

1973

In March 1973, a "bear raid" saw the BSA Group's shares slump dramatically, immediately affecting production at Meriden and hastening the deal with Norton-Villiers. Three hundred workers were laid off at Meriden, but somewhat surprisingly Triumph still made a profit that quarter. The British government agreed to inject nearly £5 million into the new company, and in July 1973 Norton Villiers Triumph (NVT) emerged from the ashes of what was the once-great BSA concern. One month later, at the age of 72, Edward Turner died, and Bert Hopwood resigned following a disagreement with Dennis Poore over the future direction of the new company. Almost immediately, Poore decided Meriden was not suitable for his master plan, envisaging selling Meriden to Jaguar and moving all motorcycle production to Small Heath and Norton's Wolverhampton. On September 14, 1973, Poore announced to shocked workers at Meriden that by November 30 the workforce would be reduced by a third, by a further third on January 1, 1974, and with closure of the factory on February 1. The workers' reaction was swift: they immediately occupied and picketed the factory. The siege lasted 18 months, and Meriden would never be the same again. But despite these troubled times, Triumph's 1973 range included several new models and was the largest it had been for several years. In America, Norton-Villiers and BSACI merged in December 1973, with the new company Triumph-Norton Inc. (TNI) headquartered in Duarte.

1973 T150V Trident 750

The Trident finally became what it should have been four years earlier. All models now had a five-speed transmission, and a Lockheed front disc brake replaced the weak conical drum. UK versions had new cigar-shaped silencers, US examples had 8-inch high-rise handlebars, and all had new colors, a polished rear conical hub, and larger chromed fenders front and rear. *Cycle* magazine pitted the new Trident against the wave of 1973 superbikes, commenting, "The Triumph Trident was generally recognized as the best-looking bike in the test." As a handling package the Trident also

1973 T150V Trident
(Differing from 1972)

Front Brake
10-inch Lockheed Disc
Wheelbase
58 inches
Color
Jet Black and red

Despite a hailstorm and extremely wet roads, Tony Jeffries took the victory in the 1973 Production TT on Les Williams' *Slippery Sam* Trident. *Slippery Sam* won this race five years in succession, from 1971 to 1975. The front brake was a four leading shoe Fontana front instead of a disc, and the silencers were the earlier "ray-gun" type.

impressed: "The Triumph offering the best combination of cornering ability and confidence." And despite new, quieter mufflers, the five-speed Trident ran considerably faster than two years earlier, running a standing quarter mile in 12.718 seconds at 106.0 miles per hour, for an overall third-place finish. Despite this good result, Triumph was still aware that despite the front disc brake and five-speed gearbox, the Trident was seen as antediluvian among the latest Japanese 750s. This year they promoted the Triumph owner in advertising as, "He's very rare. He's an expert."

By 1973 there was no longer an official racing budget, and while Percy Tait, Tony Jeffries, Dick Mann, Gene Romero, and Gary Scott still campaigned the North-framed triples, they were outclassed this year by Peter Williams' John Player Nortons in England and the new wave of two-strokes in the US. But while the triples were struggling in F750 racing, Tony Jeffries won the Production TT on *Slippery Sam*.

1973 T120R, T120RV, T140V Bonneville, TR6R, TR6RV, TR7RV Tiger 650 and 750

Considering the general uncertainty surrounding BSA and Meriden, it was surprising the twins received so many updates this year. Although the 650 T120R Bonneville and 650 TR6R Tiger continued for the

continued on page 140

CHAPTER FIVE

The TRX75 Hurricane 750

The X75 Hurricane started out as a private project in America between Don Brown and Craig Vetter which began life soon after the release of the Rocket 3/Trident, when BSA's West Coast head Don Brown loaned motorcycle-fairing manufacturer Craig Vetter a new Rocket 3 as a basis for an American-style custom sport bike. Brown drew primarily on a custom 1950 Thunderbird he had owned in the early 1950s: a slender machine with near-perfect symmetrical lines. No one at BSA Group knew of the project, and it was funded from BSA Inc. petty cash.

During 1969 Vetter prepared a prototype and presented his creation to BSACI in October. The prototype was shipped to Stephen Mettam, BSA Group's chief stylist, at Umberslade Hall. Here the prototype was carefully measured and analyzed before being sent back to America to feature on the cover of the September 1970 edition of *Cycle World*. When Rocket 3 production ended, the project transferred to Meriden. After a three-year gestation, the X75 finally made

ABOVE: Craig Vetter's X75 Hurricane was the first factory custom sports bike, and one of the most successful. The exhaust system was unique, and the one-piece fiberglass tank and seat unit were groundbreaking. Vetter's idea was more to improve the flawed original triple rather than creating a chopper-like custom.

RIGHT: Not the most practical design, the Hurricane's gas tank only held 1.66 gallons, but the curvy bodywork provided a narrow profile, accentuated by a small 1½-person seat.

LEFT TOP: As a road racer Vetter would have preferred a disc brake, but BSA didn't provide them. Vetter's prototype thus featured a massive drum brake, and as they were instructed to build it exactly as Vetter had designed it, the factory fitted an off-the-shelf conical front brake rather than the Lockheed disc of the T150 Trident this year. Also setting the Hurricane apart were light aluminum Borrani wheel rims, a custom front fender, and polished and redesigned triple clamps.

LEFT BELOW: The right-side three-into-three exhaust system with fanned-out megaphones was based on that of Jim Rice's 1969 BSA Rocket 3 flat tracker. They compromised right-side ground clearance, and as they would have trouble meeting new American noise and emissions regulations, all US Hurricanes built after January 1973 were backdated.

1973 TRX75 Hurricane
(Differing from 1973 T150V)

Horsepower
60 at 7,250 rpm
Gearbox
Four-speed (on early examples)
Front Brake
8-inch twin leading shoe
Wheelbase
60 inches
Dry Weight
458 pounds
Color
Camaro Hugger Red

it into production in September 1972, and while only 500 examples were initially envisaged, around 1,200 (reports vary between 1,154 and 1,175) were eventually built. The Hurricane caused a stir in 1973, with *Cycle* magazine commenting, "The bike took a lot of guts to build—any bike that departs from the carefully honed middle ground of stylistic normality is taking a big, hairy gamble with sales, and the Hurricane makes the largest break with normality in recent motorcycling history."

Craig Vetter's concept may have been impractical and functionally flawed, but the X75 Hurricane was the quintessential factory custom sports bike. Widely copied, it has never been bettered.

ABOVE: With its forward-slanted cylinders, the X75's engine was more BSA Rocket 3 than Triumph Trident. The engine incorporated extended cylinder head fins as on Vetter's original prototype. The frame was also a BSA-type twin downtube.

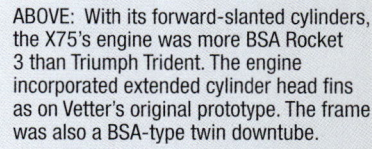

LEFT: The bare left side drew attention to the engine, making it Vetter's favorite side. Not Vetter's idea, the 2-inch-longer front fork accentuated the chopper look, and combined with the rib front tire it contributed to some Meriden testers' criticism of the handling. Another reason for backdating production was to avoid having to fit mandatory turn signals from 1973.

CHAPTER FIVE

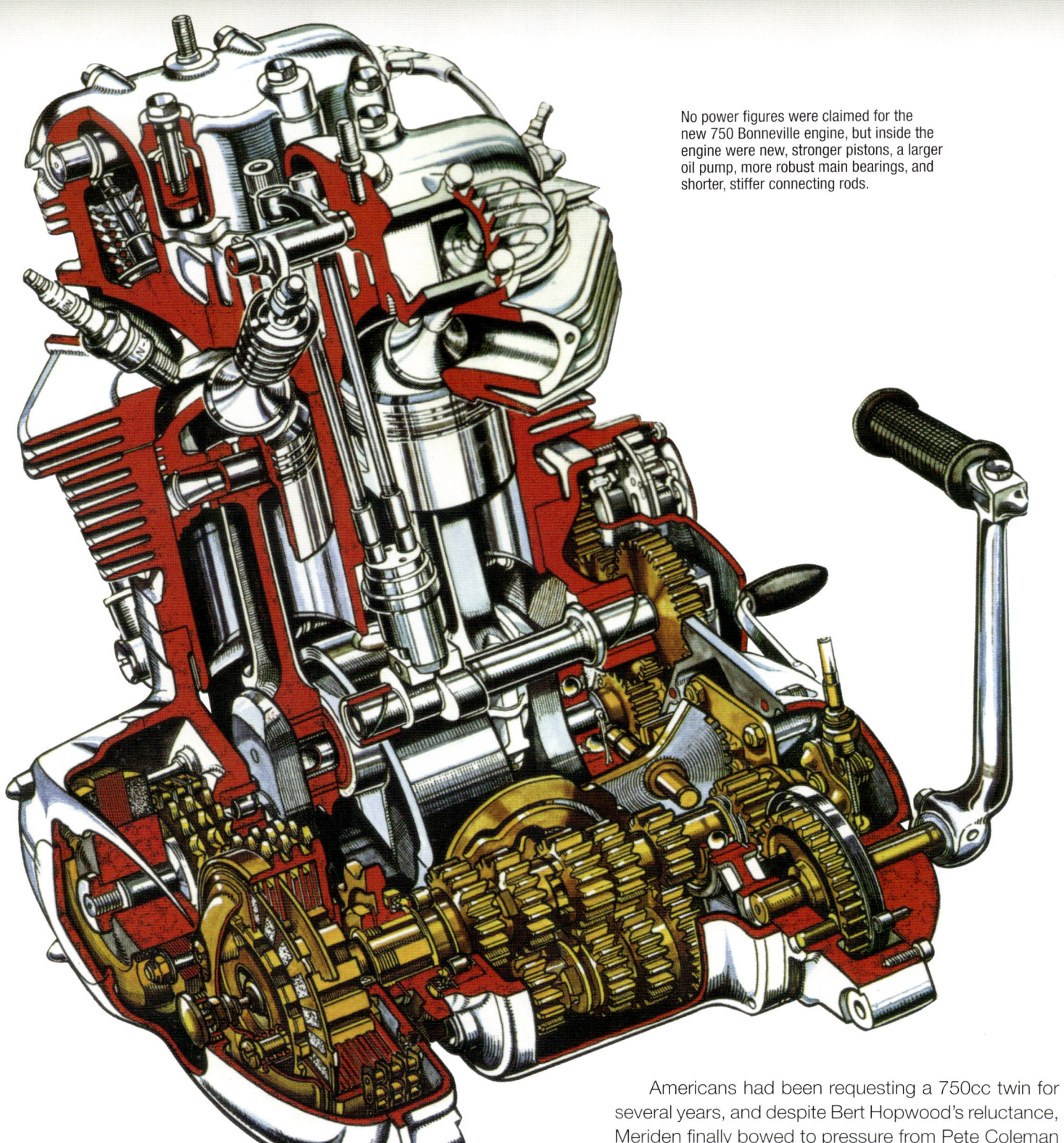

No power figures were claimed for the new 750 Bonneville engine, but inside the engine were new, stronger pistons, a larger oil pump, more robust main bearings, and shorter, stiffer connecting rods.

OPPOSITE: The 750 engine featured revised rocker inspection plates and had shorter cylinders than before, most with a 10-bolt cylinder head. This March 1973 example has shorter cylinders but not the 10-bolt cylinder head and new crankcase castings.

Continued from page 137

United Kingdom and general export in 1973, often with the earlier high-seat frame, the most important development was the introduction of the 750 T140V and TR7RV Tiger. All models were fitted with a T150 Trident front fork and Lockheed disc brake. A strengthened and improved five-speed gearbox also remained optional for the T120R and the TR6R, and after the introduction of the revised 750 in December 1972, the 650 gained the new 750 crankcases. Early-model-year 1973 UK 650s twins also continued with the 1972 colors and painted fenders.

Americans had been requesting a 750cc twin for several years, and despite Bert Hopwood's reluctance, Meriden finally bowed to pressure from Pete Coleman and the 750 Bonneville and Tiger appeared for 1973. Initially this was simply an overbored 650, displacing 724cc, but in December 1972 new cylinder castings allowed the bore to be increased, providing 744cc. The cylinder block was also shorter, as were the conrods (half an inch shorter, from the 1971 BSA racing A70), allowing the rocker boxes to be left in place during engine installation or removal, and the new cylinder head featured 10-stud fixing. Also new was a triplex primary drive drain. 750 Bonnevilles and Tigers had more substantial chromed fenders front and rear, while US and UK versions differed in the handlebar and gas tank: a 2½-gallon tank was standard in the States, with the larger 3½-gallon "bread bin" tank for Britain.

1971–1975

CHAPTER FIVE

ABOVE: Shared with the T150V Trident, the single Lockheed front disc brake was a significant improvement over the previous drum brake. The brake caliper had a chrome cover.

RIGHT: The dual seat was reprofiled for 1973, and the larger, cast-aluminum taillight bracket included a larger Lucas lens. Headlight brackets were now rubber-mounted black-painted "ears."

1973 T120R, T120RV, T140V, TR6R, TR6RV, TR7RV (Differing from 1972)

Bore	75mm (T140RV from JH15345, TR7V from KH17122); 76mm (T140RV from XH22019, TR7V from AH23727)
Stroke	85mm (T140RV and TR7V)
Capacity	724cc (T140RV from JH15345, TR7V from KH17122); 744cc (T140RV from XH22019, TR7V from AH23727)
Compression Ratio	7.9:1 (UK 750s); 8.6:1 (US 750s)
Carburetors	Amal Concentric R930/87, R930/88 30mm (T140RV); R930/89 (TR7V)
Front Brakes	10-inch Lockheed disc
Dry Weight	413 pounds (T140V, T120V); 403 pounds (TR7RV); 390 pounds (T140V, TR7RV US)
Colors	Hi Fi Vermillion and gold (T140V); Astral Blue and white (TR7RV)
Engine and Frame Numbers	T120R T120RV from JH15366; T140V from JH15435 to GH36466; TR6R, TR6RV, TR7RV JH15475 to GH35466

Cycle magazine reviewers were less than enamored with their TR7RV, noting, "Vibration was evident everywhere human anatomy came in contact with the motorcycle." *Cycle World* announced the T140V as "Triumph's best Bonneville." The handling impressed everyone, and the T140V was the quickest Bonneville yet tested, although surprisingly not quite as fast as *Cycle*'s TR7RV, which went through the standing-start quarter mile in 13.513 seconds at 95.64 miles per hour. Earthy and solid, the unrefined 750 was created at America's request, but it still struggled against increasingly sophisticated Japanese competition.

1973 TR5T Trophy Trail 500 (Adventurer) and T100R Daytona

This year saw the return of the 500 twin trail bike as the single-carburetor TR5T Trophy Trail, or Adventurer. A "parts bin special" mixture of Triumph and BSA, the TR5T shared little with its T100C predecessor and featured a BSA Victor MX oil-bearing frame, Rickman aluminum conical hub brakes, an aluminum front fork, and a 2¼-gallon BSA B50 aluminum gas tank. A first

1971–1975

ABOVE: Little had changed since 1968: the 1973 T100R Daytona still featured the earlier single downtube frame, traditional fork with gaiters, and earlier cast-iron twin leading shoe drum brake. This brake anchor plate was painted black this year. The famous old-style teardrop mufflers also found their final resting place on the 1973 Daytona.

LEFT: For 1973 the T100R Daytona received chrome-plated fenders and a larger Lucas taillight. Dealers converted a few, such as this example, into street versions from the unpopular single carburetor TR5T Trophy. The Daytona may have been obsolete, but it was still extremely good looking and aesthetically balanced.

CHAPTER FIVE

1973 T100R and TR5T
(Differing from 1972 T100R)

Compression Ratio	7.5:1 (UK)
Carburetor	Single Amal 928/21 Concentric 28mm (TR5T)
Front Brake	6-inch single leading shoe (TR5T)
Front wheel	WM1x21-inch front (TR5T)
Front tire	3.00x21-inch (TR5T)
Wheelbase	54 inches (TR5T); 54.75 inches (TR5T US)
Dry Weight	322 pounds (TR5T); 320 pounds (TR5T US)
Colors	Vermillion or Astral Blue and white (T100R); Hunting Yellow with polished aluminum (TR5T)
Engine and Frame Numbers	T100R JH15597 to DH31460; TR5T KH16597 to GH34399

The TR5T Trophy Trail was a very attractive machine but was hampered by limited suspension travel, a weak front brake, and excessive weight. Although not many were built, it was the only model that wasn't in short supply in America in 1973.

for Triumph was the use of Nipon Denso instruments. Although on paper the TR5T looked heavy and its low exhaust unsuitable for serious off-road duty, six TR5Ts were supplied by Meriden and prepared at the BSACI facility in Duarte for American and British teams in the 1973 ISDT, held in the Berkshire area of Massachusetts. The United Kingdom managed second place behind a crack Czech team, with TR5T-mounted riders gaining three gold medals and one silver. *Motorcyclist* magazine found the performance of the TR5T disappointing, and their test machine was also plagued by problems caused by excessive vibration. They summed it up by saying, "The Triumph is a compromise machine. It is the best handling dual purpose machine available today but performance is far below the competition."

For the T100R Daytona, now the oldest model in Triumph's lineup, it was very much business as usual. Virtually unchanged since 1968, the Daytona looked like a dinosaur compared to new double overhead camshaft eight-valve offerings from Japan, such as Yamaha's TX500. While the B-range 650s had evolved into reasonably modern motorcycles with new frames, forks, brakes, and five-speed gearboxes, the 500 retained the earlier frame, still with sidecar lug, cast-iron drum brake and hubs, and old-style iron front fork. Minor weaknesses, such as the screw-on-fall-off rocker caps, also remained on the T100R. To an outsider new to the Triumph experience, it was almost as if Triumph was using up its stock of obsolete parts to build the Daytona. *Cycle* magazine could find no compelling reason for anyone to buy the T100R over the Yamaha TX500, commenting, "Everything the Triumph does the Yamaha does better—except be a Triumph."

1974

Production problems at Meriden during 1973 resulted in a stock shortage, particularly of twin-cylinder roadsters in the United States, and the 1974 range was announced slightly earlier than usual, in July. The Meriden blockade prematurely halted supply of 1974 models, and after the Labour Party surprisingly won the general election in February, Meriden's workers found a sympathetic ally in the new industrial secretary, Tony Benn. NVT and Meriden were at an impasse, but in the meantime Dennis Poore managed to retrieve some of the Trident tooling from Meriden, and eventually the release of around 1,000 twins built before the blockage. Benn, on the other hand, clearly supported the workers and in July arranged a government loan and grant for the establishment of a workers' cooperative at Meriden. The pickets were lifted briefly, only to return in November 1974 while the financial negotiations with NVT were concluded. With virtually no supply from Meriden for over a year, the situation was disastrous in America. In February US Norton's Roger Stange replaced Felix Kalinsky, and the distribution was clearly Norton-biased. Although new Tridents were readily available, without Bonnevilles, 68 US dealers relinquished their franchises in November 1974. Triumph's reputation in the United States was virtually wiped out overnight, and it would take a lot of rebuilding.

1974 T150V Trident 750

Although hundreds of Trident engines were already stockpiled at Small Heath, due to the blockade they couldn't be shipped to Meriden for completion. In November 1973 NVT made plans to set up a new T150

1974 T150V Trident
(Differing from 1973)

Compression Ratio	8.25:1 (UK)
Dry Weight	462 pounds (UK)
Color	Black and gold

1971–1975

Norton Villiers Triumph placed this advertisement in magazines during 1974 and allocated dealers in the United States stock based on past sales. But no one clamored for the Trident: they wanted the unobtainable Bonneville.

Triumphs are rolling again!

Thanks to the success of the Norton Villiers Triumph Birmingham factory in re-tooling to counteract the Meriden factory blockade, Triumphs-the experts' bikes-are here again. At first, demand may be hard to fill. *Reserve yours now.*

*Triumph Trident
3-cylinder 750 cc superbike,
5-speed gearbox. Lockheed
hydraulic disc brake.*

TRIUMPH

Write or phone to find your nearest dealer in Great Britain and Europe Norton Triumph Europe Limited, North Way, Andover, Hampshire. Tel: Andover 61411

BIKE BRITISH TRIUMPH WITH NORTON IN '74.

production line at Small Heath, but this proved difficult and expensive. With many drawings unavailable, tools and jigs had to be refabricated, at a cost rumored to be £500,000. But by April 1974 Tridents were rolling off the makeshift production line, and except for yellow gas tank panels, they were ostensibly identical to the 1973 version. *Motor Cycle World* tested one of the new Small Heath Tridents and came away surprised, summarizing the Trident with, "A machine that oozes an aura of power and sleekness. In short, we loved it."

At the Isle of Man this year *Slippery Sam* continued its winning ways, this time in the hands of TT legend Mick Grant. The capacity limit was now 1,000cc, and now with disc brakes for the first time, Grant won at an

CHAPTER FIVE

This Trident, built in May 1974 on the makeshift Small Heath production line, was ostensibly identical to the 1973 version.

Yellow gas tank stripes distinguished the 1974 Trident. US examples had high handlebars.

1974 T120V, T140V, TR7RV
(Differing from 1973)

Colors
Red and white (T140V); purple and white (T120V); green and white (TR7RV)

Engine and Frame Numbers
T120V GJ55101 to KJ59067; T140V GJ55101 to NJ60032; TR7RV JJ58039 to JJ58064

average speed of 99.72 miles per hour, beating a field of 900cc BMWs, Kawasaki Z1s, and Laverda 1000s.

1974 T120V, T140V Bonneville, TR7RV Tiger 750

Although the few 1974 650 and 750cc twins available this year were built prior to the blockade (or during the blockade's hiatus in June and July 1974), they did incorporate a few updates. Most of these were designed to counter oil leaks, but as the number of larger twins were produced this year was so small, it was largely academic. Of the stock released to NVT after July 1974, most were T120Vs. These were all five-speed, with chromed fenders and quieter reverse cone mufflers. As the TR7RV Tiger finished in August

1971–1975

In Cherokee Red and gold-lined white, the 1974 750cc Bonneville looked stunning—but this brochure picture was about as close as most came to it.

1973, only a small number were available, and this year there was no TR6 Tiger.

1974 T100R, T100D Daytona, and TR5T Trophy Trail Adventurer 500

As the 1974 TR5T Trophy Trail Adventurer was only built during July 1973, the numbers were very small, and, but for the gas tank color, they were identical to the 1973 version. Unloved and a poor seller, while the stock of all other Triumphs remained in limbo, this was the model NTI provided to magazines for testing. *Cycle Guide* was not impressed, their TR5T proving extremely unreliable. They said, "Every day we rode the TR5T something broke or stopped working. If this were 1950s you might expect this, but this is 1974. Triumph technology appears to be in the 1950s."

1974 T100R, T100D, and TR5T *(Differing from 1973)*

Front Brake
10-inch disc (T100D)
Dry Weight
378 pounds (T100D)
Color
Blue and white (T100R, T100D); red with polished aluminum (TR5T)
Engine and Frame Numbers
T100R, T100D HJ56408 to JJ58103; TR5T HJ56408 to HJ57336

Apart from longer, cigar-shaped mufflers, the US T100R Daytona continued as before, but the 1974 UK T100 was initially intended to ship with the T140's aluminum front fork and single front disc brake. This was the T100D, but only a few were built before the 1973 blockade.

1974 TR5MX Avenger 500

While still under BSA control, a new addition to Triumph's 1974 range was the TR5MX Avenger. A competition motocross machine aimed at the United States, this was very much BSA B50MX with a few

The TR5T was almost unchanged from 1973. Only produced for a month prior to the Meriden blockade, this was one of the few models available in the United States during the early part of 1974.

147

CHAPTER FIVE

ABOVE: Light and effective but very much a BSA B50MX, the TR5MX Avenger was the last of the big thumper British trail bikes. The exhaust system was definitely not street legal.

RIGHT: The BSA-derived engine displaced a full 500cc and featured a USDA-approved spark arrestor.

1974 TR5MX Avenger

Bore	84mm (89mm option)
Stroke	90mm
Capacity	499cc (560cc option)
Compression Ratio	9:1
Carburetor	Amal 932 Concentric 32mm
Tires	3.00x21-inch front and 4.00x18-inch rear
Brakes	6-inch front, 7-inch rear
Dry Weight	260 pounds
Color	Blue with polished aluminum

minor updates. The engine's origins were in the BSA 441cc Victor, now stretched to 499cc, but with a caged roller big end and additional ball race on the crank drive side. Also available was a high-performance 560cc kit (with 89mm piston).

The chassis was the same oil-in-frame type as the 1971 T25T Trail Blazer, but with a polished aluminum gas tank, a lightweight conical front hub, and a pair of racing Girling shock absorbers. There was no battery, but an alternator-powered lighting kit was available as an extra. The TR5MX Avenger was perfect as a fire-roader, hill-climber, boonie-basher, or motocross mount, but it was the end of a breed of big-bore, four-stroke thumper dirt bikes.

1975

While the Meriden factory was blockaded, NVT at Small Heath continued creating the next-generation electric-start Trident. The idea began back in 1969, but development became more intensive after the racing program was suspended in 1972. The original intention was to increase the capacity of 830cc, calling it the Thunderbird III, but the Ford association complicated the name, and the larger capacity became too difficult to implement. At the same time, the Meriden picketers were adapting the Bonneville with a left-foot gear change and right-foot brake to meet new US legislation. Labour won the October 1974 general election, and with Tony Benn supporting the workers' co-op, Dennis Poore reluctantly agreed to the disposal of the Meriden factory, which was finally ratified in March 1975. Meriden paid NVT £3.9 million for the factory and rights to produce the Bonneville, and while Bonneville production commenced almost immediately, NVT was now in serious trouble. The British government refused to renew £4 million of export credit guarantees, and in August Norton's Wolverhampton plant was placed into receivership. Then in September NVT announced the closure of the Small Heath plant. NVT was effectively dead, and the final Tridents were special white Saudi Arabia fleet versions known as the Cardinal, built until February 1976, after which the factory closed. In order to distribute Meriden's Triumphs, Dennis Poore created a new company within NVT in December, NVT Motor Cycles Ltd. This enabled Norton-Triumph Inc. (NTI) to distribute the Bonneville, and they immediately dumped the existing stock of Tridents (and Commandos) through savage price-cutting. It was an ignominious end to a great lineage.

1975 T150V and T160 Trident 750

Due to a delay in the release of the electric-start T160 Trident, the earlier T150V continued for several months basically unchanged. T160 production eventually commenced in November 1974, initially for export, and included more than 200 updates over the T150.

Doug Hele headed the engineering for the T160 project, with Brian Jones leading the design team and styling by Jack Wickes. Wickes adapted

TOP: With a longer and lower profile than the T150, the T160 had perfect proportions. And it provided substance to match the style.

ABOVE: New was a rubber-mounted instrument console with standardized instrument lights and ignition switch. The speedometer and tachometer faces bore the NVT "wiggly worm."

149

CHAPTER FIVE

The T160's exhaust arrangement was completely new, the center cylinder featuring a Y-shaped manifold. The front fork was also shorter, but the Lockheed front brake was unchanged. Although it was now fitted with an electric start, the T160 still retained a kick-start lever. The rear brake was now a Lockheed disc with an underslung caliper.

many styling components of Craig Vetter's stillborn Bonneville TT, including the beautiful traditional teardrop tank, tilted instruments, headlamp, and grab rail. To allow for the Lucas M3 electric-start motor, the three-cylinder engine featured the BSA Rocket 3 and Hurricane's inclined cylinders, and as a reduction in noise was required, the exhaust system included four headers and Norton-style "black cap" annular discharge mufflers. Derived from the production racer *Slippery Sam*, the frame's lower frame tubes were raised to improve ground clearance, and handling improved with a shorter front fork, pivoting in tapered roller steering head bearings, and a longer swingarm. The engine was moved higher and forward, improving weight distribution, and the only negative feature was a marked increase in weight. Other updates included a federally mandated left-side gearshift, with the rear brake lever on the right now operating a 10-inch disc brake. And with either the small 3-gallon or larger 5¼-gallon gas tank, the T160 was undoubtedly one of the best-looking motorcycles of the day. The main downside was the price, at $2,870 until they were later discounted. *Cycle World* praised the handling, saying, "It's nothing short of fantastic." *Motorcyclist*'s venerable Bob Greene also agreed that the handling was superb, but found "It's not a lightweight, nor does it feel like one. But in the Triumph's favor, the weight is low, and on a bending road it inspires confidence above and beyond." The one area where the new Trident disappointed was in engine performance. Although *Bike* in the United Kingdom managed a top speed of 126.05

1975 T160 Trident
(Differing from 1974)

Compression Ratio	9.5:1 (8.25:1 UK)
Rear brakes	10-inch disc
Dry Weight	503 pounds
Color	Sunflower Yellow and Ice White; Cherokee Red and Ice White

miles per hour, *Cycle World*'s standing quarter time of 13.954 seconds at 95.64 miles per hour was way below par for a superbike in 1975.

This year *Slippery Sam*, painted in John Player colors, managed a fifth successive win in the TT production race. Backed by Poore and NVT, Dave Croxford and Alex George won the 10-lap race at 99.60 miles per hour. This was the end of the road for *Slippery Sam*, as rules limiting production machines to five years rendered the bike obsolete.

1975 T120V, T140V Bonneville, TR7RV Tiger 750

Although Meriden recommenced production in March, the first Meriden Triumph rolled off the production line in April. Meriden's first task was the completion of around 1,500 partially assembled 1974 models, still with a right-side gearshift and rear drum brake. Production continued until June, including alongside the T140V a number of 650cc T120Vs, a five TR7RV Tigers, and even some leftover Daytonas. But the factory was marking time until they could introduce the US-legal Bonneville for 1976.

1975 T120V, T140V, TR7RV
(Differing from 1974)

Engine and Frame Numbers
T120V DK61000 to NJ60070; T140V DK61000 to GK62239; TR7RV GK62244 to GK62248

Quiet but restrictive, the T160 mufflers were the Norton Commando "black cap" annular discharge type.

It's Here!

It was known as the Bonnie. The production hot rod of the motorcycle industry. You couldn't mistake that lean look, its powerful vertical twin engine and its distinctive tank. The Bonneville had class. It was light, stable, with incredible acceleration. And the guy who rode one knew he had the most responsive, best handling motorcycle ever built. Others have tried to copy the Bonneville for 20 years, but haven't come close. Today, you can still enjoy the classic styling, the raw power and the unmatched handling that sets the legendary 750cc Bonnie apart. Make your own legends. See your Triumph dealer, now.

The Bonnie

THEY NEVER FORGOT IT.

Also available...single carburetor model, the Triumph Tiger.

NVT AMERICA

1261 State College Parkway

6

1976–1987

The Last Hurrah

Meriden Cooperative and Les Harris Bonnevilles

With the door dramatically and unexpectedly closing on Norton Villiers Triumph, the Meriden Cooperative (where every worker was effectively a co-owner) was left as the only remnant of the once-great British motorcycle industry. But while Meriden's workforce was generally dedicated to producing a quality product, the fundamental problem was that the Bonneville was now an eccentric oddity: a light, simple, enthusiast's motorcycle with lots of character but lacking broad mainstream appeal. Co-op leader Dennis Johnson likened the modern-day appeal of Triumph's twin to the MGB sports car. Meriden was now faced with adapting the aging twin for the modern world and creating a niche market. In the United States, dealer John McCoy commented, "Triumph was still appreciated by a dedicated group of enthusiasts who were content with a good looking, good handling, simpler bike from an earlier era. Sales would be lower, but steady."

1976

As the Meriden Cooperative embarked on introducing a US-spec Bonneville, they were inadvertently caught up in NVT's troubles. The co-op's deal involved selling their entire production to NVT, and agreeing on a suitable price was problematic. There were also serious production problems because the 400-strong labor force was unable to build the required 230 motorcycles a week to break even. Financial troubles continued throughout 1976, and there was simply no capital available for the development of future models. By May 1976 Meriden posted a £1 million loss, but the situation improved slightly later in the year with the introduction of a coproduction venture with Moto Guzzi. Still, Meriden built 9,782 twins between June 1975 and June 1976 and managed to secure a better price for each from Dennis Poore. But they weren't out of the woods yet. By the end of 1976 the Meriden workforce numbered 700, and they would need to build an unrealistic 350 Bonnevilles a week to stay afloat.

OPPOSITE: NVT, and later TMA, indulged in massive magazine advertising. Almost a saturation campaign, these advertisements emphasized the Bonneville's tradition.

CHAPTER SIX

ABOVE: Also new for 1976 was a rear disc brake. This further complicated rear wheel removal.

ABOVE RIGHT: The 1976 Bonneville looked very similar to the 1975 version but now had a federally mandated left-side gearshift.

US distribution was still pivotal for Triumph, but Norton-Triumph Inc., under Roger Stange, struggled. After nearly two years without Bonnevilles, US dealers rejoiced, only to be dismayed at the $1,895 sticker price.

1976 T140V Bonneville, TR7RV Tiger 750

The T120V was dropped this year, and the existing T140V and TR7RV were adapted with a left-side gearshift and right-side rear brake. At the same time, the rear brake became a 10-inch Lockheed disc, matching the front, with an underslung rear caliper. The gearshift modification was achieved via a crossover shaft at the back of the gearbox, and as some of the shifting precision was lost, the left-shift Bonnevilles met with some resistance from traditional conservative buyers. In other respects the 1976 Bonneville and single-carb TR7RV were the same as for 1975, with the same colors.

1977

By 1977 the co-op was close to becoming a viable concern, but the familiar financial crises continued. Meriden wanted to buy the Triumph name and marketing rights from NVT, but this time the government refused financial aid. Many workers were laid off for several weeks as Meriden was forced to reduce stock. Fortunately the giant company GEC came to rescue, purchasing 2,000 motorcycles and providing much-needed liquidity. In return, unused factory space would be used to build Puch Maxi mopeds. Bonneville production recommenced in March at a rate of 250 motorcycles a week, and one of the bonuses of GEC's involvement was managerial support, including unpaid assistance from ex-British Leyland boss Lord Stokes. Lord Stokes assisted co-op director Brenda Price with establishing a new US subsidiary, Triumph Motorcycles America (TMA), running out of Placentia, California. Despite the factory closure, Meriden managed to build 11,931 motorcycles for the year to June 1977. Dennis Johnson, chairman of the co-op since its inception, then decided he wanted "to do more things than simply eat and sleep Meriden motorcycles."

1977 T140V Bonneville, TR7RV Tiger 750

Few technical updates featured for the 1977 twins, but the front fender was new, no longer with a forward

1976 T140V, TR7RV
(Differing from 1975)

Compression Ratio
7.9:1 (TR7RV)
Rear Brake
10-inch Lockheed disc
Dry Weight
390 pounds
Engine and Frame Numbers
T140V HN62501 to GN72283; TR7RV DN70186 to EN71867

1976–1987

LEFT: But for a new front fender, the Bonneville changed little for 1977. Unfortunately, the revised fender wasn't as well supported, and many split. The familiar 750cc twin still had Amal Concentric carburetors at this stage.

BELOW: Reliability problems, particularly electrical, persisted into 1977. Though the bikes were charismatic and expensive, vibration remained a bugbear as always. Clinging onto the past, the Bonneville represented a continuation of tradition but had limited appeal.

1977 T140V, TR7RV
(Differing from 1976)

Compression Ratio
7.9:1 (all versions)
Colors
Pacific Blue (T140V); Signal Red (TR7RV
Engine and Frame Numbers
T140V GP75000 to JP84931; TR7RV HP74444 to HP84475

mount. With technical development centered on meeting the 1978 Environmental Protection Agency (EPA) regulations, Meriden resorted to offering the Bonneville in a new color, Polychromatic or Pacific Blue, and the Tiger in Signal Red. The earlier colors of red and white (T140V) and green and white (TR7RV) were also available.

Amazingly, Triumph continued to set records at Bonneville, and in 1977 four more records were set, bringing the unbroken total to 37. Three of these were set by 50-year-old Jack Wilson, taking his record total to nine.

1978
Following Dennis Johnson's resignation, previous service manager John Nelson returned to the co-op as managing director, with Brian Jones as chief engineer. Confidence at Meriden was high, and 1978 was looking like its best year yet. Production was up to almost 12,000 motorcycles, peaking at 320 a week, but circumstances conspired against Meriden this year. The bulk of production headed to the United States, where it joined a huge inventory of unsold 1977 stock. As 1977 sales to dealers were way below budget, TMA was running out of operating cash. The

continued on page 158

155

CHAPTER SIX

1977 T140J Bonneville Silver Jubilee Limited Edition 750

During 1977 Meriden needed a quick lifesaver, and GEC Sales Executive Brian Reilly suggested capitalizing on the Queen's Silver Jubilee (25th anniversary). Meriden was awash with stockpiled motorcycles, and Reilly suggested these be converted and reworked into a special commemorative model. Thanks to Lord Stokes, the Palace approved this limited-edition model and, launched in August, each came with a special certificate. Although it was only a cosmetic adaptation, it was a brilliant marketing move and provided Meriden with a lot of publicity, particularly in the United Kingdom. They also represented 25 percent of the year's production.

Ostensibly the Jubilee was a standard 750 Bonneville, but it was set apart by some dubious gaudy details, including shiny chrome-plated engine covers, fork covers, and taillight bracket. Blue and red striping highlighted the rather undistinguished silver paint, and striping accents continued on the 19- and 18-inch wheel rims and special Dunlop K91 "Red Arrow" tires. An unusual styling feature was the blue seat with red piping, continuing to the silver chain guard, accented in three colors. For some of the final Jubilees, Triumph fitted a pair of Girling gas-filled upside-down shocks, with the spring preload at the top, that would feature on later Bonnevilles.

Initially it was envisaged that only 1,000 Jubilees would be built; hence every machine carried the proud boast of "One in a Thousand." However, demand persisted, so Meriden produced another 1,000 and finally a further 400 for general export. For these final examples Triumph replaced the emblem with "Limited Edition." Obviously "One in about Two and Half Thousand" didn't have quite the same ring. One thousand were destined for the United States, but they sold very slowly as Americans couldn't really identify with the royal event. Even in America the extravagant looks were considered over the top, and Jubilees sat in US showrooms, gathering dust well into the 1980s. For many the Jubilee was considered an investment, and Jubilees were squirrelled away with few or no miles put on them—some even remaining in their shipping crates. As Meriden's Chief Engineer Brian Jones said later, "Fifty percent of owners put them away and never ran them, so the warranty call back was minimal!"

TOP: The Silver Jubilee Bonneville for the United States had the traditional 2½-gallon gas tank. *Triumph Motorcycles*

ABOVE: UK and general-export examples had a larger, "bread bin"-style gas tank and fork gaiters.

LEFT: Early examples had this side cover decal, but it was a misnomer, as a second 1,000 were produced, and later a further 400. The second series had a less-exclusive "Limited Edition" decal. *Triumph Motorcycles*

1977 T140J *(Differing from T140V)*

Color — Silver and Royal Blue

ABOVE: Although early Jubilees had NVT instruments, later examples such as this had new Smiths instruments.

LEFT: The royal treatment extended to the highlighted chain guard, and the primary drive cover was chrome-plated.

BELOW: The Jubilee Bonneville was a great marketing move, and many were put away as investments.
INSET: The Jubilee's blue seat was highlighted with red piping and special graphics. *Triumph Motorcycles*

1976–1987

CHAPTER SIX

1978 T140V, T140E, TR7RV (Differing from 1977)

Carburetors
Amal R2930 30mm Mark II (T140E)
Horsepower
49 at 7,000 rpm (T140E)
Dry Weight
413 pounds (T140V, TR7RV); 395 pounds (T140V T140E US)
Colors
Black and crimson, Chocolate Brown and gold, blue and silver (T140V, T140E); crimson and silver (TR7RV)
Engine and Frame Numbers
T140V, T140E HX00100 to JX10747; TR7RV PX02117 to HX10584

TOP: The UK Tiger for 1978 was Crimson Red with silver, featuring an unusual U-shaped band on the front fender. These early model year examples don't yet have the upside-down Girling shocks.

ABOVE: Strange colors were introduced for 1978, including this US Bonneville in Chocolate and gold with a brown seat. US models also received Western handlebars.

Continued from page 155

appointment of Jack Hawthorne, from Honda, as the new TMA president headed off a planned dealer revolt, and while the new 1978 models were well received, the weakening of the US dollar saw a massive drop in revenue. Because 1978 prices were fixed at a higher exchange rate, there would be no profit, but rather a £700,000 loss.

1978 T140V, T140E Bonneville, TR7RV Tiger 750

As Meriden was preoccupied with the US market, they were forced to adapt the Bonneville to meet the EPA regulations coming into effect from January 1978. In the meantime the existing Bonneville and Tiger continued with startling new colors and minor updates. A new composite cylinder head gasket replaced the troublesome copper type, and during the year the Jubilee's upside-down Girling shocks appeared.

To meet the new US emissions standards, Brian Jones and Jock Copeland designed a new cylinder head with parallel inlet ports to accommodate a pair of square-bodied Amal Mark II Concentric carburetors. These were mounted on rubber stubs and no longer included the messy float ticklers. Also updated was the engine breathing arrangement, with the breather now venting into the airbox instead of the hose running down the rear fender. The new engine was placed in the T140E (*E* for EPA), with sufficient testing of 9,500 endurance miles completed by December 1977. These were sent to the United States starting in March 1978, but they weren't available elsewhere until 1979.

With the Bonneville back in the US, *Cycle World* was impressed enough to rate it a Heavyweight Roadster honorable mention in their 10 Best Bikes of 1978, noting, "The Twin still sounds as if human beings built it and we can forgive it for not leaking oil." Very few Tigers made it to the States, but *Motor Cycle World* found the TR7RV Tiger "nice but somewhat overpriced. And as Pink Floyd was fond of saying, 'Desperation is the English way.'"

1979

At the end of 1978, former Jaguar Car Chairman and Labour MP Geoffrey Robinson became Meriden's unpaid chief executive, and he would supervise the co-op's shrinkage in the wake of Margaret Thatcher's Tory government's rise to power in the spring of 1979. Meriden was hoping to waive more than a million pounds of interest due in July 1979, and they were faced with a serious decline in sales, particularly in the United Kingdom, and unsold stock of around 8,000 motorcycles. Robinson wanted 200 redundancies, with production reduced to three days and 200 bikes a week. This created considerable bitterness among the workforce, but Meriden received no reprieve from Thatcher's ideologically opposed government. The co-op now sought a long-term solution in a partnership with a larger company and began a Third World

1976–1987

sales drive, including a highly profitable sale of 1,300 military models to Nigeria.

Amid this commotion, Meriden knew they had to drag the Bonneville into the modern era if they were to survive. The first of these more modern designs was the T140D Bonneville Special, created by co-op stylist Tom Hyam. The T140D demonstrated a certain blinkered optimism, and with the unfavorable exchange rate it had to be sold at an unsustainably high price ($3,225) in America. US dealers simply didn't order many bikes for 1979, and stockpiles continued. Ultimately this became unsustainable.

1979 T140D Bonneville Special, T140E Bonneville, TR7RV Tiger 750

All Bonnevilles this year incorporated the 1978 T140E cylinder head and carburetor updates, although the Tiger retained the earlier single Amal Concentric carb and was no longer sold in America. A special 27-horsepower Tiger was also built this year for the German market. The other significant new feature was Lucas Rita electronic ignition, and a useful improvement was the replacement of Phillips-head engine screws with an Allen-head type. Smiths supply problems saw Veglia instruments become standard, in addition to a neutral indicator light, new Lucas handlebar switches, chrome-plated fenders for all models, a lockable seat, and a rear rack. The 1979 Bonneville proved quite successful, in the United Kingdom at least, with patriotism ensuring that one in three 750cc motorcycles sold was a Triumph. Considering the competition included the new Suzuki GS750 and Yamaha's XS750, this was quite impressive. *Bike* magazine praised the 1979 Bonneville for its light weight and excellent handling but feared for its future, saying, "If Triumph are murdered by an unnecessarily pedantic bureaucracy then a lot of people are going to be sorry."

ABOVE: With its cast-aluminum wheels, stylish colors, and TT-style exhaust system, the T140D looked impressive but didn't appeal to the traditionalist.

LEFT: The T140D finned ignition cover was highlighted in black, and the gas tank featured a simpler Triumph tank badge. All Bonnevilles this year included the new cylinder head with Mark II Amal carburetors.

159

CHAPTER SIX

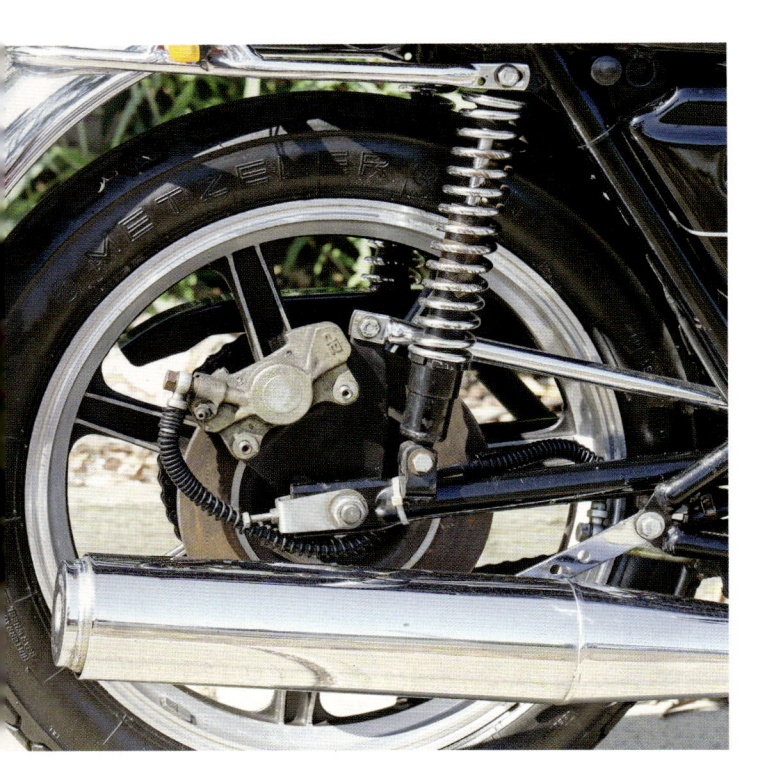

TOP LEFT: The rear brake arrangement was altered for the T140D, with the brake caliper above the swingarm. The single annular discharge muffler was reminiscent of the T160 Trident.

TOP RIGHT: A truncated front fender also distinguished the T140D. The seat on this example is nonstandard. Originally the Girling shock absorbers were inverted, but as this looked odd many owners reversed them.

ABOVE: One of the color schemes for the 1979 T140E was this stunning black and silver. With electronic ignition and Amal Mark II carbs, it met new EPA regulations and appealed to the traditionalist. The 1979 price was a competitive $2,615, but it wasn't enough to ensure profitability.

1979 T140D, T140E, TR7RV
(Differing from 1978)

Ignition
Lucas Rita electronic
Wheels
Lester cast-aluminum 19- and 18-inch (T140D)
Rear Tire
4.25x18-inch TT100 (T140D)
Dry Weight
400 pounds (T140D)
Colors
Black and red, blue and silver, black and silver (T140E, TR7RV); black and gold (T140D)
Engine and Frame Numbers
T140D, T140E HA11001 to KA24999; TR7RV HA1109 to XA24608

The success of the Japanese "factory custom," particularly Yamaha's XS650 Specials in America, prompted the T140D Bonneville Special. Released in March 1979, this was ostensibly a cosmetic variation of the T140E, with US-made, seven-spoke, cast-aluminum Lester wheels, a stepped seat, a unique two-into-one TT Special–style exhaust system, and stylish colors. The swingarm was strengthened (this would appear on all models for 1980), with the rear brake caliper now located above it. While the T140D looked impressive, it had more show than go. It also suffered a confused identity, being neither truly sporty nor a real custom. *Cycle Guide* was also unimpressed. They "hated the cast wheels and single muffler, wishing Triumph had left the Bonneville alone." *Cycle Guide* wasn't alone, and large numbers of unsold T140Ds piled up in TMA's Placentia warehouse. The T140D was also the only Triumph to suffer a general recall; this was due to the fat rear tire not clearing the rear fender on some carelessly assembled bikes. Triumph's service bulletin caused some bewilderment with dealers because the official repair called for a block of wood and a common C-clamp. Triumph also commissioned local exhaust manufacturer Jardine to produce an official replacement for the quiet standard muffler.

1980

Early in 1980 Meriden was in discussion with several companies regarding a partnership, including Honda, Kawasaki, and Marubeni/Suzuki in Japan, and exhaust and suspension manufacturer Armstrong Equipment. Stockpiles in the United States resulted in a drastic reduction in the US operation, with Brenda Price temporarily replacing Jack Hawthorne at TMA. Production

1976–1987

T140E Bonneville European

at Meriden had to be cut back to 100 motorcycles a week, initially with no US build. The Bonneville was struggling to meet new emissions regulations, and 1980 models couldn't be sold in California (where the distribution was based).

Suzuki was extremely interested in a partnership but desired complete control, with production increased to 30,000 motorcycles, including 250, 500, and 750cc. Following a massive devaluation in the yen in June, this all fell through, leaving Armstrong as the only realistic buyer. Armstrong boss Harry Hooper wanted the workforce cut to 150, part of the factory sold off, production reduced to 100 motorcycles a week, and the interest debt waived.

While awaiting the outcome of negotiations between Armstrong and the government, a two-day week was introduced and production reduced to 50 bikes a week. Ultimately the government refused to support Armstrong's takeover, and the co-op was forced to implement 300 redundancies. Meriden's future was still uncertain, but as the Export Credit Guarantee Fund was looking at ways of recouping capital from unsold stock in the United States, they saw Meriden, with their existing distribution network, as their best solution to secure a financial result. In return, the government agreed to write off the co-op's outstanding loan and accrued interest. Meriden had survived by the skin of its teeth, but all this disruption resulted in a disastrous production year, with only 2,572 motorcycles built for the 1980 model year.

With the co-op's future secured, at least in the short term, Geoffrey Robinson resigned as chief executive, and John Rosamond was confirmed as board chairman with ex-Ford and Jaguar man Bob Lindsay as managing director. Public support for the co-op remained strong, and the Triumph Bonneville won the prestigious *Motor Cycle News* Machine of the Year award.

1980 T140D, T140E Bonneville, TR7RV Tiger 750

During this difficult period Brian Jones managed to incorporate a number of significant updates to the Bonneville for 1980, including a more efficient four-valve oil pump, improved primary chain adjustment, a strengthened swingarm with improved chain adjuster,

Only UK-spec Bonnevilles were built for 1980, either European versions such as this one or American versions with high handlebars and a small gas tank. Both versions featured fork gaiters.

161

1980 T140D, T140E, TR7RV (Differing from 1979)

Colors
Black and red, gray and red (T140E); Silver Blue and gold (TR7RV)

Engine and Frame Numbers
T140D, T140E PB25001 to KB27500; TR7RV PB25193 to JB27513

a rear brake caliper mounted above the swingarm, stands relocated to improve ground clearance, and rubber-mounted rear turn signals. With considerable T140D stock available, this was offered in the United Kingdom for 1980 with a "bread bin" gas tank and a choice of standard or two-into-one exhaust systems. The T140E Bonneville continued as two versions, the T140E Bonneville European and T140E Bonneville American. British market features included fork gaiters, rear number plate board, and gas tank emblem.

The electric-start Bonneville and Executive finally appeared this year, lavishly launched in London on May 23, 1980, with pop star David Essex and former F1 champion James Hunt in attendance. Although orders for the electric-start Bonneville were strong, component supply and casting problems saw this delayed until the 1981 model year.

1981

A combination of low US demand and a British recession resulted in considerable reduction in both production and sales of motorcycles and spare parts in the final three months of 1980. These disastrous business conditions continued into 1981, and production over the winter months was reduced to three days a week.

Emissions smog laws were constantly changing, and smaller European companies, including Triumph, struggled to maintain the pace of change. The US Triumph distributor was located in California, but the 1980 and 1981 models could not be sold there because they did not meet the clean-air standards. Enthusiasm had already passed by the time that certification was achieved, for the bike was effectively obsolete. California dealers were kept supplied from 1980 until 1982 with brand-new 1979½ models while we waited for something new. The T140D Special excess inventory was largely cleared out with sales in California through the end of 1982.

As the 1981 Bonneville was still not EPA certified, TMA in America was still selling new 1979½ models while struggling to dispose of T140D stock as required by the agreement with the Export Credits Guarantee Department. Experienced industry figure Wayne Moulton was officially appointed TMA president in February, coming from Kawasaki, where he had been involved in the creation of the LTD series and the Z1-R. In response to the failure of the T140D, Moulton began work on a project he believed to be in tune with American tastes. This was later sent back to Meriden and would become the TSX. By March the orders were coming in again, the co-op was working five days a week, and to appease creditors Meriden released their 1982 model range early, at the National Exhibition Centre (NEC) Show in April 1981.

This had a positive effect on rebuilding Triumph, in Britain at least, but demand was still weak. There was also good news for the co-op in June regarding the sale of stockpiled motorcycles in the United States, as the Conservative government granted an extension for payment to the Export Credit Guarantee Fund until April 1982 and wrote off the company's loan and accrued interest. But as the poor economic situation in England ensured a depressed home market, Triumph's market share continued to slide. And although TMA's orders were increasing, they were nowhere near earlier levels. A year after the government's debt write-off, the co-op was in the same precarious position. Several times during 1981 Triumph faced closure because both motorcycle and spare parts production were way below budget forecast, and at the annual general meeting in November, Managing Director Bob Lindsay was dismissed. The co-op would now be run collectively, and during this upheaval, development of the new eight-valve TSS continued, with the design office also beginning work on Wayne Moulton's TSX Custom.

1981 T140D, T140E, T140ES Bonneville, Bonneville Executive Touring, TR7RV, TR7RVS Tiger 750

Only 1,423 motorcycles were built for the 1981 model year, with sales of 1981 models virtually nonexistent in America, mainly due to the $3,995 price. The model range this year included the UK and US Bonneville Electro, Tiger TR7RV, UK Bonneville Executive Touring, and the previous US T140D. Component supply was always problematic, and this year saw the gas tanks sourced from Italy and some models with Marzocchi suspension, while the electric-start motor was made

All T140Es until 1981 had a cylinder head with parallel intake ports and a pair of Amal Mark II carburetors mounted on rubber intakes. Instead of the problematic push-in type introduced back in 1971, the exhaust header pipes were now a threaded stub.

1981 T140D, T140E, T140ES, TR7RV, TR7RVS
(Differing from 1980)

Starting
Lucas M3 electric-start (T140ES, Executive ES, TR7RVS)
Dry Weight
430 pounds (T140ES); 482 pounds (Executive ES)
Color
Smoked Red or blue (T140ES, Executive)
Engine and Frame Numbers
T140E, T140ES KDA28001 to DDA29427;
TR7RV KDA28097 to DDA29398

under license in India. Persistent problems with the push-in exhaust header pipes saw a return to the pre-1971 threaded stub exhaust, and the right-side main bearing was upgraded to a twin lip roller from a troublesome ball type. For the electric-start models there was a larger, 14 AH battery and a three-phase 14.5-amp alternator with three-pack Zener diode. The Bonneville Executive came with a beautiful Brearley Smith fork-mounted fiberglass fairing and 3.9 cubic feet of dedicated luggage, and US versions were all electric-start.

LEFT: The traditional kick-start-only Bonneville reached its zenith in 1981, the black and red colors still the most popular.

BELOW: One of the features setting the 1981 Bonneville apart from the 1980 version were the rectangular, black, German-made ULO turn signals, which replaced the traditional Lucas type. Most 1981 Bonnevilles had a T140D-style stepped seat.

CHAPTER SIX

RIGHT: A new engine cover distinguished the T140ES, and the electric-start motor fitted very neatly behind the cylinders in the old magneto position. A reduction gear linked to a Borg Warner sprag clutch driving an intermediate timing gear. A kick-start lever was still fitted in 1981 but later became an option.

BELOW: In Smoke Blue, the 1981 T140ES was a very handsome machine. The fenders were now stainless steel instead of chrome. The small electric-start motor didn't compromise the aesthetics; it was only just visible underneath the Amal carbs. If only it had appeared 10 years earlier.

1981 T140LE Bonneville Royal Limited Edition

As he favored limited-edition models as a way of stimulating sales, Co-op Chairman John Rosamond sought approval from the Prince of Wales for a Commemorative Bonneville Royal Limited Edition celebrating the royal wedding on July 29, 1981. Rosamond also put it to Wayne Moulton, who exhibited little enthusiasm, the memories of the slow-selling Jubilee still strong. Released early in July, Bonneville Royals were ostensibly T140ESs updated to the new 1982 specification with Bing carburetors, an ungaitered front fork, and Marzocchi Strada rear shock absorbers. UK versions received Morris cast-aluminum wheels and triple Lockheed disc brakes. The Royal was the first Triumph with a chrome-plated gas tank since the TR5 Trophy and Edward Turner's ban, and rather strikingly styled. UK examples had a silver-painted frame and a matte-black engine with polished fin edges. The US versions were quite different, ostentatiously styled with a slimline gas tank, black frame, highly polished engine, and "King and Queen" seat. Unlike the UK versions, these had wire wheels and the usual single Lockheed front disc brake. All came with Royal-badged side covers and a certificate of authenticity. A total of 250 Royals were built, 125 UK and 125 US versions, and the popularity of the Royal in England saved Meriden from a potentially disastrous year. Demand also continued on the other side of the Atlantic, and as the first series wasn't certified for sale in California, in August 1981 a second series Royal (without the "Wedding") was produced for the United States. This was now similar to the UK Royal Wedding, with Morris wheels, triple disc brakes, larger Euro-style gas tank, and a black frame, and it was sold as a 1982 model.

The US version of the Royal was extremely garish and had an odd two-tone "King and Queen" seat. Not a great styling success. *Triumph Motorcycles*

1982

Due to many component supply difficulties, in particular the US-supplied Morris cast wheels and Italian shock absorbers and gas tanks, Meriden continued to struggle to meet production targets and early in 1982 suffered a severe cash-flow crisis. Three years on, several hundred T140Ds still remained unsold in America, and Meriden was endeavoring to introduce new models, in particular the eight-valve TSS. Money was so tight that dealers even had to purchase the new 1982–1983 sales brochures, and to reduce overhead Triumph started to look at selling part of the factory and relocating. UK dealers were also asked to pay cash on delivery for completed motorcycles, and when the enlarged 11-model 1982–1983 range was launched at the NEC Show in April 1982, Triumph was hoping for increased orders. Shown for the first time were the TSS and TSX Custom, as well as the controversial TS8-1. This was optimistically listed at £3,149 but wouldn't make production. But when fewer than 100 orders materialized from the show, morale at Meriden slumped yet again. In the United States, TMA managed to qualify for dealer financing, agreeing to buy back new unsold inventory from the finance companies in the event of a dealer default, but it came too late. The delayed arrival of 1982 models in the middle of the winter instead of September was disastrous, virtually ensuring the selling season was lost.

A short reprieve came in November 1982, when the West Midlands Enterprise Board agreed to finance the export of motorcycles to the United States and development costs totaling £465,000. The workforce now numbered only 188, producing a mere 80 motorcycles a week, but a total of 1,692 for the 1981–1982 period represented a slight improvement over the previous year. In the meantime plans to sell the Meriden site were well underway, but by the end of 1982 the money finally ran out and production ground to a halt. Most of the workforce was laid off, leaving a skeleton staff working around the clock to keep the co-op afloat.

1982–1983 T140E, T140ES Bonneville, T140AV Bonneville Executive Touring, TR7RV, TR7RVS Tiger, TR7T Tiger Trail 750

With motorcycles built to order during this period, there was a much wider range of variants and colors available than before, with many more options. The 1982 model range was originally displayed in April 1981 and included the T140E and ES (UK and US), Executive Touring, TR7RV, and TR7T, and a new 650, the TR65 (in UK and US versions). As in 1980, the designation *US* described the specification and didn't actually mean it was intended for the US market, and no TR650s were exported to the States. The Bonneville US continued with the small gas tank, wire wheels, and single front disc brake, while the Euro Bonneville included Morris cast wheels, triple disc brakes, and the new, larger, Italian-made gas tank. (As the British gas tank supplier had discontinued production at short notice, from September 1981 the new-style Italian gas tanks appeared on all Bonnevilles.) With Girling ceasing production during 1981, Marzocchi units

CHAPTER SIX

New for 1982 were the black-painted forks, 60-watt sealed beam headlight with new rubber mounting, and Bing carburetors. This is the Bonneville T140 ES electric-start USA.

1976–1987

gradually replaced the Girlings. The cast-iron brake discs were also reduced slightly in diameter, to 9.8 inches (250mm) and no longer chrome-plated, and Dunlop sintered metal brake pads were fitted. Initially only fitted to the Executive, Bing constant velocity carburetors gradually replaced the Amal Mark IIs on twin-carb models, with a new air filter and side cover arrangement. The Bing carbs were already certified for some older BMWs, and this allowed quicker and less costly implementation for Triumph. This year saw a 60/45-watt Lucas sealed beam headlight, matte-black flexible headlight brackets, and a cold-core insert in the dual seat to improve rider comfort.

For 1982 the models sold in the US were the three Bonnevilles: T140ES Bonneville, T140LE Royal, and T140EX Executive. These all had the new-style Italian-made gas tank with hinged flip-up filler, but they could be optioned with the earlier teardrop tank while stocks lasted. The new US T140ES Bonneville was one of Meriden's most successful models this year, with 389 orders secured within two weeks of it going on sale. A promotional bonus at this difficult time also came when Richard Gere used a T140E in the 1982 Paramount film *An Officer and a Gentleman*. Gere was a Triumph enthusiast, having previously purchased a Bonneville while on tour in the United Kingdom with the stage production *Grease*. By the time the film appeared, Gere's earlier model T140E looked decidedly vintage, but it still didn't hurt Triumph's sales.

Initially shown in October 1980, the bright yellow TR7T Tiger Trail appeared in 1981 and was Meriden's short-lived attempt at creating a dual-purpose bike in the mold of the BMW R80 G/S. Flawed and unattractive, the TR7T retained the Tiger's Amal Mark I Concentric carburetor but had a lower compression ratio and a new inlet camshaft, as well as off-road equipment including a 21-inch Radaelli steel front rim, high-mounted sprung plastic front fender, and vulnerable black exhaust system. A bright green TR65T Trail was also available briefly later in 1982, basically a TR7T with the 650cc engine. Both models were unpopular and only sold in small numbers. Other new models built in very small numbers for 1982 were the anti-vibration T140AV Bonneville and TR7AV Tiger, with a rubber-mounted engine to reduce vibration. This anti-vibration frame was ostensibly a prototype for the projected new T2000 Phoenix double overhead camshaft 900cc twin.

A second series of 1982 models was displayed at the NEC Show in April, now described as 1982–1983 models. Updates included dogleg brake and clutch levers and an improved electronic ignition, and production toward the end of 1982 was very much mix and match, using whatever components were available at the time. While most models received Marzocchi shock absorbers, Paiolis from the TSX were fitted to the TR7RV and TR7RVS Tiger, and the tank colors for all models were inconsistent.

1982/83 T140E, T140ES, T140AV Executive, TR7RV, TR7RVS, TR7T
(Differing from 1981)

Compression Ratio	7.4:1 (TR7T)
Carburetors	Bing Type 94 CV 32mm (Executive)
Front Tire	3.00x21-inch (TR7T)
Rear Brake	7-inch drum (TR7T)
Dry Weight	383 pounds (TR7T)
Color	Yellow (TR7T)
Engine and Frame Numbers	T140E, T140ES, T140AV EDA30001 to BDA31693 (1982); T140E, T140ES, T140AV BEA33001 to AEA34393 (1983); TR7RV, TR7RVS, TR7T JDA29428 to BDA31917 (1982); TR7RV, TR7RVS, TR7T BEA33009 to AEA34386 (1983)

1982–1983 TR65 Thunderbird 650

At the end of June 1981, production of the 1982 TR65 Thunderbird commenced. Essentially a budget version of the TR7RV Tiger, the 750 stroke was shortened and, with a new crank, a cylinder with one less fin, and a Tiger cylinder head, Meriden created the Thunderbird with minimal outlay. This economy model also included a single-phase alternator, points ignition, a drum rear brake, and a simpler dash without a tachometer. Replacing the tachometer was a very ugly warning light binnacle. The engine featured a matte-black finish with a black siamesed exhaust system.

LEFT: The Bonneville Executive was handsome, if somewhat heavy, the weight detracting from its sport prowess.

LEFT BELOW: The 1982 Bonneville T140ES Euro with the new-style gas tank, Morris wheels, and double front Lockheed discs.

167

CHAPTER SIX

1982/83 TR65 Thunderbird

Bore	76mm
Stroke	71.5mm
Capacity	649cc
Carburetor	Single Amal
Front tire	3.25x19-inch
Rear tire	4.00x18-inch
Front Brake	9.8-inch Lockheed disc
Rear brake	7-inch drum
Dry Weight	395 pounds
Color	Smoked Burgundy
Engine and Frame Numbers	TR65 EDA30001 to BDA31693 (1982); TR65, TR65T BEA33009 to AEA34386 (1983)

Two versions were available, the Euro with the larger gas tank and the US with a smaller tank.

1982–1983 T140 TSS 750

Weslake eight-valve cylinder heads had been a popular conversion for 650 twins during the 1960s, and it was Weslake Meriden turned to for cylinder heads for the TSS 750. The electric-start crankcase and gearbox were stock, but inside the engine was a stiffer crankshaft with narrower, fatter big-end journals. This reduced vibration, raising the safe ceiling to 10,000 rpm, previously unheard of for a Triumph twin. The new cylinder head included four valves set at a 30 degree included angle, the conventional pushrods operating a pair of forked rockers. The aluminum head featured integral rocker boxes, eliminating oil leaks, with new flat-topped pistons and aluminum cylinders contributing to higher horsepower and cooler running. UK versions were fitted with Amal Mark II carburetors, while US examples had Bing carbs. This engine was installed in a T140 chassis, all having a dual-disc Lockheed front brake and Marzocchi Strada shock absorbers, with Morris wheels an option.

Unfortunately, following the TSS launch in February 1982 the delivery of cylinders and heads from Weslake was slow, and by the time of the official release in April only half the orders were fulfilled. And the design was insufficiently developed, with oil leaks a problem, and many warranty claims due to cylinder head porosity, blown Cooper ring head gaskets, and flawed camshaft design. Meriden had plans to offer the TSS in the AV anti-vibration frame for 1983, the crankcase, cylinders and head no longer black, but the factory closed before this happened. In many respects the TSS delivered what many enthusiasts were demanding: a smoother engine and real performance. The TSS was fast, now a 120-mile-per-hour motorcycle, but it was still expensive ($3,895) and virtually unobtainable (only 112 were imported into the United States). The TSS certainly showed a lot of promise; Jon Minonno won the modified production class in the 1982 Daytona Battle of the Twins Race on Jack Wilson's Big D Triumph TSS at 155 miles per hour. But for lack of development the TSS could have been what Triumph needed to stay alive, but this wasn't to be.

1982–1983 T140 TSX Custom 750

Released at the same time as the TSS, the TSX closely followed Wayne Moulton's factory custom concept of a street cruiser equally comfortable winding down British country roads or cruising the boulevard. The basic architecture was a T140ES with a black-highlighted engine, shorter megaphone-style mufflers and fenders, Morris cast wheels (the rear a squat 16-inch), 3.6-gallon Italian-made gas tank, and a stepped twin seat. The TSX also had Paioli shock absorbers and a wider swingarm (to accommodate the fat tire), and the workers at Meriden raided the obsolete spare parts supply to fit the earlier Lucas turn signals and a chrome-plated headlamp. The ostentatious styling never really struck a chord, and functionally the change in geometry and small rear wheel had a negative effect on the TSX' handling. Early TSX production was also troubled by an unexpected rear brake problem. Traditionalists asked the question, why? And the $3,695 price and initial reliability issues didn't endear it to the custom crowd. For 1983 Meriden planned two versions of the TSX, the TSX 8 Custom, with the TSS eight-valve engine (still in black), and the TSX 4 Custom. Although displayed at the 1983 NEC Show, neither made it to production.

1983

At the beginning of 1983 Meriden's 41 remaining workers faced a daunting prospect. The last Meriden Bonneville and Tiger were built on January 7, but the co-op still ambitiously planned for future models and hoped to relocate to a new factory in a nearby vacant Dunlop building. With the council offering £1 million if the private sector could match it, Triumph made a huge effort for the NEC Show in March, showing a sleeved-down 599cc Daytona and TSX-styled Thunderbird, TSX 8 and TSX 4, TSS in AV frame, and

1982/83 T140 TSS

Bore	76mm
Stroke	82mm
Capacity	744cc
Horsepower	57 at 6,500 rpm
Compression Ratio	9.5:1
Carburetors	Amal Mark II 34mm (UK); Bing CV 32mm (US)
Front Tire	4.10x19-inch
Rear Tire	4.10x18-inch
Front Brake	Twin 9.8-inch Lockheed discs
Rear Brake	9.8-inch Lockheed disc
Dry Weight	410 pounds
Color	Black with gold stripes, black and red
Engine and Frame Numbers	T140TSS from CEA33027

1976–1987

ABOVE: Designed as an introductory model, ideal for the first-time big-bike rider, the Thunderbird TR65 had a black engine and exhaust system. This is the TR65 USA, with a smaller gas tank, but it wasn't sold in the US.

LEFT: Handsome in black and gold, the TSS was fast but insufficiently developed. All TSSs had twin Lockheed front disc brakes and remote reservoir Marzocchi Strada shock absorbers. *Australian Motorcycle News*

BELOW: Not particularly attractive, the TSS engine featured a pair of centrally located 12mm spark plugs. The pushrod arrangement was similar to before, with two tunnels, but improved sealing. This general-export version has Amal Mark II carburetors, with US examples receiving Bing CV. *Australian Motorcycle News*

CHAPTER SIX

1982/83 T140 TSX Custom

Compression Ratio	7.4:1 (after GEA33965)
Carburetors	Bing 32mm CV
Front Tire	MJ90x19-inch
Rear Tire	MT90x16-inch
Front Brake	9.8-inch Lockheed disc
Rear Brake	9.8-inch Lockheed disc
Dry Weight	415 pounds
Color	Gypsy Red or Midnight Black
Engine and Frame Numbers	T140TSX from GEA33528

new Bonnevilles in UK and US trim. Both of the latter had a drum rear brake, with the US Bonneville featuring TSX styling and a 16-inch wire-spoked rear wheel and the UK version rear-set footpegs. The T2000 Phoenix was meant to usher in a new era, but Britain was in an economic crisis and the co-op's door finally closed on August 26, 1983, when the workers accepted voluntary liquidation. Eventually 40-year-old Staffordshire businessman John Bloor acquired the name, patents, and manufacturing rights, and the Meriden factory was sold to make way for a housing development.

In the United States TMA closed with dignity, paying out warranty claims and continuing to support dealers after selling the spare parts inventory to JRC Engineering. This goodwill would stand Triumph in good stead when they reentered the US market more than a decade later.

The Harris Bonnevilles, 1985–1988

Soon after acquiring the name and rights, John Bloor leased the rights to build the existing T140 Bonneville for five years to one of the failed bidders, Les Harris' Racing Spares in Devon. Harris intended to improve the Bonneville with TSS crankshafts and aluminum barrels, but Bloor's agreement specified that the bikes must be the same specification as Meriden's pre-closure. Harris invested considerably in tooling, initially planning a 1984 release, but component supply problems saw this delayed. The first Harris Bonneville rolled out of the Newton Abbot factory on June 25, 1985.

The Harris Bonneville's engine was much as before, but it included a TSX Custom crankshaft with machined bob weights and earlier Amal Concentric carburetors, which were described as Mark 1.5 as they incorporated the Mark-II cold-start slides. All Harris Bonnevilles were kick-start only, and although the frame was still British-made, by 1984 most other British suppliers were out of business. Harris thus turned to Italy for much of the chassis equipment, including Veglia instruments, 38mm Paioli forks and shocks, Brembo brakes, Radaelli wheel rims, and

Not really hitting the mark, the 1983 T140 TSX cruiser emphasized style over function. Meriden raided the parts bin to find earlier Lucas turn signals and a chrome-plated headlamp. *Australian Motorcycle News*

170

Lafranconi mufflers. The switches and front master cylinder were German Magura, so this was truly a European Bonneville. Two versions were produced, the UK model with a 4.5-gallon tank and the US with 2.5 gallons.

From the outset the Harris operation was doomed. Due to the high cost of liability insurance the Harris Bonneville couldn't be sold in the United States, and as it was expensive to make, it was difficult to sell in the United Kingdom. By 1988 the existing dies for crankcases, cylinder heads, and rocker boxes were also worn out and needing replacement. With less than two years left for the license, Harris couldn't justify the expense of replacing the dies and the last Bonneville was produced in March 1988.

So, more than 50 years after the release of Edward Turner's Speed Twin, his venerable design went out with a whimper. It was a rather inglorious end for one of the most influential motorcycle engine designs of all time, but for a prewar design to continue in production through until the 1980s was remarkable. Fortunately, a new era for the British motorcycle industry was already underway, one in which the lessons from the failure of the previous generation were learned.

ABOVE: Harris Bonnevilles were kick-start only, and many components were sourced from Italy, including the large Lafranconi silencers. *Australian Motorcycle News*

LEFT: The rear brake was a Brembo disc and the suspension Paioli. *Australian Motorcycle News*

1985–1988 Harris T140 Bonneville

Carburetors	Amal Mark 1.5
Front Tire	100/90H19
Rear Tire	110/90H18
Front Brakes	Twin 260mm Brembo
Rear Brake	260mm Brembo
Dry Weight	410 pounds
Colors	Black and red, black and gold, silver and black
Engine and Frame Numbers	EN000001 to SN001258

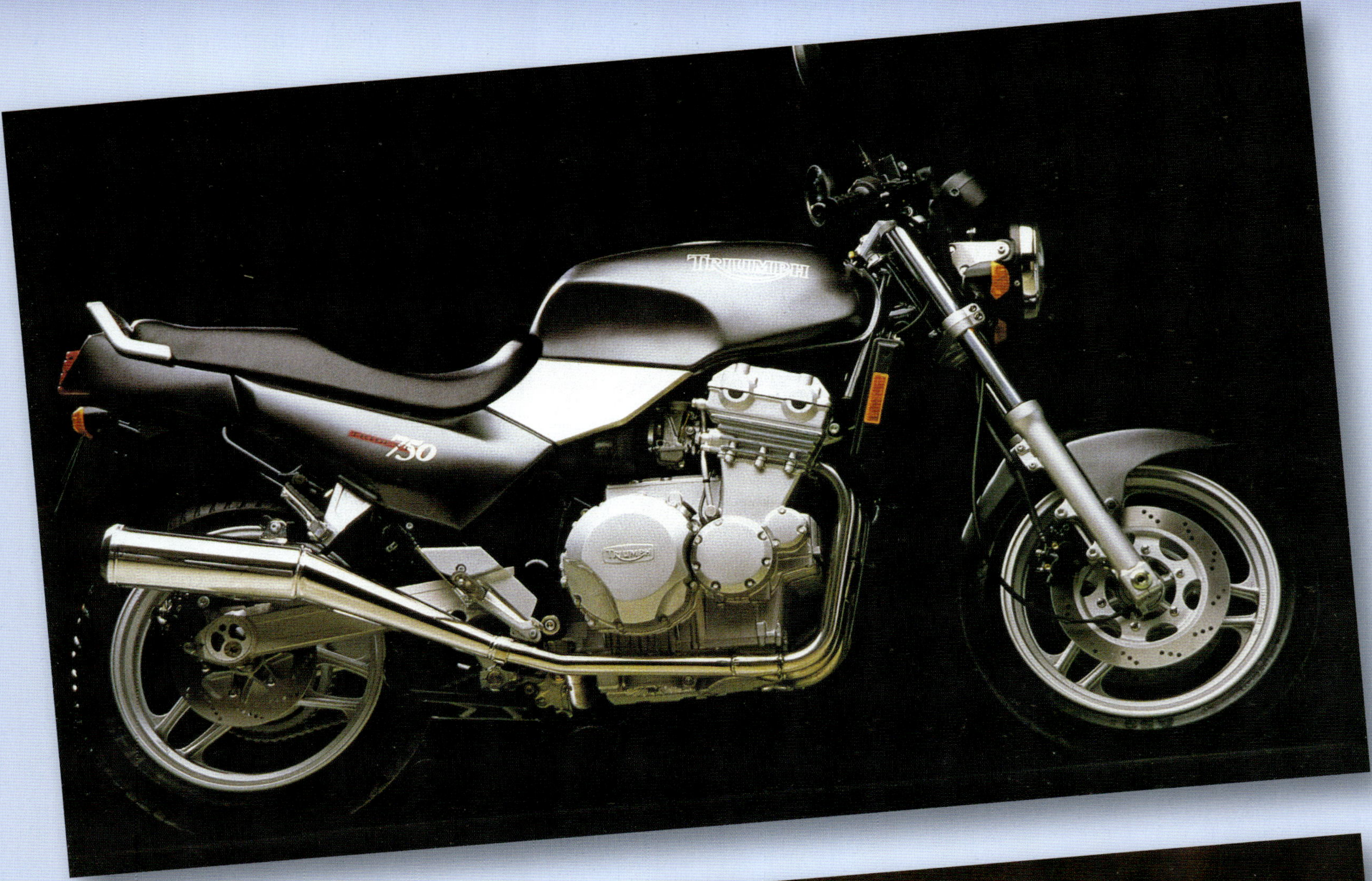
TRIUMPH
750

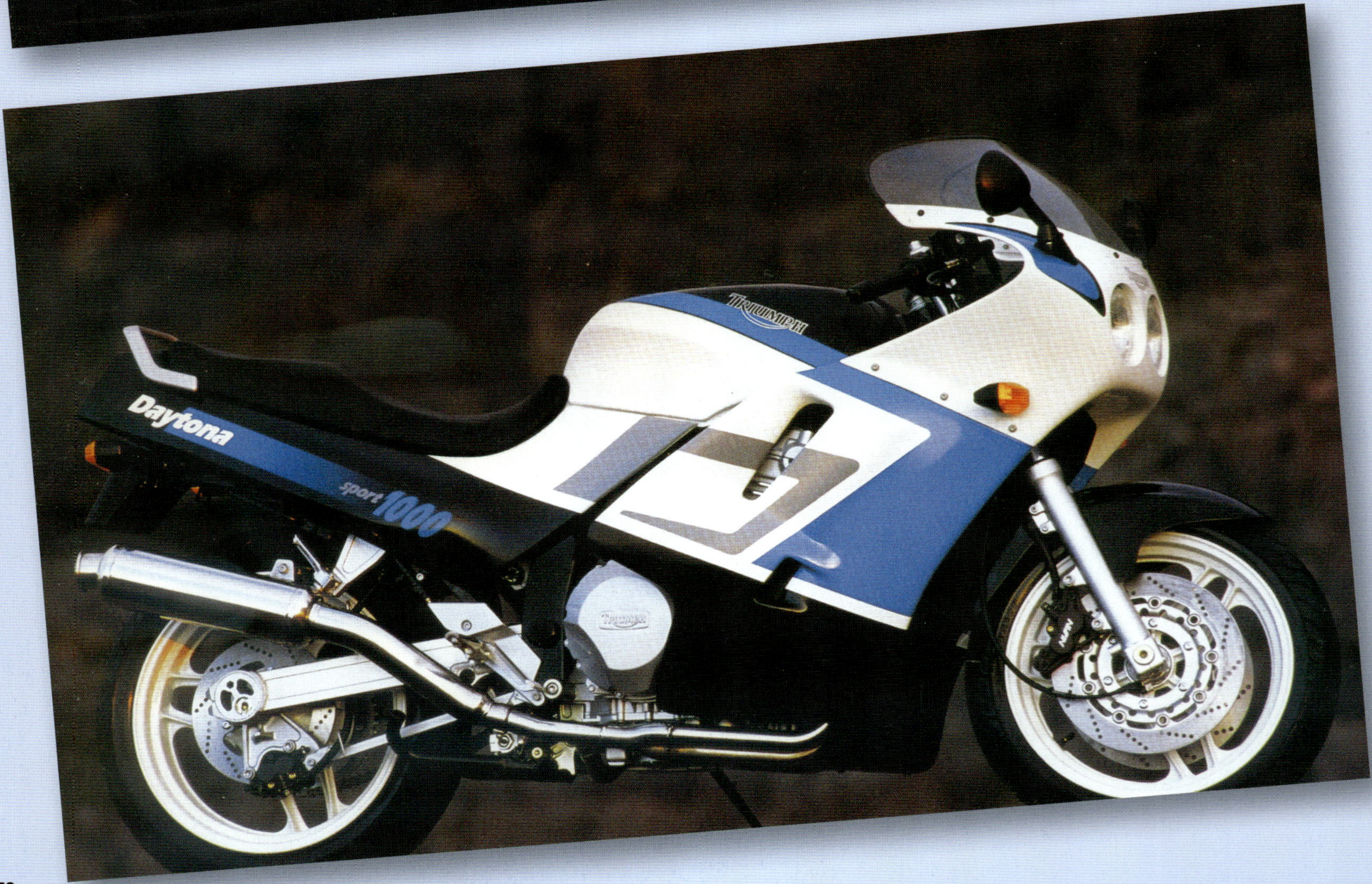
TRIUMPH
Daytona
sport 1000

7

1990–1996

Resurrection at Hinckley

The Modular Approach; Spine-Frame Triples and Fours

After acquiring Triumph's name and rights in 1983 and subsequently leasing these to Les Harris, John Bloor quietly set about designing a new range of motorcycles that bore no relationship to the previous Bonneville. Determined not to repeat the series of "false dawns" that had epitomized many British motorcycle industry revivals (notably Hesketh and Norton), Bloor set up a research and development unit and began looking for a suitable factory site. It soon became evident that Bloor's business philosophy differed substantially from Edward Turner's earlier model, which had characterized Triumph in the previous era. Rather than pursue immediate short-term profits with underdeveloped products, Bloor sanctioned all-new designs, underwriting the development costs for several years. Although the Triumph name was well respected, it also came with some baggage, particularly in regard to electrical and general reliability. Bloor thus decided on a conservative approach, promoting the new range as modern, not retro, and launching Triumph as small, high-quality producer with the intention of eventually growing into a large-volume mainstream manufacturer competing directly with the Japanese.

One of the projects Bloor acquired from Meriden was the twin-cylinder Phoenix, codenamed the Diana, but after comparing this to current Japanese offerings the idea of developing this was quickly abandoned. In 1985 Bloor's predominantly ex-Meriden team decided on an economically expedient modular approach, the full range sharing many parts. This was built around a common spine frame and three- and four-cylinder, liquid-cooled engines in four capacities, all with a 76mm bore. Common components on early versions even extended to the wiring looms. The first four-cylinder prototype was running by

ABOVE: Preproduction testing took place with professional racers at the former RAF and US Air Force base at Bruntingthorpe in Leicestershire. Flanking long-time Triumph employee Gary McDonnell (who joined Meriden in 1969 as an apprentice fitter) are Mark Phillips, Keith Huewen, and Steve Tonkin.

OPPOSITE TOP: Although very similar, the Trident 750 wasn't as popular as the 900. The 1991 version was available in Metallic Gunmetal Gray and black. Silver highlighting included the engine cases, swingarm, fork legs, and six-spoke wheels.

OPPOSITE BOTTOM: The Daytona 1000 was available in white, blue, and black for 1991. The Daytona was too tall and heavy for a true sport bike, and as the short-stroke engine provided little power under 8,000 rpm, they didn't prove popular. Compared to the Trident, the rear brake caliper was situated underneath the swingarm, and the larger front brakes included four-piston brake calipers.

CHAPTER SEVEN

John Bloor

Reclusive and enigmatic, John Bloor epitomizes the rags-to-riches story. The son of a coal miner, Bloor was born in 1944 in a small Derbyshire village, but after suffering poor health and long periods of absence he left school at 15 to become a plasterer. Within two years he had set up his own company, building his first house by the age of 20. Bloor Homes gradually expanded to become one of the largest privately owned house builders in the United Kingdom, contributing significantly to the East Midlands' revival. It was his quest to buy the Meriden factory site for a housing development that resulted in Bloor's acquisition of the Triumph name, but his track record showed there was no place for complacency and sentiment in his agenda. Triumph was revived as a business, not an exercise in nostalgia. In 1995 Bloor was awarded the Order of the British Empire for services to motorcycling, and after surviving the building slump of the late 2000s, Bloor subsequently benefited from new government "Help to buy" incentives for house buyers. In 2002 Bloor Homes sold 1,870 houses, the largest number by a sole-owner company. Triumph Motorcycles has also continued to prosper under Bloor, with production increasing to more than 50,000 and Triumph claiming 20 percent of the British over-500cc market. By 2014 Bloor was seventh in Birmingham's Rich List with an estimated net worth of £520 million. Modest and private, Bloor rarely gives interviews and maintains a hands-on role at Hinckley. His approach has always erred on the conservative side, and it has paid dividends.

Hinckley's modular approach extended to the new three- and four-cylinder engines, all sharing a 76mm bore but with varying strokes creating four capacities.

1987, and in 1988 Bloor purchased a 10-acre site in Hinckley, Leicestershire, and proceeded to build a 150,000-square-foot modern factory known as T1. The factory opened in 1990, with about one-third of components manufactured in-house to ensure superior quality control. Triumph's renaissance was shrouded in mystery, and when Triumph displayed the new range to the press on June 29, 1990, journalists were stunned. Privately financed to the tune of £80 million, Bloor had built the state-of-the-art plant and developed the bikes in secret. After the initial range of preproduction models launched at the Cologne Show in September, a six-model lineup was displayed in December at the NEC in Birmingham, with production scheduled for the 1991 model year.

1991 Model Year

Central to the resurrected Triumph lineup was the association with famous past model names, the range consisting of two naked Tridents (900 and 750cc), two sport Daytonas (1000 and 750cc), and two sports touring Trophys (1,200 and 900cc). The engine design emphasized durability, the valves operated by twin overhead camshafts running directly in the head, these driven by a Hy-Vo chain on the right side of the engine. The cylinder head design was the same for all engines, with two 32mm inlet valves and two 28mm exhaust valves inclined at 39 degrees. Valve adjustment was by bucket and shim, and a one-piece crankshaft was coupled to balancer shafts, one for the triples and two for the fours. An example of Triumph's quest for reliability was the crankshaft manufacturing process, in which the crankshaft was heat-treated for 30 hours in a plasma nitride furnace and machined to tolerances of 5 microns. The horizontally split crankcase included a wet-liner block, and as all engines shared the same mounting points, the frames and gas tanks were interchangeable between all models. The spine frame and aluminum swingarms were also manufactured on a robotic production line in-house, the swingarm MIG instead of TIG welded to minimize distortion.

With no UK-based suspension or carburetor suppliers left, Kayaba supplied the front fork and shock absorbers, Shin Nippon the wheels, and Mikuni the carburetors. Emphasis was placed on build quality, each machine undergoing a cold engine test, culminating in a rolling road analysis, and resulting in a range of British motorcycles with genuine character. Practicality was also a priority, with major service intervals at 6,000 miles and only 11 service items across the entire range.

Production commenced in February 1991 at only five bikes a day, with the first shipment to Germany in March. Germany was selected as a proving ground to assess the overall quality, and the first examples were four-cylinder Trophy 1200s. By June three-cylinder production commenced, and 750 Daytonas, 900 Trophys, and new Triumphs were appearing in UK dealers' showrooms. Deliveries began to the Netherlands, Australia, and France later in the year. A modest total of 2,414 motorcycles left the Hinckley factory during 1991, but it was a very promising beginning. As venerable journalist Alan Cathcart commented after riding the Trophy 1200, "This Lazarus-like rebirth of one of the most famous marques in motorcycle history is a new seventh force in world motorcycle manufacture."

1991 Trident 900 and 750

As a basic, naked roadster with an upright riding position, the Trident continued the Meriden Bonneville's style, but with considerably more power, vastly improved brakes and suspension, and substantially superior strength and reliability. With more torque, the 900 was much more popular than the 750, but both were extremely competent first efforts, the tall and heavy spine frame suited to the naked style.

1991 Daytona 1000 and 750

The sport model in the range was the Daytona, produced in three- and four-cylinder versions, both with a short stroke. As a result they lacked the torque of the 900 triple and 1200 four and received a lukewarm reception. The spine frame also contributed to a heavy, tall bike that was too cumbersome to be considered a true sport bike. Designed to compete with Japanese superbikes like the Yamaha FZR and Suzuki GSX-R, the Daytona was simply too heavy, with unsuitable power characteristics.

While the result was confused, the Daytona did come with some quality equipment, including Nissin four-piston front brake calipers, a chrome Motad exhaust system, and fully adjustable Kayaba suspension. Triumph claimed a top speed of around 160 miles per hour for the Daytona 1000. The plastic bodywork was also Japanese-inspired. Developed by ICI and noted for its durability and flexibility, the fairing included a pair of 5.5-inch headlights.

1991 Trophy 1200 and 900

Another famous name carryover, the first Hinckley Trophy was the 1200 four, followed by a more popular 900 triple. These 1991 Trophys were "Sports Tourers" with a full fairing and clock, low screen, low handlebars, and no standard luggage. The engine had an oil cooler, and while the Kayaba front fork was non-adjustable, the rear unit had provision for adjustable preload and damping. Unlike the Daytona, with its twin headlights, the Trophy's large fairing had a rectangular headlight; like the Daytona, however, the big Trophy had silver engine cases and swingarm, and a chrome exhaust system. Although large and heavy, the Trophy received favorable initial reviews and would become the longest living of the spine-frame models.

1992 Model Year

With the establishment of new export markets in Japan, Italy, Spain, and Switzerland, Triumph continued to expand. But for cosmetics and colors the model range was unchanged, the color range increasing during the year, as did production numbers, to 3,098 for 1992.

1991 Trident 900 and 750

Model	TC338 and TC333
Type	Inline three-cylinder DOHC, 4 valves per cylinder liquid-cooled
Bore	76mm
Stroke	65mm (900); 55mm (750)
Capacity	885cc (900); 749cc (750)
Horsepower	97 at 9,000 rpm (900); 89 at 10,000 rpm (750)
Compression Ratio	10.6:1 (900); 11:1 (750)
Carburetors	Triple Mikuni BST 36mm flat slide CV
Ignition	Digital electronic
Gearbox	Six-speed
Frame	High-tensile 600MPa micro-alloyed steel
Swingarm	Aluminum with eccentric chain adjuster
Front Fork	43mm Telescopic
Rear Suspension	Gas-charged rising rate
Brakes	2x296mm disc 2x2 calipers front, 255mm disc rear
Wheels	17x3.5-inch front and 18x4.5-inch rear
Tires	120/70x17-inch front and 160/60x18-inch rear
Wheelbase	58.7 inches
Dry Weight	467 pounds
Colors	Gun Metal Gray and black, British Racing Green and black

1991 Daytona 1000 and 750

Model	TC343 and TC330
Type	Inline three- and four-cylinder DOHC, 4 valves per cylinder liquid-cooled
Bore	76mm
Stroke	55mm
Capacity	998cc (1000); 749cc (750)
Horsepower	120 at 10,500 rpm (1000); 89 at 10,000 rpm (750)
Compression Ratio	11:1
Carburetors	3 or 4 Mikuni BST 36mm flat slide CV
Ignition	Digital electronic
Gearbox	Six-speed
Frame	High-tensile 600MPa micro-alloyed steel
Swingarm	Aluminum with eccentric chain adjuster
Front Fork	43mm Adjustable Telescopic
Rear Suspension	Gas-charged adjustable rising rate
Brakes	2x310mm disc 2x4 calipers front, 255mm disc rear
Wheels	17x3.5-inch front and 18x4.5-inch rear
Tires	120/60x17-inch front and 160/60x18-inch rear
Wheelbase	58.7 inches
Dry Weight	518 pounds (1000); 480 pounds (750)
Colors	Blue, white, and black; black, Gun Metal Gray, and black; red, white, and black

The first of the new Hinckley Triumphs was the Trophy 1200, initially a sports tourer. Unusual and distinctive colors with bold stripes distinguished the 1991 versions.

CHAPTER SEVEN

1991 Trophy 1200 and 900

Model	TC340 and TC336
Type	Inline three- and four-cylinder DOHC, 4 valves per cylinder liquid-cooled
Bore	76mm
Stroke	65mm
Capacity	1,180cc (1200); 885cc (900)
Horsepower	140 at 9,000 rpm (1200); 99 at 9,000 rpm (900)
Compression Ratio	10.6:1
Carburetors	3 or 4 Mikuni BST 36mm flat slide CV
Ignition	Digital electronic
Gearbox	Six-speed
Frame	High-tensile 600MPa micro-alloyed steel
Swingarm	Aluminum with eccentric chain adjuster
Front Fork	43mm Telescopic
Rear Suspension	Gas-charged adjustable rising rate
Brakes	2x296mm disc 2x2 calipers front, 255mm disc rear
Wheels	17x3.5-inch front and 18x4.5-inch rear
Tires	120/70x17-inch front and 160/60x18-inch rear
Wheelbase	58.7 inches
Dry Weight	529 pounds (1200); 489 pounds (900)
Colors	Gun Metal Gray and silver; black and red; Gun Metal Gray, black, and silver; Gun Metal Gray, black, and red

1992 Trident 900 and 750 (Differing from 1991)

Colors
British Racing Green, Cherry Red, Cherry Black

1992 Daytona 1000 and 750 (Differing from 1991)

Colors
Lancaster Red, Assam Black, Radiant Red

1992 Trident 900 and 750
By 1992 it was clear the market preferred the torque of the 900cc triple, the 900 outselling the 750 four to one.

1992 Daytona 1000 and 750
One of Bloor's positive initiatives was to heed market forces and consumer feedback, and as the two short-stroke Daytonas weren't proving popular with

BELOW: Except for colors, the Trident was unchanged for 1992. Some examples had new side cover decals that did not state the engine capacity.

1990–1996

1992 Trophy 1200 and 900 (Differing from 1991)

Colors
Lancaster Red, Caribbean Blue, Charcoal Gray, burgundy, Oxford Blue (later)

journalists or customers, this was to be their final year. As with the Trident, both Daytonas were unchanged but for colors and graphics.

1992 Trophy 1200 and 900
Still available in 1200 and 900cc versions, the Trophy initially featured bold color stripes like the Daytona, but during the model year the stripes were dropped for a single color. At this time Oxford Blue and burgundy were added to range of color options, with Lancaster Red dropped.

1993 Model Year
By October 1992 Hinckley had built 5,000 bikes, which coincided with the launch of the new 1993 range at the Cologne Show. While retaining the modular methodology, the Daytona was revamped to include the long-stroke 900 and 1,200, a faired Trident Sprint was introduced, and, in response to German and French demand for an adventure BMW GS-style bike, the Tiger was announced. The lineup expanded to eight models, but now with only three engines (the 750 and 900 triples and 1,200 four). In July 1993 Hinckley's T1 factory celebrated building 10,000 motorcycles, and production increased to 6,512 this year, with additional export markets including Sweden.

1993 Trident 900 and 750, Trident 900 Sprint
Cosmetic updates on the Trident 900 and 750 saw a black wrinkle engine finish replace the silver, black extending to the wheels and exhaust downpipes.

TOP: The 1992 Daytona 1000 in Lancaster Red with white and black stripes. The short-stroke Daytona wasn't popular and lasted only two years.

ABOVE: Also largely unchanged for 1992 was the Trophy. Initially these had stripes like the Daytona's, but later in the model year they became a single color. *Triumph Motorcycles*

177

CHAPTER SEVEN

RIGHT: New for 1993, and filling a gap between the Trophy and unfaired Trident, was the Trident Sprint with a wind tunnel–developed half fairing. The instruments were frame-mounted, and unlike the Trophy, the Sprint fairing included a pair of quartz halogen headlamps. This is the 1994 version with four-piston brake calipers. *Triumph Motorcycles*

OPPOSITE: Another new model for 1993 was the Enduro-style Tiger 900. With an integrated fairing and tank, the Tiger was the first Triumph with a plastic gas tank. This is also a 1994 version, with black engine cases. *Triumph Motorcycles*

1993 Trident 900 and 750, 900 Sprint (Differing from 1992)

Model
TC362 (Sprint)
Horsepower
100 at 9,500 rpm (900); 97 at 11,000 rpm (750)
Dry Weight
474 pounds (Sprint)
Colors
British Racing Green, Cinnabar Red, Diablo Black, Caspian Blue (Sprint)

1993 Daytona 1200 and 900 (Differing from 1992)

Type
TC354 and TC357
Bore
76mm
Stroke
65mm
Capacity
1,180cc (1200); 885cc (900)
Horsepower
145 at 9,500 rpm (1200); 100 at 9,500 rpm (900)
Compression Ratio
12:1 (1200); 10.6:1 (900)
Dry Weight
503 pounds (1200); 476 pounds (900)
Colors
Racing Yellow, Barracuda Blue, Pimento Red

Crinkle black stretched to the handlebars, contrasting with polished alloy levers and chromed mirrors. The Trident wheels also received polished aluminum rims, a new embossed seat, a seat height lowered to 30.5 inches, lower footpegs, and redesigned engine hoses to improve the cosmetics. New this year was the Trident Sprint 900, ostensibly a Trophy 900 but with a top half fairing and twin headlights. The result of customer feedback requesting some weather protection for the Trident, the Sprint was a fusion of disparate components but soon established its own identity and was very successful. All Tridents had black engine cases, wheels, and exhaust downpipes this year, and for 1993 the Sprint 900 still featured Trident side cover decals. This year only the Sprint's front brake calipers were the dual-piston type.

1993 Daytona 1200 and 900

In response to criticism, the Daytona was considerably updated this year. Not only did the long-stroke 1200 four and 900 triple replace the short-stroke 1000 and 750, but the Trophy 1200's four-cylinder engine received hotter cams, higher-compression pistons, and cylinder head porting. Both had a standard oil cooler, and John Mockett designed a new fairing with a lower screen and revised rear panels. The black engine, frame, front fork, upswept exhaust, and wheels contrasted with bold single colors. All Daytonas could be specified with an optional rear seat cover, and the new model looked a lot sportier, particularly with the new graphics. It was still heavy, but with the Japanese voluntarily restricting their large-capacity sports bikes to 125 horsepower at the time (in the United Kingdom, at least), the big Daytona 1200 was the most powerful bike available, with its top speed an impressive 160 miles per hour. Still designed primarily for the German market, new Daytonas shipped to nearly 40 German dealers in February 1993, riding back to Germany through snow and ice to prove their robustness.

1993 Trophy 1200 and 900

While the Trophy's profile, still with a rectangular headlamp, remained similar to that of 1992, updates this year comprised a black finish for the engine, exhaust headers, and wheels. The Trophy's revised riding position featured higher handlebars and lower footpegs. Instrumentation now included a digital clock, and the Trophy 1200 shared the Daytona's front four-piston Nissin brake calipers and floating discs.

1993 Trophy 1200 and 900 (Differing from 1992)

Horsepower
110 at 9,000 rpm (1200)
Brakes
2x310mm disc 2x4 calipers front (1200)
Dry Weight
518 pounds (1200); 485 pounds (900)
Colors
Caspian Blue, Candy Apple Red, British Racing Green

1993 Tiger 900

Although the large-capacity Enduro style was not particularly popular in the United Kingdom, Triumph's emphasis on expanding into Europe saw the first new bike to appear since the original six-model lineup, the Tiger 900. Designed by John Mockett, to reduce tooling costs this featured a plastic gas tank, at the time not legal in the United Kingdom. Triumph subsequently had the tank homologated, but the real market was Germany, where the Tiger was the most popular model in 1993.

While maintaining the modular approach, the Tiger's frame was strengthened with a longer wheelbase, wire-spoked wheels, a longer seat, an upswept three-into-two exhaust system, and a longer leading axle front fork. The front brake calipers were a floating twin-piston type, and the rear shock absorber was fully adjustable and included a remote reservoir. Off-road features extended to an engine guard and lever protectors, and the engine was retuned with new camshafts for less power and more torque. At the request of German dealers the main engine covers were painted the same color as the bodywork, but as paint durability was a problem, this only appeared on 1993 Tigers.

1994 Model Year

This year saw the introduction of a new paint facility and the expansion of the Hinckley works with planning permission for a second factory, T2, on a 40-acre site. Still following an evolutionary approach, new models included the Speed Triple café racer and Daytona Super III sports bike, taking the lineup to 10 models. The Daytona Super III featured new Cosworth-designed crankcases and cylinder head castings that would soon be incorporated across the range. With the number of sprag clutch teeth increasing from 51 to 53, Triumph assumed the earlier sprag clutch problem had been overcome and the new crankcases no longer included an access lid in the crankcase for the sprag clutch. Unfortunately this wasn't the case, and changing the sprag clutch on post-1994 models now involved splitting the crankcases. Production at

1993 Tiger 900

Type	Inline three-cylinder DOHC, 4 valves per cylinder liquid-cooled
Bore	76mm
Stroke	65mm
Capacity	885cc
Horsepower	84 at 8,500 rpm
Compression Ratio	10.6:1
Carburetors	Triple Mikuni BST 36mm flat slide CV
Ignition	Digital electronic
Gearbox	Six-speed
Frame	High-tensile 600MPa micro-alloyed steel
Swingarm	Aluminum with eccentric chain adjuster
Front Fork	43mm Telescopic
Rear Suspension	Monoshock remote reservoir fully adjustable
Brakes	2x276mm disc 2x2 calipers front, 255mm disc rear
Wheels	19x2.5-inch front and 17x3.0-inch rear
Tires	110/80x19-inch or 100/90x19-inch front and 140/80x17-inch rear
Wheelbase	61.4 inches
Dry Weight	461 pounds
Colors	Sandstone, Caspian Blue, Cinnabar Red

CHAPTER SEVEN

Apart from a crinkle black finish, the Trident 900 was essentially unchanged for 1994. The black and red colors were also the same, as would continue for 1995 and 1996. Triumph Motorcycles

After receiving a new fairing, longer-stroke engine, and new graphics for 1993, for 1994 the Daytona was updated with new 17-inch wheels and wider tires.

1994 Trident 900 and 750, Sprint 900
(Differing from 1993)

Rear Suspension
Monoshock with adjustable damping (900)
Brakes
2x310mm disc 2x4 calipers front (Sprint)
Wheelbase
59.4 inches (1,510mm)

Hinckley continued to rise, now up to 10,407 motorcycles during 1994, with exports beginning to Canada, taking the number of export markets to 33.

1994 Trident 900 and 750, Sprint 900

As usual, most updates to these models for 1994 were cosmetic, and this year the Sprint 900 was no longer termed a Trident. The Tridents received a reprofiled seat, crinkle black finish on the engine, headlight and triple clamps, and silver-highlighted wheels. Additionally the Trident 900's rear suspension unit was now adjustable, as were the clutch and brake levers. The Sprint 900's front brakes were updated to 310mm floating discs with Nissin four-piston calipers, and the instrumentation included a digital clock. Apart from red being no longer offered for the Trident 900, the colors were unchanged.

1994 Daytona 1200 and 900

The most significant update to the Daytona 1200 and 900 for 1994 was a pair of 17-inch three-spoke Brembo

180

1994 Daytona 1200 and 900 (Differing from 1993)

Horsepower
97 at 9,000 rpm (900)
Wheels
17x3.5-inch front and 17x5.5-inch rear three-spoke
Tires
120/70xZR17 front and 180/55xZR17 rear
Dry Weight
496 pounds (1200); 470 pounds (900)

wheels replacing the six-spoke type. These had hollow spokes and wider rims, and a rear wheel hugger incorporated the chain guard. Colors remained unchanged.

1994 Trophy 1200 and 900

As the two Trophy models were both well established by 1994, updates concentrated on wheels and brakes. Both the 1200 and 900 received the new Brembo wheels, and the Trophy 900 got the 1200's improved front brakes. Colors were also unchanged, although the red was deleted.

1994 Tiger 900

Updates for 1994 were cosmetic only, with the outer engine case now in a black crinkle finish instead of bodywork-matched.

1994 Daytona Super III

Determined to attempt to create a real sports bike out of the spine-frame Daytona, Triumph released

1994 Trophy 1200 and 900 (Differing from 1993)

Horsepower
107 at 9,000 rpm (1200); 97 at 9,000 rpm (900)
Brakes
2x310mm disc 2x4 calipers front (900)
Wheels
17x3.5-inch front and 17x5.5-inch rear three-spoke
Tires
120/70xZR17 front and 170/60xZR17 rear
Dry Weight
511 pounds (1200); 478 pounds (900)
Colors
Caspian Blue, British Racing Green

1994 Tiger 900 (Differing from 1993)

Colors
Caspian Blue, Pimento Red

the Daytona Super III during 1994. The Super III engine featured new cylinder heads and crankcases designed by Cosworth Engineering in Worcester. Cosworth also helped develop new high-lift camshafts and a gas-flowed cylinder head. With higher-compression pistons and carburetor and exhaust updates, the power increased by 15 percent. Along with carbon-fiber fenders and silencers, the Super III was the first Triumph with new Alcon-produced six-piston brake calipers, which radically improved the braking performance. Alcon specialized in Indy car braking systems. Although the new crankcases and carbon-fiber components saved around 10 pounds, the Super III was still much heavier than comparable Japanese superbikes, notably the 407-pound Honda Fireblade. The Super III was also extremely expensive, selling in the United Kingdom for £9,699, compared to £8,199 for a standard 900 Daytona.

TOP: Red was also a popular color for the 1994 Daytona. *Triumph Motorcycles*

ABOVE: The Trophy 1200 received new wheels and brakes for 1994. *Triumph Motorcycles*

CHAPTER SEVEN

ABOVE: All Daytona Super IIIs were Racing Yellow with a black lower fairing. The fenders and mufflers were carbon fiber.

RIGHT: The front brakes included floating discs and six-piston Alcon "Tr-6" calipers.

1994 Speed Triple 900

While the Daytona Super III didn't really succeed as a state-of-the-art sports bike, the other new 1994 model, the niche-market Speed Triple, hit the nail on the head. Triumph described the Speed Triple as "a pure enthusiasts machine," and with its low bars, rear-set footpegs, and single round headlight, the naked café racer style pioneered a new market. Inspired by a modified Trident special of Italian distributor Carlo Talamo's, *Numero Tre*, this was the Triumph triple in its rawest form, and it was an instant hit. More Daytona than Trident, the Speed Triple wasn't only a styling exercise; it also featured the Daytona's wider wheels and black upswept mufflers. The front fork was fully adjustable, with the rear brake caliper mounted below the black swingarm. While the 885cc triple was much the same as the Daytona, with the new Cosworth crankcases, the exposed plumbing was tidied, and inside was only a five-speed gearbox. The Speed Triple's steering was also quicker than the Daytona and Trident, with an additional 5mm of triple clamp offset, increasing from 35 to 40mm.

1990–1996

1994 Daytona Super III

Model	TC310
Type	Inline three-cylinder DOHC, 4 valves per cylinder liquid-cooled
Bore	76mm
Stroke	65mm
Capacity	885cc
Horsepower	115 at 10,500 rpm
Compression Ratio	12:1
Carburetors	3 Mikuni BST 36mm flat slide CV
Ignition	Digital electronic
Gearbox	Six-speed
Frame	High-tensile 600MPa micro-alloyed steel
Swingarm	Aluminum with eccentric chain adjuster
Front Fork	43mm Adjustable Telescopic
Rear Suspension	Monoshock adjustable
Brakes	2x310mm disc 2x6 calipers front, 255mm disc rear
Wheels	17x3.5-inch front and 17x3.5-inch rear
Tires	120/70xZR17 front and 180/55xZR17 rear
Wheelbase	58.7 inches
Dry Weight	465 pounds
Color	Racing Yellow

1994 Speed Triple 900
(Differing from Daytona 900)

Model	TC301
Gearbox	Five-speed
Dry Weight	461 pounds
Colors	Racing Yellow, Diablo Black

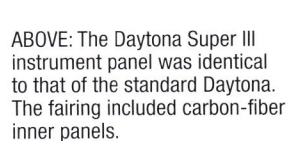

ABOVE: The Daytona Super III instrument panel was identical to that of the standard Daytona. The fairing included carbon-fiber inner panels.

LEFT: The Daytona Super III had new Cosworth crankcases, the semicircular (rather than flat) sprag clutch cover no longer allowing access to the sprag clutch without opening the crankcases. Unlike the regular Daytona 900, the swingarm was black.

In July Triumph returned to racing with a one-off Speed Triple Challenge support race for the British Grand Prix at Donington Park Circuit. This was a precursor for a one-make UK Speed Triple series in 1995 and 1996. The Speed Triple was arguably Hinckley's most significant model yet, and it accounted for more than 1,500 of the total model year production of 7,800 bikes.

1995 Model Year

At the 1994 Cologne Show Triumph released the Thunderbird, this coinciding with a return to the US market in October out of a new operation based in Peachtree City, Georgia. Aimed squarely at America, the Thunderbird drew heavily on Triumph's heritage and, with expected strong demand, was allocated 25

183

CHAPTER SEVEN

Also new for 1994 was the Speed Triple. The name was a clever reference to Edward Turner's inspirational Speed Twin, and this bare-bones café racer–style motorcycle proved extremely popular. Most sold in 1994 were black. Triumph Motorcycles

percent of production capacity. Triumph continued to invest in manufacturing equipment, installing a further seven computer numerical control (CNC) machines, this allowing the machining of engine cases to be reduced from 18 to 7 minutes. Technical updates across the range this year extended to a revised sprag clutch, one of the most persistent problems on early Hinckley Triumphs, and a new igniter box. After the successful Speed Triple Challenge at Donington, Speed Triple Challenge series continued in the United Kingdom and were established in the United States, France, and Germany during 1995. Dealerships were also established in South Africa and Singapore, and in August 1995 the 30,000th Hinckley Triumph was delivered (a Thunderbird to Australia). A total of 13,525 bikes was built in 1995, and also introduced was the range of Triple Connection clothing.

1995 Trident 900 and 750, Sprint 900, Tiger 900, Daytona 1200 and 900, Daytona Super III, Trophy 1200 and 900, Speed Triple 900

While the base Trident 900 and 750 continued unchanged, still with the same color options, the Sprint 900 now received Charcoal Gray three-spoke Brembo wheels, with Triumph-branded four-piston brake calipers. The engine finish was Charcoal Gray, and detail updates extended to twin passenger grab rails and a more rounded tail unit and taillight. But for a black swingarm, black-faced instruments, and the addition of black as a color, the Tiger 900 was unchanged, while the Daytona 1200 and 900 were also as for 1994,

1995 Sprint, Tiger, Daytona, Trophy, Speed Triple *(Differing from 1994)*

Wheels
17x3.5-inch front and 17x5.5-inch rear three-spoke (Sprint)

Tires
120/70xZR17 front and 170/60xZR17 rear (Sprint)

Colors
Nightshade, Candy Apple Red, British Racing Green (Sprint); Diablo Black (Tiger, Daytona); British Racing Green, Nightshade (Trophy); Fireball Orange (Speed Triple)

with the addition of black and a black swingarm. The bright yellow Daytona Super III was as before, but with only 300 produced, and with the Trophy 1200 and 900 about to be updated for 1996, they were offered in two new colors only, one a dramatic and rather unusual purple. But for orange replacing yellow, the five-speed Speed Triple was also unchanged this year. All models were available in the United States for the first time, and *Cycle World* tested a Daytona 900, finding, "Triumph has polished fundamentally mature design and technology, breaking no new ground, to produce a very appealing motorcycle with its own special character." The Sprint 900 met with a more favorable reception:

1990–1996

For 1995 the Speed Triple was offered in bright orange, the bold color becoming iconic and contributing to the Speed Triple's cult status. This example has an aftermarket three-into-one exhaust system, providing even more noise and further emphasizing the visceral engine performance.

"Simple, unpretentious, fast, not perfect, but a lot of fun to ride." It was also the first reborn Triumph to win *Cycle World*'s Best Bikes Award, the Sprint taking Best Open-Class Streetbike.

1995 Thunderbird

Twenty-nine years after the demise of the iconic Meriden Thunderbird, Triumph released their new, retro-styled Thunderbird. Although still part of Hinckley's modular approach, the Thunderbird included design elements strongly evocative of earlier Triumphs, reaching the American enthusiast by chiming a chord of recognition with the past. The Thunderbird's engine included new crankcase covers, a cylinder block with fins cast all the way down (cast by Cosworth Engineering and replicating the air-cooled 650 twin), and a restyled cylinder head. Engine styling drew heavily on earlier cues, the polished balancer and crankshaft cover joined as one and the clutch cover with TRIUMPH cast into it. With a five-speed gearbox, the engine replicated the relaxed nature of the original Thunderbird, producing less power and more torque. Other retro features included plenty of chrome, the 1957-style chrome-grill Mouth Organ gas tank badge, classic flat seat, and classic "peashooter"-style mufflers. A low (29.5-inch) seat height, stretched-out wheelbase, and wire-spoked Spanish Akront rims with a single disc and floating brake caliper widened the Thunderbird's appeal to the cruiser market, and it was also one of the first Triumphs offered with a range of custom accessories.

As the Thunderbird's release coincided with Triumph's return to the American market, it was Triumph's best-selling model in 1995 at around 3,000 units. Celebrities such as Bruce Springsteen took to it, and *Motorcyclist* magazine named the Thunderbird Cruiser of the Year, describing it as "A masterpiece of polished aluminum and glistening chrome." *Cycle World* was more guarded in their enthusiasm, saying, "The Thunderbird works well for what it is. It represents

With plenty of chrome, and bristling with retro cosmetic touches, the Thunderbird was extremely successful and subsequently initiated a range of derivatives. The specification was basic, not even including a center stand, and underneath the retro styling was a spine-frame triple not unlike the rest of the lineup. This year only the Thunderbird received an embossed clutch cover. *Triumph Motorcycles*

185

CHAPTER SEVEN

1995 Thunderbird

Model	TC339
Type	Inline three-cylinder DOHC, 4 valves per cylinder liquid-cooled
Bore	76mm
Stroke	65mm
Capacity	885cc
Horsepower	69 at 8,000 rpm
Compression Ratio	10:1
Carburetors	3 Mikuni 36mm flat slide CV
Ignition	Digital electronic
Gearbox	Five-speed
Frame	High-tensile 600MPa micro-alloyed steel
Swingarm	Aluminum with eccentric chain adjuster
Front Fork	43mm Telescopic
Rear Suspension	Monoshock
Brakes	1x320mm disc 2-piston caliper front, 285mm disc rear
Wheels	18x2.5-inch front and 16x3.5-inch rear
Tires	110/80x18-inch front and 160/80x16-inch rear
Wheelbase	61 inches
Dry Weight	485 pounds
Colors	Cherry Red and cream, Aegean Blue, Diablo Black

the outlook of Triumph's corporate rearview mirror, a nod to heritage and to a perceived, and apparently real, American appetite for nostalgia."

1996 Model Year

This was the final year with the entire range based around the spine frame, and although total Hinckley manufacture now exceeded 40,000 motorcycles, production was down slightly, to 11,041. New markets were found in Malaysia and Thailand, and new models included an updated Trophy and the Thunderbird-based Adventurer. In preparation for the second generation to be presented for 1997, runout models were offered to use up spare parts, particularly stocks of the short-stroke 750 triple engine.

While the Trident was largely unchanged for 1996, the revamped Trophy now featured a larger, curvaceous fairing and dedicated luggage rack.
Triumph Motorcycles

1996 Sprint, Tiger, Daytona, Speed Triple
(Differing from 1995)

Gearbox
Six-speed (Speed Triple)

1996 Trident 900 and 750, Sprint 900, Tiger 900, Daytona 1200 and 900, Daytona Super III, Speed Triple 900 and 750

Updates for the Trident and Sprint this year included a new aluminum nitrogen-charged shock absorber, featuring a hard anodized internal finish for increased service life, with the 900s receiving a lighter (6-pound) shock absorber with 12 damping settings and the Sprint a lighter exhaust system. Apart from a black swingarm, Trident and Sprint colors were unchanged. With only 125 sold in the United States during 1995, the Tiger 900 wasn't proving exceptionally popular, and apart from a new luggage rack it remained as before.

For this year the Daytona, Super III, and Speed Triple also received the lighter, nitrogen-charged rear shock absorber and now featured a shorter revalved Kayaba front fork that sat flush in the top triple clamp. The Daytona and Speed Triple's brake calipers were gold rather than black, and this year the Super III was only available in a limited number of 150. Super III total production over three years was only 805, and while not the most contemporary sports bike, it provided a foundation for the next sport Daytona. Still carrying Triumph's high-performance flag, *Cycle World* had this to say of the Daytona 1200: "The 1200 Four has an addictive midrange and top-end punch, truly an intoxicating effect."

Also finishing this year was the Speed Triple 900, now with a six-speed gearbox. A small number of Speed Triple 750s (rumored to be between 140 and

1996 Trophy 1200 and 900 (Differing from 1995)

Dry Weight
518 pounds (1200); 485 pounds (900)
Colors
British Racing Green, Merlot Red, Pacific Blue

250) were also built and included the Trident's six-spoke wheels (the rear an 18-inch) and smaller section tires. Only available in Diablo Black, the Speed Triple 750 was almost apologetic. Lacking the 900's power and charisma, the 750 didn't offer the same riding experience and didn't appear in any brochures.

1996 Trophy 1200 and 900

As far back as 1992 Triumph decided the Trophy required a major redesign, positioning it more as a touring motorcycle and distancing it from the sport Daytona. The style centered on a new, twin-headlamp John Mockett–designed fairing, tested in the William F1 team wind tunnel, and improved ergonomics. With higher, pulled-back handlebars, a slightly taller seat and screen, and integrated luggage, the Trophy finally came of age as a touring motorcycle. New instruments included chrome bezels, a fuel gauge and an analog clock, and options extended to fairing hand extensions, fully enclosed chain guard, and top box. The resulting new Trophy wasn't exactly light or particularly good looking, but it was effective. The British press loved it, *Bike* magazine claiming, "The new Trophy has truly entered the premier league of motorcycling. This bike is a worthy challenger to the best."

1996 Thunderbird and Adventurer

This year the Thunderbird received minor updates, the classic style enhanced with the swingarm now constructed of oval-section aluminum with traditional screw and lock-nut chain adjusters. The polished embossed clutch cover was dropped, replaced by a slimmer cover, and the Thunderbird's success prompted the release of the cruiser-style Adventurer. Still inspired by earlier Triumphs, the Adventurer's gas tank included traditional embossed rubber kneepads and 1968-style Eyebrow badges. And while very much a Thunderbird underneath, the Adventurer's high-rise handlebar, single seat, megaphone exhaust, and ducktail rear fender created quite a different image. Also offered with a variety of accessories, the Adventurer was another success, with over 2,000 sold during 1996.

Toward the end of 1996 Hinckley's initial renaissance came to an end. Successfully launched utilizing an economically expedient modular approach, the time had come for a more pragmatic modus. In an age of increased specialization and extremely competent competition, the strong, effective, but heavy spine frame had its limitations, particularly for a sports bike. While the spine-frame models would overlap for a few more years, Hinckley's next episode would embrace new tubular aluminum frames and fuel-injected engines.

1996 Thunderbird and Adventurer (Differing from 1995)

Model
TC399 (Adventurer)
Tires
150/80x16-inch rear (optional)
Dry Weight
496 pounds (Adventurer)
Colors
British Racing Green and cream (Thunderbird); Heritage Gold and ivory, violet and ivory (Adventurer)

With its high handlebars and swooping ducktail rear fender, the Adventurer was unashamedly intended for the US cruiser market but didn't really hit the mark. This is the 1997 version, identical to 1996 but for the color.
Triumph Motorcycles

The 1996 Trophy's new face included a pair of very distinctively shaped headlights.
Triumph Motorcycles

1990–1996

187

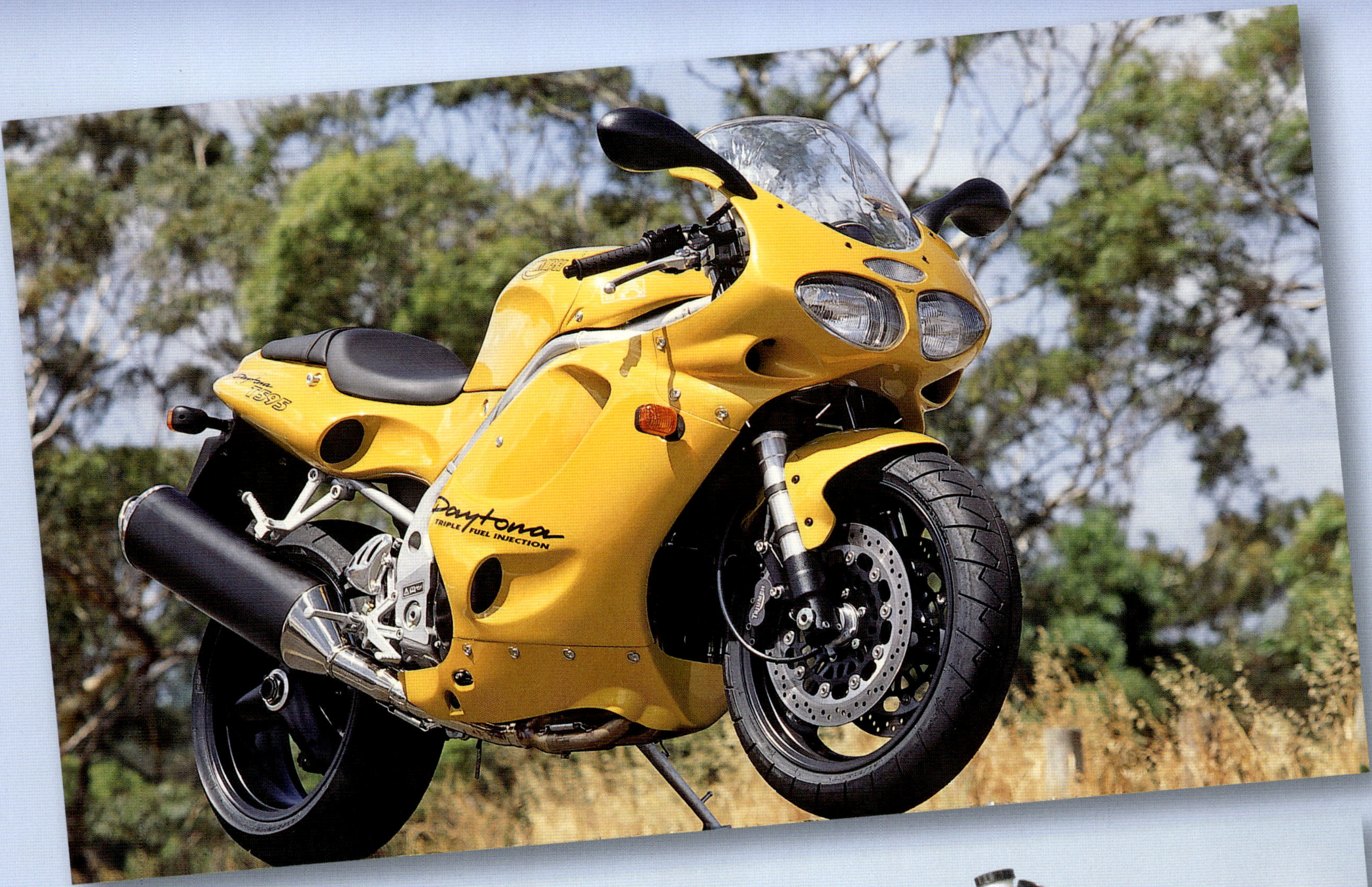
Daytona
TRIPLE FUEL INJECTION

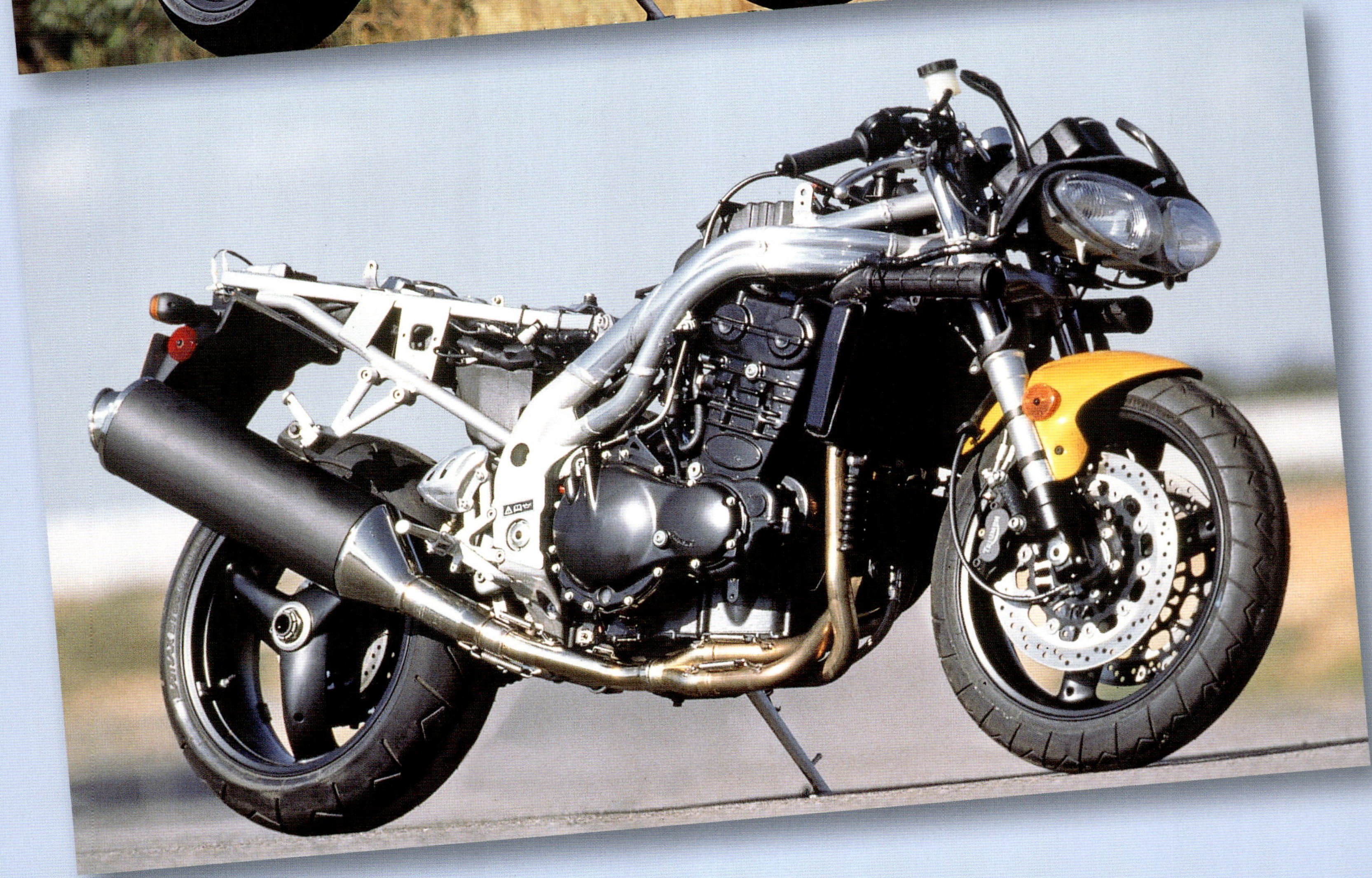

1997–2002
Consolidation

Second Generation: New Frames and Fuel Injection

After celebrating a highly successful initial phase with their first-generation spine-frame models, Triumph embarked on their next era with the unveiling of the new Daytona T595 and Speed Triple T509 at the Cologne Show in October 1996. These much more modern designs marked a milestone for Triumph, and unlike the previous Daytona, the T595 could compete on equal terms with other new-generation sports bikes. Whereas the long, heavy, trucklike, but admittedly powerful, spine-frame Daytona emphasized durability, it was effectively a dinosaur from another age. Replacing the earlier Daytona 900, the considerably more lithe and modern T595 ushered Triumph into a new domain. The T509 replaced the Speed Triple, but other spine-frame models continued in the short term, most finishing in 1998, as the fuel-injected range gradually expanded.

1997 Model Year

Worldwide distribution expanded with the addition of the Czech Republic and Turkey this year, and Hinckley celebrated their 50,000-unit production landmark in March 1997 (with a UK Daytona T595). Due to the demand for the new T595 and T509, motorcycle manufacture increased 37 percent in 1997, to 15,160, while Triumph continued to expand the range of accessories for all models. Triumph's reputation for reliability and longevity was further cemented in July when Nick Sanders set a new world record, traveling around the world in an astonishing 31 days and 17 hours on a Daytona 900.

1997 Daytona T595

Back in July 1993 Triumph had decided to expand on their modular concept by creating two new models to replace the Daytona 900 and Speed Triple. At the time the Super Sport segment was the fastest growing, and this was expected to continue over the next five years. Selecting the Honda Fireblade and Ducati 916 as benchmarks, by August 1993 Harris Performance had provided a prototype rolling chassis. John Mockett began work on the styling in March 1994, with the final

OPPOSITE TOP: The Daytona T595 ushered in Triumph's second generation and represented a significant departure from the previous modular approach. *Triumph Motorcycles*

OPPOSITE BOTTOM: Central to the T595 were a new aluminum frame and single-sided swingarm. *Triumph Motorcycles*

CHAPTER EIGHT

1997 Daytona T595

Type	Inline three-cylinder DOHC, 4 valves per cylinder liquid-cooled
Bore	79mm
Stroke	65mm
Capacity	955cc
Horsepower	128 at 10,200 rpm
Compression Ratio	11.2:1
Fuel Injection	Sagem 41mm throttle bodies
Ignition	Digital inductive
Gearbox	Six-speed
Frame	Aluminum alloy
Swingarm	Aluminum single-sided with eccentric chain adjuster
Front Fork	45mm Adjustable Telescopic
Rear Suspension	Monoshock adjustable rising rate
Brakes	2x320mm disc 2x4 calipers front, 220mm disc rear
Wheels	17x3.5-inch front and 17x6.0-inch rear
Tires	120/70 ZR17 front and 190/50 ZR17 rear
Wheelbase	56.7 inches (1,440mm)
Dry Weight	436 pounds (198kg)
Colors	Strontium Yellow, Jet Black

model completed by March 1995. In the meantime work began on the engine and chassis, the first of these running by August 1995.

Central to the T500 series was a new frame, a perimeter aluminum twin-spar design with oval extrusions and a single-sided swingarm. The frame's total weight was only 24.3 pounds, and to further save weight upside-down forks were eschewed in favor of a conventional 45mm Showa. The wheels were three-spoke Brembos with Nissin brakes. With a target of 130 horsepower, providing a maximum speed of 165 miles per hour, the three-cylinder engine was completely redesigned. The cylinder bore was increased, now Nikasil-coated aluminum with semi-forged pistons, and Lotus Engineering consulted in the port and combustion chamber redesign. Updates included larger valves, shorter inlet guides, new cam profiles, and a three-into-one stainless steel exhaust system, and by carefully shaping the throttle body bores the intake and exhaust ports were effectively extended beyond their tracts. Magnesium engine covers contributed to the engine weighing 25 pounds less than before. The sophisticated Anglo-French Sagem MC2000 electronic engine-management system incorporated a Siemens SAHC microprocessor, with dual 41mm injectors feeding through a pressurized airbox, and to enhance engine life a two-way cooling system was developed in a climatic wind tunnel.

Representing a great leap forward, the new Daytona was small, light, and built for speed. And for an all-new effort it was surprisingly polished, though not perfect. *Cycle World* said, "Gorgeous motorcycle, a giant leap forward for the revitalized British firm, shame about the flawed fuel injection, spongy suspension, and exhaust header that grounds in right-hand turns."

1997 Speed Triple T509

Released at the same time as the Daytona T595, the Speed Triple T509 shared the Daytona's chassis, but with a smaller-capacity, lower-horsepower engine. Patterned after the stripped-down streetfighters that were all the rage in Britain at the time, the radical styling raised a few eyebrows. With its Buell-like mini-fairing, funky, ocular discordant bug-eyed headlamps, and unceremoniously tacked-on radiator-mounted turn signals, the T509 screamed weird. Visually bizarre, it was certainly an acquired taste. Options included high or low handlebars and a fly screen. Functionally the T509 met most criteria, but it suffered for lack of Daytona horsepower, and *Cycle World* wasn't convinced, saying, "It really does have the look of a home-brew." *Motorcyclist*, on the other hand, crowned the T509 Best Hooligan Machine of 1997.

The new Speed Triple T509 shared the Daytona chassis, but the style was controversial. *Triumph Motorcycles*

1997 Spine-Frame Models

But for the usual color variations, the existing Trident, Tiger, Daytona 1200, Thunderbird, and Adventurer continued much as before. The Trophy's colors remained the same, but it received stainless steel exhaust header pipes. While the standard Sprint was unchanged and in the same colors, a limited run of budget black Sprint Sports was also produced toward the end of 1996 and then returned two years later.

1998 Model Year

With a large number of updates scheduled for 1999, this year saw a hiatus in lineup expansion. The only new model was the Thunderbird Sport, while the Daytona 1200 ended with a limited edition. Other spine-frame triples were restyled, and for some this was their final year. It was obvious Triumph was looking to create a more modern, fuel-injected lineup, gradually phasing out the less popular spine-frame models. Production increased slightly during 1998, to 16,831.

1998 Daytona T595 and Speed Triple T509

For the 1998 T595, Triumph addressed earlier criticisms by remapping the Sagem fuel injection, installing a stiffer shock absorber spring, and rotating the exhaust collector, tucking the pipes in closer. The revised mapping (#9822) rectified the midrange flat spot and surging, while a 14kg shock spring replaced the soft 13kg unit. Cornering clearance at maximum lean increased 0.4 inch with the altered exhaust, and the combined result was a more effective sport bike. The only other change this year was the addition of red alongside yellow and black. The Speed Triple T509 also continued much as before, but with the previous optional high handlebar as standard equipment and two new colors, with orange no longer available. Carrie-Anne Moss rode one in the 1999 release *The Matrix*.

1997 Speed Triple T509
(Differing from the T595)

Bore	76mm
Capacity	885cc
Horsepower	106 at 9,100 rpm
Compression Ratio	11:1
Wheelbase	56.6 inches (1,437mm)
Dry Weight	432 pounds (196kg)
Colors	Lucifer Orange, Jet Black

1997 Trident, Sprint, Tiger, Daytona 1200, Trophy, Thunderbird, Adventurer
(Differing from 1996)

Colors	Quicksilver (Trident 750); Quicksilver and blue, black and red, green and cream (Trident 900); Chili Red, Khaki Green, Jet Black (Tiger); Jet Black (Daytona 1200); Cardinal Red, Imperial Green, Jet Black (Thunderbird); amber and copper, turquoise, aubergine (Adventurer)

ABOVE: With the Tiger offered with optional luggage for 1997, its focus moved towards touring rather than off-road. *Triumph Motorcycles*

LEFT: With more than 7,000 sales worldwide, by 1997 the Thunderbird was Triumph's most successful model. *Triumph Motorcycles*

CHAPTER EIGHT

ABOVE: For 1998 the Speed Triple had higher handlebars as standard equipment. *Triumph Motorcycles*

OPPOSITE TOP: The Trident received Sprint-style rear bodywork for 1998, with passenger grab rails and a black panel near the airbox. *Triumph Motorcrcycles*

OPPOSITE BOTTOM: New for 1998, the Sprint Executive came standard with integrated luggage. *Triumph Motorcycles*

1998 Daytona T595, Speed Triple T509
(Differing from 1997)

Colors
Tornado Red (T595, T509); Roulette Green (T509)

1998 Trident 900 and 750, Sprint Sport and Sprint Executive, Tiger, Daytona 1200, Trophy 1200 and 900, Thunderbird, and Adventurer

Continuing as an entry-level model for one more year, the Trident received the Sprint's rear bodywork for 1998, along with two passenger grab handles and new colors. Also in its final year, the Sprint evolved into two distinct versions, Sport and Executive. The Sprint Sport received lower handlebars, firmer suspension, and exhaust raised for improved ground clearance, while the Executive came with standard integrated luggage. Both had a black engine, frame, and wheels, and fully adjustable rear suspension. This year also saw the end of the spine-frame Tiger, now with more accessories and two new colors.

In response to demand for the 145-horsepower Daytona 1200 after it effectively finished in 1997, a

1998 Trident, Sprint Sport and Executive, Tiger, Daytona 1200, Trophy, Thunderbird, Adventurer
(Differing from 1997)

Wheelbase
62.2 inches (1,580mm) (Thunderbird and Adventurer)
Dry Weight
481 pounds (Sprint Executive); 485 pounds (Tiger)
Colors
Pacific Blue (Trident 750); Pacific Blue and silver (Trident 900); Pacific Blue, Jet Black, platinum (Sprint); Volcanic Red, British Racing Green (Tiger); platinum, turquoise (Trophy); Jet Black and red (Thunderbird); Jet Black and silver (Adventurer)

limited edition of 250 1200SEs became available during 1998. These featured gold (rather than white) lettering on the black tank and fairing, gold wheels, the Super III's Alcon six-piston front brake calipers, and a numbered plaque. The United States also received

150 examples, called the Daytona 1200SP, which sold as a 1999 model year version.

One of the most popular spine-frame models, the Trophy continued largely unchanged, but the rear suspension unit was remotely adjustable and the engine, frame, and wheels black. The successful Thunderbird received new colors and more chrome, while the rather extreme Adventurer was toned down with lower handlebars, leather panniers, and a chrome sissy bar.

1998 Thunderbird Sport

Adding another dimension to the successful Thunderbird range, the Thunderbird Sport blended traditional styling with a higher-performance package. More café racer than cruiser, the Thunderbird Sport's 885cc engine produced more power, with an additional sixth gear added to the Thunderbird's five close ratios. Carburetion was by Keihin rather than Mikuni. Chassis updates included fully adjustable longer travel suspension front and rear, twin front disc brakes, and wide-rim, spoked 17-inch wheels. Classic-inspired styling extended to the twin right-side reverse cone mufflers, echoing the seminal X75 Hurricane, and to perforated chrome airbox covers imitating the 1960s and 1970s individual type. Combining retro styling with a modern sport chassis proved a successful marriage,

CHAPTER EIGHT

Apart from a black finish for the engine, matching the frame and wheels, the Trophy continued unchanged for 1998. Platinum was also a new color this year. *Triumph Motorcycles*

RIGHT: The 1998 Sprint Sports had lower handlebars and firmer suspension. *Triumph Motorcycles*

OPPOSITE ABOVE: The first Thunderbird Sport had a distinctive dual-muffler exhaust system. *Triumph Motorcycles*

OPPOSITE BELOW: The Adventurer's rather extreme styling was considerably revised for 1998. Now more cruiser than custom, it had a roadster-style dual seat and standard leather panniers. *Triumph Motorcycles*

194

1998 Thunderbird Sport
(Differing from the Thunderbird)

Model	TC398
Horsepower	82 at 8,500 rpm
Carburetors	3 Keihin 36mm flat slide CV
Gearbox	Six-speed
Front Fork	43mm Adjustable Telescopic
Rear Suspension	Monoshock adjustable
Brakes	2x310mm disc front
Wheels	17x3.5-inch front and 17x4.25-inch rear
Tires	120/70 R17 front and 160/70 R17 rear
Dry Weight	494 pounds (224kg)
Colors	Racing Yellow and black, Tornado Red and black

and the Thunderbird Sport outlived all other spine-frame models and has since garnered a cult following.

1999 Model Year

Triumph continued to move away from their modular concept, introducing two more second-generation models, the Sprint ST and the new Tiger. But although it was inevitable that the spine-frame range would wither out, Triumph still made the most of the existing models that weren't particularly disadvantaged by the weight and high center of gravity of the older design. Thus the Trophy and Thunderbird offshoots continued with styling updates and the addition of the Legend TT. But the success of the new generation was really paying dividends, with total production increasing to 20,402 this year. Triumph also celebrated five years in the US market, expanding to 200 dealers, with the United States now accounting for 20 percent of sales.

CHAPTER EIGHT

ABOVE: While the single-sided swingarm wasn't new, the Sprint ST's aluminum perimeter frame was a Triumph first. *Triumph Motorcycles*

RIGHT: New for 1999, the Sprint ST was one of Triumph's best models yet, winning several awards as the best sport tourer. *Triumph Motorcycles*

1999 Sprint ST

Following the success of the Daytona T595, the Sprint ST was another all-new design. With the idea of replacing the Sprint, at one stage Triumph's third-best-selling model, initial design work began in April 1996, with the first prototypes running by April 1997. While the reworked 955cc engine with new camshafts, cast pistons, and steel liners, was similar to the Daytona, the chassis was designed specifically for sport touring. Unlike the Daytona's trademark oval aluminum, the Sprint ST frame was twin-spar aluminum, still with a single-sided swingarm. As with other recent designs, John Mockett was responsible for the styling, and the fairing was designed to combine sport style with touring ergonomics. A large, 5.6-gallon polymer gas tank provided a great range, and with optional color-coordinated luggage the Sprint ST set a new sport touring benchmark. *Cycle World* thought so too, and in a comparison test with the BMW R1100S, Ducati ST4, and Honda Interceptor, they said, "It is incredibly

1999 Sprint ST

Type	Inline three-cylinder DOHC, 4 valves per cylinder liquid- cooled
Bore	79mm
Stroke	65mm
Capacity	955cc
Horsepower	108 at 9,200 rpm
Compression Ratio	11.2:1
Fuel Injection	Sagem MC2000
Ignition	Digital inductive
Gearbox	Six-speed
Frame	Twin-spar aluminum alloy
Swingarm	Aluminum single-sided with eccentric chain adjuster
Front Fork	43mm Adjustable Telescopic
Rear Suspension	Monoshock adjustable rising rate
Brakes	2x320mm disc 2x4 calipers front, 255mm disc rear
Wheels	17x3.5-inch front and 17x6.0-inch rear
Tires	120/70 ZR17 front and 180/55 ZR17 rear
Wheelbase	57.9 inches (1,470mm)
Dry Weight	456 pounds (207kg)
Colors	Tornado Red, Jet Black

1997–2002

An all-new Tiger was introduced for 1999. This included the T509's fuel-injected 885cc engine and a new perimeter frame in steel.

comfortable, with serious grunt, and a vice-free chassis. It's been a while but we could get used to a British King." *Cycle World* also named it their best sport touring bike of 1999.

1999 Tiger

Alongside the new Sprint was an all-new Tiger for 1999. The focus was reorienting the Tiger closer to a more street-oriented touring role, based around a modified T509 engine and a new tubular steel perimeter frame. Triumph developed the engine, with different camshafts, exhaust, and engine mapping, in conjunction with Lotus; Kayaba assisted in the suspension design and Nissin the brakes. Development began in March 1996, with John Mockett again entrusted with the styling. In July 1997 Sagem received a complete bike to finalize engine management, and production commenced in December 1998. Ergonomics were a major consideration, with the seat adjustable to three different heights between 33 and 33.8 inches

1999 Tiger

Type	Inline three-cylinder DOHC, 4 valves per cylinder liquid-cooled
Bore	76mm
Stroke	65mm
Capacity	885cc
Horsepower	86 at 8,200 rpm
Compression Ratio	11.3:1
Fuel Injection	Sagem MC2000
Ignition	Digital inductive
Gearbox	Six-speed
Frame	Twin-spar steel
Swingarm	Twin aluminum box-section with eccentric chain adjuster
Front Fork	43mm Telescopic
Rear Suspension	Monoshock adjustable rising rate
Brakes	2x310mm disc 2x2 calipers front, 285mm disc rear
Wheels	19x2.5-inch front and 17x4.25-inch rear
Tires	110/80 R19 front and 150/70 R17 rear
Wheelbase	61 inches (1,550mm)
Dry Weight	474 pounds (215kg)
Colors	Lightning Yellow, Jet Black

CHAPTER EIGHT

1999 Speed Triple
(Differing from 1998)

Bore	79mm
Capacity	955cc
Horsepower	108 at 9,200 rpm
Compression Ratio	11.2:1
Wheelbase	56.7 inches (1,440mm)
Dry Weight	432 pounds (196kg)
Colors	Roulette Green

1999 Trophy, Thunderbird, Thunderbird Sport, Adventurer
(Differing from 1998)

Gearbox
Six-speed (Thunderbird)
Front Fork
43mm Adjustable Telescopic (Thunderbird)
Rear Suspension
Monoshock adjustable (Thunderbird)
Front wheel
19x2.5-inch (Adventurer)
Front Tire
110/90R19 (Adventurer)
Wheelbase
63.8 inches (1,620mm) (Adventurer)
Dry Weight
464 pounds (211kg) (Adventurer)
Colors
Cardinal Red, Aston Green (Thunderbird)

in a matter of seconds. The Tiger's styling was more minimalist than its predecessor's, the twin elliptical headlights continuing a trademark style initiated with the T595.

1999 Daytona 955i and Speed Triple

The Daytona T595 was renamed the 955i this year and fitted with new exhaust camshafts and further engine-management updates to smooth and fatten power delivery. All 1999 fuel-injected triples featured revised throttle bodies with 5-degree butterflies instead of 12-degree, as well as a new air bypass system. Apart from an updated rear suspension unit and new graphics, in other respects the Daytona continued as before. For 1999 the Speed Triple received the Daytona's 955cc engine, tuned for more midrange power. The frame finish was new, as were the decals, with T509 now dropped from the model name. Both the Daytona 955i and the Speed Triple featured in the extremely successful movie *Mission: Impossible 2*, released in 2000. The final chase scene in which Tom Cruise does elaborate stunts certainly didn't hurt Speed Triple sales.

1999 Trophy, Thunderbird, Thunderbird Sport, Adventurer

As the Trophy was getting toward the end of its model life, updates were minor, centering on a fairing and screen redesigned to improve weather protection, as well as new aluminum handlebars. The colors were as before, but turquoise was dropped. After not initially appearing in the 1999 lineup, toward the end of the 1999 model year the Thunderbird reappeared in higher specification, now with a six-speed gearbox, lower riding position and seat, powder-coated engine, tank knee pads, and fully adjustable suspension. The Thunderbird Sport continued unchanged from 1998, while the Adventurer received the Thunderbird's new lighter, lower subframe, providing a seat height of only 26.5 inches. Other Adventurer updates included the front end raked out to 27 degrees, further increasing the wheelbase, and a 19-inch front wheel. The wheel rims were now chrome-plated steel and the engine powder-coated silver.

1999 Legend TT

Introduced midway through 1998 as an early-release 1999 model, the Legend TT was a budget Thunderbird with nonadjustable suspension, a lower seat, and an entry-level price of $7,695. Although based on the Thunderbird, the styling was a little more retro 1960s, with a teardrop gas tank, tucked-in side covers, Thunderbird Sport wheels, and reverse cone mufflers. Shared with the Adventurer were the nonadjustable suspension and five-speed gearbox. As *Cycle World* said, "The Legend TT is fun to ride and easy on the eyes." As Triumph entered the 21st century, few disagreed.

After not initially included in the 1999 model lineup, during 1999 the popular Thunderbird reappeared, now with a six-speed gearbox and adjustable suspension.

The Legend TT was the budget classic model in the 1999 lineup.

2000 Model Year

For the first time, the 11-bike lineup was grouped into Sport, Touring, and Classic, the second-generation models expanding to include the Sprint RS, with Triumph also introducing their ground-up new design, the TT600. Coinciding with a decade of production at Hinckley, the TT100 also represented a change in philosophy, with modularity no longer a consideration in the design. Production at Hinckley was also up markedly, with 28,842 motorcycles produced during 2000, and in December 1999 the 100,000th Triumph (a Sprint RS for Italy) was built.

2000 Sport
TT600

In the wake of their 1990s resurrection, Triumph had carved out a niche by offering bikes that didn't directly compete with the Japanese, but all this changed with the release of the TT600 in 2000. Steering clear of design modularity, the TT600 was almost a textbook 600cc four-cylinder in the typical Japanese mode, deviating only in the use of fuel injection instead of carburetors. The concept was first discussed in early 1996, with engine development commencing in December 1996 and chassis development in January 1997. Early in 1998 the first engine was running, with final track testing taking place in August 1999.

The short-stroke engine featured expensive, high-pressure diecast crankcases and Nikasil-coated aluminum cylinders, with the oil-cooled rare earth alternator and starter drive incorporated on the crankshaft ends. The first production super sport with EFI, the Sagem MC1000 engine-management system featured a single injector per cylinder, with four 38mm throttle bodies nestling inside a large 2.2-gallon pressurized airbox.

Housing the new engine was a 28-pound, four-celled, extruded aluminum twin-spar frame with a twin-sided aluminum swingarm; further weight-saving measures extended to one-piece cast-aluminum handlebars and three-spoke cast-aluminum wheels. Kayaba provided the adjustable suspension, a conventional fork chosen over the heavier inverted type, with Sunstar brake discs and Nissin calipers. After extensive wind-tunnel testing, Triumph achieved their lowest ever claimed C_d figure, and early impressions indicated it was a match for comparable offerings from the Japanese manufacturers. Certainly, in terms of handling and braking the TT600 was impeccable, but unfortunately the engine lacked low-end power, and endemic fuel-injection glitches didn't endear the TT600 to critics.

1999 Legend TT
(Differing from The Thunderbird)

Model	TC396
Gearbox	Five-speed
Front Fork	43mm Telescopic
Rear Suspension	Monoshock
Tires	120/70 R17 front and 160/70 R17 rear
Dry Weight	474 pounds (215kg)
Colors	Cardinal Red, Obsidian Black, Imperial Green

CHAPTER EIGHT

Triumph's first attempt at a Japanese-style super sport was the TT600. While a commendable effort, early examples suffered from fuel-injection glitches. *Triumph Motorcycles*

2000 TT600

Type	Inline four-cylinder DOHC, 4 valves per cylinder liquid-cooled
Bore	68mm
Stroke	41.3mm
Capacity	599cc
Horsepower	108 at 12,750 rpm
Compression Ratio	12:1
Fuel Injection	Sagem MC1000
Ignition	Digital inductive
Gearbox	Six-speed
Frame	Aluminum beam perimeter
Swingarm	Aluminum twin-sided
Front Fork	43mm Adjustable Telescopic
Rear Suspension	Monoshock adjustable rising rate
Brakes	2x310mm disc 2x4 calipers front, 220mm disc rear
Wheels	17x3.5-inch front and 17x5.5-inch rear
Tires	120/70 ZR17 front and 180/55 ZR17 rear
Wheelbase	54.9 inches (1,395mm)
Dry Weight	374 pounds (170kg)
Colors	Jet Black and yellow, Tornado Red and silver

Sprint RS

Designed to bridge the gap between the Daytona 955i and Sprint ST was the half-faired Sprint RS. Evolving out of the Sprint ST, and sharing many of its components, including the 955cc fuel-injected engine and aluminum twin-spar frame, the RS featured a cheaper twin-sided swingarm, a slightly narrower rear wheel, and a 5mm-longer Showa rear shock absorber. This steepened the steering head angle by 0.5 degrees, to 24.5 degrees, and with a wide range of optional equipment and a sticker price of only $9,499, the Sprint RS appealed to those leaning toward the sport side of sport touring.

Daytona 955i and Speed Triple

Updates to the Daytona 955i for 2000 saw the wheelbase reduced by allowing full rotation of the eccentric axle adjuster, gearbox modifications, new graphics, reworked engine management, and a circular-section brushed stainless steel muffler. The Speed Triple was unchanged this year.

2000 Touring
Sprint ST, Tiger, Trophy 1200 and 900,
Except for some additional colors, all Touring models continued unchanged for 2000.

2000 Classic
Thunderbird, Thunderbird Sport, Adventurer, and Legend TT
Designed to evoke nostalgic memories with modern sport ability, the Thunderbird Sport received a new exhaust system with single reverse cone mufflers on each side instead of the Hurricane-inspired right-side dual muffler. The colors were intentionally reminiscent of the 1966 T120 Bonneville. Other Classic models were unchanged this year, the six-speed Thunderbird remaining slightly higher-spec than the five-speed Adventurer and Legend TT.

2001 Model Year
The big news for Triumph this year was the release of the much-awaited new Bonneville, heralding a range that would continue to expand over the next few years. Also for 2001 the Tiger received the 955cc engine, while during the model year the Daytona was significantly updated. Production increased moderately during 2001, to 29,082 motorcycles.

2001 Sport
TT600
As the TT600 entered its second year it received new paint and graphics, and more notably, refined fuel-injection mapping and new camshafts to smooth the previously flawed power delivery. A balancer pipe was added to help boost low-speed torque, and while it all added up to an improvement, the TT600's reputation was already irreparably damaged. Trying to beat the Japanese at their game was all very well, but along the way the character was lost.

Daytona 955i
Initially the Daytona 955i appeared unchanged for the model year except for Caspian Blue replacing Tornado Red, but early in 2001 it received a major ground-up update. Inheriting many features from the TT600, the third-generation 955 included new, high-pressure, diecast crankcases, a crankshaft-mounted alternator and starter, lighter con-rods, and a higher compression ratio. The new cylinder head featured 1mm-larger intake valves, 1mm-smaller exhaust valves, and a reduction in the included valve angle from 39 to 23 degrees. The Sagem closed-loop fuel-injection system now included smaller injectors and larger throttle bodies, and with a larger, reshaped airbox, the new engine boasted a power increase of nearly 13 percent. One significant improvement was the relocation of the sprag clutch to the end of the crankshaft, improving both reliability and ease of servicing.

2000 Sprint RS (Differing from Sprint ST)

Swingarm	Aluminum twin-sided
Rear wheel	17x5.5-inch
Dry Weight	438 pounds (199kg)
Colors	Racing Yellow, Lucifer Orange, Eclipse Blue

2000 Daytona 955i and Speed Triple (Differing from 1999)

Wheelbase	53.6 inches (1,431mm) (955i)
Colors	Aluminum Silver (955i)

2000 Sprint ST, Tiger, Trophy (Differing from 1999)

Colors	Sapphire Blue (Sprint ST); Eclipse Blue (Tiger)

2000 Thunderbird Sport, Adventurer (Differing from 1999)

Colors	Opal White and Tangerine (Thunderbird Sport); Jet Black and Lucifer Orange (Adventurer)

2001 TT600 (Differing from 2000)

Compression Ratio	12.5:1
Colors	Caspian Blue, Racing Yellow, Caspian Blue and silver

Resplendent in colors evoking the 1966 Bonneville, the 2000 Thunderbird Sport also had a new exhaust system with individual mufflers on each side.

CHAPTER EIGHT

RIGHT: Released midway through 2001, the new Daytona 955i received an updated engine with more power and a redesigned fairing. *Triumph Motorcycles*

BELOW: Along with a slightly narrower rear wheel and small tire, the Daytona's swingarm was now dual-sided. This would only last through 2002. *Triumph Motorcycles*

the wheelbase. Other updates included a new aluminum-bodies rear shock with lighter spring, a narrower rear wheel and tire, and a new fairing with a sharper nose and less distinctive, plastic-covered dual headlamps. Although the new Daytona was lighter, quicker, and faster than before, it was still outclassed by comparable Japanese superbikes.

Speed Triple and Sprint RS
Except for colors, both the Speed Triple and Sprint RS continued unchanged for 2001.

2001 Speed Triple, Sprint RS *(Differing from 2000)*

Colors
Neon Blue, Nuclear Red (Speed Triple); Jet Black (Sprint RS)

2001 Daytona 955i *(Differing from 2000)*

Model	T595NS
Horsepower	147 at 10,700 rpm
Compression Ratio	12:1
Fuel Injection	Sagem 46mm throttle bodies
Swingarm	Aluminum twin-sided
Rear Wheel	17x5.5-inch
Rear Tire	180/55 ZR17
Wheelbase	55.8 inches (1,417mm)
Dry Weight	414 pounds (188kg)
Colors	Strontium Yellow, Jet Black

While retaining the trademark tubular aluminum frame, the steering head angle was steeper (22.8 degrees instead of 24 degrees), and a stiffer and lighter double-sided swingarm replaced the (admittedly more attractive) single-sided type, also reducing

2001 Touring
Tiger
After two years as an 885cc triple, the Tiger received a retuned version of the Daytona 955cc engine for 2001. This now shared many features with the new Daytona 955i, including high-pressure diecast crankcases and the alternator and starter moved to the end of the crankshaft. Other updates included a chain-driven (rather than gear) oil pump and closed-loop fuel-injection system. Apart from revised graphics the chassis was much as before, and still more street than off-road focused.

Sprint ST, Trophy 1200 and 900,
Except for some additional colors, all Touring models continued largely unchanged for 2001. The Trophy received an Adventurer-style clutch cover and a new breather system.

1997–2002

The 2001 Speed Triple in Neon Blue. *Triumph Motorcycles*

2001 Classic
Bonneville

After reintroducing most of the famous names in Triumph's history, it was only a matter of time before the most famous of all, the Bonneville, would reappear. As 308,000 were sold in America, the Bonneville was an icon colored with nostalgia, and Triumph decided on one of their greatest models, the 1969 T120 Bonneville, as the inspiration for the new version. The project commenced in July 1997, with the team importing a restored 1969 T120R from the United States and using it as a template for styling and ergonomics. The key brief was that the new Bonneville had to be light and agile, and it had to corner well. No longer was it to be the Bonneville of old that bespake speed and performance. Traditional capacities, 750 and even 650cc, were considered, before settling on a 790cc double overhead camshaft air/oil-cooled parallel twin with the camshaft drive between the cylinders. In keeping with tradition the crank was a 360-degree type, with two balance shafts to quell vibration. The front oil drain tube was styled to resemble the original's pushrod tube, and reversing the gearbox allowed the traditional triangular engine cover on the right and larger clutch case on the left. The pair of 36mm Keihin carbs included electric heaters to combat icing, and a Sagem digital ignition. But the retro look only carried so far, with Triumph believing

The Tiger received the 955cc engine for 2001 but was quite similar to the previous version. Jet Black was a perennial favorite. *Triumph Motorcycles*

2001 Tiger (Differing from 2000)

Model	T795NE
Bore	79mm
Capacity	955cc
Horsepower	104 at 9,500 rpm
Compression Ratio	11.65:1
Colors	Roulette Green, Jet Black

2001 Sprint ST, Tiger Trophy
(Differing from 2000)

Colors	British Racing Green (Sprint ST); Sunset Red (Trophy)

203

CHAPTER EIGHT

2001 Bonneville

Type	360-degree Parallel Twin DOHC, 4 valves per cylinder air-cooled
Bore	86mm
Stroke	68mm
Capacity	790cc
Horsepower	61 at 7,400 rpm
Compression Ratio	9.2:1
Carburetors	2 Keihin 36mm
Ignition	Digital inductive
Gearbox	Five-speed
Frame	Tubular steel cradle
Swingarm	Twin-sided tubular steel
Front Fork	41mm Kayaba telescopic
Rear Suspension	Twin Kayaba shock absorbers
Brakes	310mm disc 2-piston caliper front, 255mm disc rear
Wheels	19x2.5-inch front and 17x3.5-inch rear
Tires	100/90R19 front and 130/80R17 rear
Wheelbase	58.8 inches (1,493mm)
Dry Weight	451 pounds (205kg)
Colors	Scarlet Red and silver, Forest Green and silver

The new Bonneville was intentionally styled to replicate the iconic 1969 T120R. *Triumph Motorcycles*

their new customers wanted a modern motorcycle in period dress, so the lack of a kick-start provided convenience over quirkiness.

Also traditional in layout, the chassis included a tubular steel cradle frame with twin downtubes and a spine (reminiscent of the older triples). Steering was intentionally slow, with a raked-out 29-degree head angle. The box-section swingarm pivoted in the crankcases, with disc brakes front and rear and a traditional 19-inch front wheel. As usual, John Mockett provided the styling model, the style and Scarlet Red option replicating the iconic 1969 model. With the new Bonneville Triumph had created a compact but somewhat heavy retro bike with comfortable suspension and excellent handling. They also provided a wide range of options, including a louder set of mufflers said to add 10 horsepower.

Thunderbird, Adventurer, and Legend TT
With the introduction of the Bonneville, Triumph rationalized the range of classic spine-frame triples. This year the Thunderbird Sport was dropped, and the existing Keihin-carbureted Thunderbird and Mikuni-carbureted Adventurer and Legend TT continued unchanged, the latter two in their final year.

2002 Model Year
This year Triumph celebrated the centenary of their first motorcycle by releasing two centennial models. Although a full 12-model range was anticipated for the 2002 model year, with planned production of 37,000 motorcycles, on March 15 a fire totally engulfed the T1 factory. Taking five hours to control, the fire destroyed the chassis assembly line and stores and coated much of rest of the site in a layer of soot. No motorcycles were built during the five-month rebuilding period, and as a result of the disruption, only 14,174 motorcycles were constructed this year.

2002 Sports
TT600, Daytona 955i, 955i Centennial Edition, Speed Triple, Sprint RS

While the TT600 continued with the usual new colors and injection tweaks, all the other sports models received the third-generation 955cc engine. As the Daytona 955i had been updated significantly during 2001, it continued unchanged, now joined by a limited-edition Aston Green Centennial Edition. A fusion of old and new, this included the previous single-sided swingarm and wider wheel with larger-section tire, increasing both the weight and wheelbase over the standard 955i. Also with the third-generation engine, the Sprint RS received a power boost but was otherwise unchanged. Most updates this year centered on the Speed Triple, now with the more powerful third-generation engine and less weight. The Speed Triple also retained the single-sided swingarm and wider rear wheel, but with quicker steering geometry (a 23.5-degree steering head angle) and a shorter wheelbase it set a new standard for agility. Other updates included restyled bodywork, digital instrumentation, and a more compact twin-headlight arrangement. A brilliant all-round motorcycle, to the cognoscenti the Speed Triple was the star in the lineup, and Triumph's best model. It was also a hit as a film extra, with Natalie Imbruglia riding one in the 2003 film *Johnny English*.

1997–2002

Although the Speed Triple looked similar to earlier models, the headlight arrangement was slightly revised and restyled. *Triumph Motorcycles*

2002 TT600, Daytona 955i Centennial Edition, Speed Triple, Sprint RS
(Differing from 2001)

Horsepower
118 at 9,100 rpm (Speed Triple, Sprint RS)
Compression Ratio
12:1 (Speed Triple, Sprint RS)
Rear Wheel
17x6.5-inch (955i Centennial Edition)
Rear Tire
190/50 ZR17 (955i Centennial Edition)
Wheelbase
56.1 inches (1,426mm) (955i Centennial Edition);
56.2 inches (1,429mm) (Speed Triple)
Dry Weight
420 pounds (191kg) (955i Centennial Edition);
416 pounds (189kg) (Speed Triple)
Colors
Jet Black (TT600); Acidic Yellow (955i, Sprint RS); Aston Green (Centennial Edition)

2002 Touring
Sprint ST, Trophy 1200, Tiger

With the Sprint ST continuing as one of Triumph's recent success stories, for 2002 it received the more powerful third-generation 955i engine. The more modern Sprint ST effectively hastened the demise of the Trophy 900, which disappeared with the only modular spine-frame model, the Trophy 1200. As it was still a new release, the Tiger continued unchanged.

2002 Classic
Bonneville America

Continuing where the unloved Adventurer left off, Triumph hoped to conquer the American cruiser market with a chopped retro Bonneville, the Bonneville

TOP: With more power and less weight, the 2002 Speed Triple was an improved sport motorcycle and arguably Triumph's best model. The extroverted lipstick pink (Nuclear Red) was a love-it-or-hate-it color. *Australian Motorcycle News*

ABOVE: Never as popular as the Sprint ST, the sportier Sprint RS was an underrated motorcycle. For 2002 it too received more power, but its days were numbered. *Triumph Motorcycles*

2002 Sprint ST, Tiger, Trophy
(Differing from 2001)

Horsepower	118 at 9,100 rpm (Sprint ST)
Compression Ratio	12:1 (Sprint ST)
Colors	Azure Blue, Emerald Green (Trophy)

205

CHAPTER EIGHT

With more power, the Sprint ST continued to lead the way in the Sport Touring category. *Australian Motorcycle News*

America. The idea for the project began late in 1998, and with master stylist John Mockett again responsible for the design, 120 drawing were presented to focus groups to gauge the response. The outcome suggested combining the distinctive parallel twin engine with heavier and bulkier styling. Believed to be more suitable for an American-style cruiser, the America had a rephrased 270-degree crankshaft, new engine covers, and dual walled exhausts, imparting a heavier look than the standard Bonneville. The chassis was revamped to provide more cruiser attitude, with a low, 28.3-inch seat height, stretched-out rake (33.3 degrees), and a longer wheelbase. And with three screen, three seat, and nine luggage options available among an increased range of accessories, the Bonneville America could be tailored to suit just about anyone. Triumph certainly succeeded with the America, and in its namesake country it was their best-selling bike in 2002.

Bonneville, Bonneville T100 Centennial, Thunderbird

The only Classic spine-frame triple this year was the Thunderbird, unchanged but for colors. While the Bonneville was also unchanged, joining it this year was the Bonneville T100 Centennial. This was slightly higher in specification, including a tachometer alongside the standard speedometer.

The second stage in Triumph's revival initially centered on a range of new sport bikes, but by 2002 the emphasis was shifting more toward the classic and cruiser styles. Although the factory fire seriously affected production, development continued, and the best-selling bikes were now the America and Speedmaster cruisers. And with the market for cruisers now accounting for 40 percent of worldwide motorcycle sales, it was inevitable that Triumph would concentrate on this path during their next phase.

2002 Bonneville America
(Differing from the Bonneville)

Type
270-degree parallel twin DOHC, 4 valves per cylinder air-cooled
Rear Brake
285mm disc
Wheels
18x2.5-inch front and 15x3.5-inch rear
Tires
110/80R18 front and 170/80R15 rear
Wheelbase
65.2 inches (1,655mm)
Dry Weight
497 pounds (226kg)
Colors
Jet Black and silver, Cardinal Red and silver

2002 Bonneville, Bonneville T100 Centennial, Thunderbird
(Differing from 2001)

Colors
Sky Blue and silver (Bonneville); Lucifer Orange and silver (T100 Centennial); Sunset Red and white, Jet Black and gray (Thunderbird)

1997–2002

ABOVE: With its fat rear tire, low seat, and raked-out front end, the Bonneville America was more cruiser than retro. *Triumph Motorcycles*

FAR LEFT: The twin-cylinder air-and-oil-cooled engine was similar to the Bonneville but featured a 270-degree crankshaft, providing a different low-rpm and idle note. *Triumph Motorcycles*

LEFT: With new engine covers and dual-walled exhausts, the Bonneville America imparted a heavier look. *Triumph Motorcycles*

www.triumph.co.uk

The Millennium Eye, an extraordinary new way to look at London. The Triumph Speed Four, an extraordinary new take on the naked middleweight bike.

A highly specified, fuel injected 600cc super sports engine complemented by a lightweight aluminium frame and fully adjustable suspension. True sports bike handling wrapped in streetfighter guise.

The Triumph Speed Four. In a class of its own.

The English Revolution

A FEELING OF TRIUMPH

2003–2009

After the Fire— Bigger Cruisers and Classics

The Rocket III, Thruxton, Daytona 675, Scrambler, and Street Triple

RIGHT: Marking an adventurous shift in focus, Triumph entered a new era with the astonishing Rocket III. More than any other modern Triumph, the Rocket III was built for America. *Triumph Motorcycles*

OPPOSITE: Triumph promoted the new Speed Four as a revolutionary new take on the naked middleweight bike.

Reconstruction after the devastating fire began on April 16, 2002, with production eventually recommencing in September 2002 for the 2003 model year. This year saw the introduction of the TT600 offshoot Speed Four, originally intended for 2002 release, and a Bonneville America derivative, the Speedmaster. The previous Thunderbird Sport was reintroduced, but the end of the road finally came for the venerable spine-frame Trophy and Thunderbird. During 2003 Triumph also released the much-awaited Daytona 600 Supersport, and production at Hinckley rose to pre-fire levels of 150 units a day, with 30,309 motorcycles manufactured in 2003. Also during 2003 Triumph expanded their traditional horizons and began building a subassembly manufacturing factory in Thailand.

2003 Model Year
2003 Sports
Speed Four

As the Daytona offshoot Speed Triple's success continued, it was inevitable that a Speed Four would evolve out of the much-maligned TT600. By 2003 the fuel injection was finally in order, and with new camshafts and mapping, the four-cylinder

CHAPTER NINE

2003 Speed Four (Differing from the TT600)

Horsepower	97 at 11,750 rpm
Colors	Jet Black, Roulette Green, Tangerine Orange

engine was tuned for more midrange power. Throttle response was improved over the TT600, and the chassis, identical to its sportier sibling's, provided exceptional handling and braking. With its pair of chromed headlights, compact instrument cluster, fly screen, twin air scoops, and satin-black frame, engine, and swingarm finish, the Speed Four continued the Speed Triple's effective style.

Daytona 600

Although it was a commendable effort at competing head-to-head with the Japanese, the TT600 was understyled, underpowered, initially underdeveloped, and didn't strike a chord with customers. Launched in April 2003, the Daytona 600 was Triumph's attempt at addressing the TT600's shortcomings, and while the basic architecture was carried over from the TT600, the four-cylinder engine featured CNC-machined combustion chambers, a lighter crank and starter motor, and magnesium cam covers. The most significant technical update was a Keihin fuel-injection system replacing the much-criticized Sagem. The triple-cell (rather than four-cell) twin-spar aluminum frame also saved weight, while the Kayaba front fork now included aluminum internals. With its totally new, hard-edged styling, the Daytona 600 stated that Triumph was no longer concentrating on niche models. But accomplishment in this highly competitive sector needed to be backed up by race success, and 34-year-old New Zealander Bruce Anstey provided Triumph their first Isle of Man TT win since 1975. On his ValMoto Daytona 600 Anstey won the Junior (Supersport 600) TT at 120.36 miles per hour, the fastest ever, and 10.96 seconds ahead of second place.

Underneath the Daytona 600's edgy styling was a completely revamped and vastly improved four-cylinder 600 supersport motorcycle, more than capable of taking the Japanese head-on. Triumph Motorcycles

2003 Daytona 600
(Differing from the TT600)

Horsepower
110 at 12,750 rpm
Fuel Injection
Keihin twin butterfly 38mm throttle bodies
Front Brakes
2x308mm discs
Wheelbase
54.7 inches (1,390mm)
Dry Weight
363 pounds (165kg)
Colors
Racing Yellow, silver

Daytona 955i, Speed Triple, TT600, Sprint RS

Due to its disappointing reception, after barely more than a year with the dual-sided swingarm, the Daytona 955i was now similar to the 2002 Centennial version, reverting back to the original single-sided type with a wider rear wheel. A silver example was featured in the 2004 film *Torque*, ridden by Ice Cube. Other sport models continued unchanged; the TT600 was about to be replaced by the new Daytona 600. To the disappointment of many, the Speed Triple was no longer available in outlandish lipstick Nuclear Red, while the Sprint RS, still struggling to find its niche, was now in its final incarnation.

2003 Touring
Sprint ST, Trophy 1200, Tiger

The Touring range continued as before, with this the final year for the venerable Trophy 1200, which was already no longer available in California.

2003 Classic
Speedmaster

Recreating another historical transatlantic link and inspired by the unofficially tagged 1964 T120R "Speedmaster," the Speedmaster filled a niche for a more performance-oriented cruiser. Built around the America, the Speedmaster's ethos was minimalist, with flat 'drag' handlebars and forward-set footpegs providing a laid-back riding position. The tachometer

2003 Speedmaster
(Differing from the America)

Front Brake
2x310mm discs with 2-piston calipers
Wheels
Cast-alloy 18x2.5-inch front and 15x3.5i-inch rear
Dry Weight
504 pounds (229kg)
Colors
Jet Black and Tornado Red, Jet Black and Racing Yellow

2003 Daytona 955i, Speed Triple, TT600, Sprint RS
(Differing from 2002)

Rear Wheel	17x6.5-inch (955i)
Rear Tire	190/50 ZR17 (955i)
Wheelbase	56.1 inches (1,426mm) (955i)
Dry Weight	420 pounds (191kg)
Colors	Jet Black, Tornado Red (955i); Roulette Green, Aluminum Silver (Speed Triple); Roulette Green, Tangerine Orange (TT600); Caspian Blue (Sprint RS)

2003 Sprint ST, Tiger, Trophy
(Differing from 2002)

Colors	Aluminum Silver (Sprint ST); Lucifer Orange, Aluminum Silver (Tiger); Graphite (Trophy)

was mounted in the gas tank nacelle, with a large, white-faced speedometer dominating the rider's view. Apart from dual front disc brakes, cast-alloy wheels, and blacked-out motor, the Speedmaster was pretty much the same as the America—just a bit toughed up.

Expanding into the niche for a more performance-oriented cruiser, the Speedmaster was ostensibly the same as an America, but incorporated a feet-forward riding position, cast-alloy wheels, and dual front discs. *Triumph Motorcycles*

CHAPTER NINE

2003 Bonneville, Bonneville T100, America, Thunderbird, Thunderbird Sport
(Differing from 2002)

Colors
Jet Black, Cardinal Red (Bonneville); Sapphire Blue and white, Goodwood Green and gold (T100); Caspian Blue and silver (America); Cardinal Red and silver, British Racing Green and cream (Thunderbird); Tangerine and Jet Black (Thunderbird Sport)

Bonneville, Bonneville T100, America, Thunderbird, Thunderbird Sport

Due to popular demand, the Thunderbird Sport was reintroduced, after a one-year hiatus, alongside the regular Thunderbird (now in its final year). Available with a wide range of accessories, apart from new colors, they were unchanged, as were the Bonneville and America. The Bonneville T100 continued as an up-spec Bonneville, with a tachometer, two-tone colors, and the tank with kneepads. This year the Bonneville's colors were single scheme, without kneepads.

2004 Model Year

With the factory fire well behind them, Triumph astonished everyone at the EICMA Milan Motorcycle Show in September 2003 with the release of the astounding Rocket III Cruiser, the first production motorcycle to break the 2-liter barrier. Cruisers now accounted for 63 percent of all over-600cc street bikes sold in America, and Triumph wanted a bigger slice of Uncle Sam's apple pie. Production was down slightly this year, to 26,125 motorcycles, but with a new purpose-built R&D building and 100 design staff, Triumph was looking to the future.

Rocket III

The impetus for the Rocket III went back to 1998, when Triumph had ascertained that there was a gap in the 1,200-to-1,500cc power-cruiser market. After considering 1,500 and 1,600cc, Triumph decided on a 2,000cc inline three, as at the time the largest-displacement cruiser was the 1,500cc Honda Valkyrie. But suspecting this could rise to 1,800cc and not wanting to be outdone, they pushed the displacement up to 2,300cc. Engine design began in September 2000, with the first engine running in May 2002, soon after the factory fire, and complete bikes were built by the end of 2003.

The Rocket III initiated a number of firsts for Triumph. It had the first longitudinally mounted engine, it was the first new Triumph with dry sump lubrication, and it was the first with shaft final drive, developed in collaboration with the Italian company Graziano. The pistons were the same size as those of a V10 Dodge Viper, the 8-inch con-rods longer than a small-block Chevy's, and a counter-rotating balance shaft nullified the monster torque reaction. The massive 120-degree, 39-pound forged crankshaft ran in four bearings, and as it was only 8 inches above the ground, it contributed to a low center of gravity. With two spark plugs per cylinder, Keihin developed a fuel-injection system with 52mm dual butterfly injectors. With space at a premium and the need for a large fuel tank, the airbox and filter were under the rider's seat.

The solidly mounted engine was a stressed member in the tubular steel spine frame, while Kayaba provided the suspension (an upside-down fork on the front), Nissin the front brakes, and Brembo the rear brake. Along with the massive engine, the huge 240-section rear tire was also an important visual component. As for the name, the original Rocket III may have been a BSA, but it fulfilled all the marketing criteria, emphasizing the British heritage, three cylinders, and high performance. This was enough to secure 1,500 preorders in 2003.

2004 Sports and Sports Touring
Daytona 600, Speed Four, Daytona 955i, Daytona SE, Speed Triple, Speed Triple SE, Sprint ST, Tiger

As the Daytona 600 was already available as an early-release 2004 model, the only update for the rest of the year was an additional color, Tornado Red. It was also similar for the relatively new Speed Four, now

2004 Rocket III

Type	Inline three-cylinder DOHC, 4 valves per cylinder liquid-cooled
Bore	101.6mm
Stroke	94.3mm
Capacity	2,294cc
Horsepower	140 at 5,750 rpm
Compression Ratio	8.7:1
Fuel Injection	Multipoint sequential electronic fuel injection
Ignition	Digital inductive
Gearbox	Five-speed
Final Drive	Shaft
Frame	Tubular steel, twin spine
Swingarm	Twin-sided, steel
Front Fork	43mm upside-down
Rear Suspension	Twin shocks with adjustable preload
Brakes	2x320mm disc 4-piston calipers front, 316mm disc rear
Wheels	Alloy 17x3.5-inch front and 16x7.5-inch rear
Tires	150/80 V17 front and 240/50 V16 rear
Wheelbase	66.5 inches (1,690mm)
Dry Weight	704 pounds (320kg)
Colors	Cardinal Red, Jet Black

LEFT: The fat 240-section Metzeler Marathon radial tire dominated the Rocket III's rear. The color is the optional orange "tribal" custom paint version. *Triumph Motorcycles*

BELOW: This engine and drivetrain cutaway was on display at the EICMA Milan Show towards the end of 2003. The dry sump oil tank resides under the throttle bodies, and the gearbox is stacked.

available in yellow. More updates were introduced for the Daytona 955i, now with a more aerodynamic seat, restyled cockpit, revised headlamps, and new graphics. While the Speed Triple continued unchanged, a special Jet Black edition was available, also for the Daytona 955i, with a black frame, swingarm, and wheels. With the Sprint ST and Tiger about to be updated, both went unchanged for 2004.

2004 Classic
Thruxton 900

Named after the Thruxton Circuit in Hampshire, the scene of many significant production race victories and the inspiration for the limited-production 1965 Thruxton Bonneville, the new Thruxton 900 harkened back to the café racers of the sixties. Another model evolving from one of the late Carlo Talamo's *Numero Tre* concept bikes, the Thruxton 900, although strictly speaking not a 900, was based on the 360-degree Bonneville. Along with the capacity increase came new camshafts and upswept megaphone-style mufflers. Sharing the Bonneville's frame, the front fork was pulled back, the shocks longer, and the front wheel an 18-inch, steepening the steering geometry (27-degree head angle) and shortening the wheelbase slightly. Shorter fenders, polished engine covers, clip-on

2004 Daytona 600 and 955i, Speed Four, Speed Triple *(Differing from 2003)*

Colors	Tornado Red (Daytona 600, Speed Triple); Racing Yellow (Speed Four)

213

CHAPTER NINE

Now in its eighth year, the 2004 Daytona 955i included a restyled fairing and rear seat, along with new graphics. *Triumph Motorcycles*

With its clip-on handlebars, solo seat, and checkered-flag stripe, the Thruxton emulated the 1960s café racer.

2004 Thruxton 900
(Differing from the Bonneville)

Bore	90mm
Capacity	865cc
Horsepower	69 at 7,250 rpm
Compression Ratio	10.2:1
Front Brake	320mm floating disc
Front Wheels	18 x 2.5-inch
Front Tire	100/90 R18
Wheelbase	58.1 inches (1,477mm)
Dry Weight	451 pounds (205kg)
Colors	Jet Black and silver, Sunset Red and silver

handlebars and rear-set footpegs, and checkered-flag graphics all contributed to the café racer profile.

Bonneville, Bonneville T100, America, Speedmaster, Thunderbird Sport

Alongside the new Thruxton 900, the rest of the Classic range was unchanged except for colors. The Bonneville was available with either polished engine cases or black finish, while the T100 remained more traditional, with polished engine cases and hand-painted pinstripes. The America and Speedmaster were also unchanged. The Thunderbird was dropped, and the only modular-concept spine-frame model left was the black-highlighted Thunderbird Sport, now in its final year.

2003–2009

LEFT: Functionally and aesthetically effective, the Thruxton was one of the most successful of the new breed of retro classics.

BELOW: A 4mm overbore of the Bonneville's engine provided the Thruxton with more power and torque.

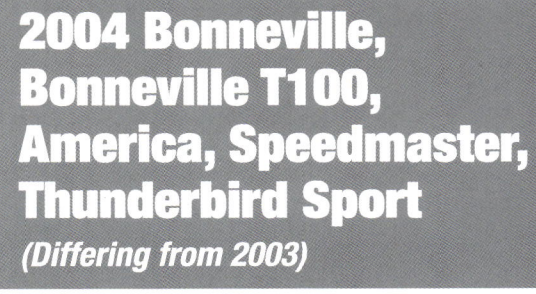

2004 Bonneville, Bonneville T100, America, Speedmaster, Thunderbird Sport
(Differing from 2003)

Colors
Jet Black (T100); Jet Black, Goodwood Green and silver (America); Jet Black, Jet Black and blue (Speedmaster); Jet Black and yellow (Thunderbird Sport)

ABOVE: With its rubber tank kneepads, polished engine cases, and hand-painted pinstripes, the Bonneville T100 was a more traditional take on the classic concept. This was its final year with the 790cc engine. *Australian Motorcycle News*

LEFT: The 2004 Thunderbird Sport colors were a reverse of those of the original 1998 version, now predominantly black with yellow highlighting. This was the last modular-concept spine-frame model, heralding the end of Hinckley's first phase. *Triumph Motorcycles*

215

CHAPTER NINE

For 2005 the Speedmaster received the larger, 865cc engine and was available in the new color combination of Jet Black and Neon Blue. *Triumph Motorcycles*

2005 Rocket III, Speedmaster, America
(Differing from 2004)

Bore	90mm (Speedmaster)
Capacity	865cc (Speedmaster)
Horsepower	54 at 6,500 rpm (Speedmaster)
Colors	Graphite (Rocket III); Jet Black and blue (Speedmaster); Mulberry Red and Graphite (America)

2005 Model Year

This year saw the highest production ever at Hinckley, with 31,600 motorcycles built, and new model grouping of Cruisers, Urban Sports, and Modern Classics was an indication of how Triumph was reorienting their range away from their traditional touring and sport lineup. The 955 grew to 1,050cc for the Speed Triple and Sprint ST, while the Speedmaster and Bonneville T100 received the 865cc engine.

2005 Cruisers
Rocket III, Speedmaster, America

As the Rocket III was ostensibly an early-release 2005 model, the only update was an additional color. The America was also unchanged, but the Speedmaster now included a detuned version of the Thruxton's larger-capacity engine, but with the Speedmaster's 270-degree crank.

2005 Urban Sports
Speed Triple, Sprint ST

Heading the new Urban Sports lineup this year were a significantly revamped Speed Triple and Sprint ST, now with an all-new, longer-stroke 1,050cc three-cylinder engine. With horizontally split, high-pressure diecast crankcases, a four-bearing nitride 120-degree crank, and new profile camshafts chain-driven on the right side of the engine, even the water hoses were more neatly tucked away. The new cylinder head breathed through 33.5mm and 27mm valves set at a 23-degree included angle, and a new Keihin injection system hosted 46mm throttle bodies. The Speed Triple's engine was tuned for more peak power than the Sprint ST's.

Although similar to earlier models, with identical geometry, the Speed Triple's black aluminum chassis was also new, as was the Showa suspension, a 45mm upside-down fork on the front and remote reservoir rear shock. A first for Triumph was the radially mounted four-piston Nissin front brake calipers, and while retaining the single-sided swingarm, the rear wheel was reduced to 5.50 inches. New features included a digital instrument display with analog tachometer and a high-level underseat exhaust system. As a badass naked street bike, the Speed Triple was as good as it got, *Cycle World* voting it their 2005 Best Open-Class Streetbike.

The Sprint ST also received a new perimeter-style aluminum frame, with sharper steering and a reduced

ABOVE: Also receiving the larger engine, the Sprint ST's featured edgy new styling as well. The fairing included triple headlights. *Australian Motorcycle News*

LEFT: The Speed Triple received a significant makeover for 2005, including a larger-capacity engine and an upside-down front fork with radial brake calipers. This is a similar 2007 version in Roulette Green. *Triumph Motorcycles*

wheelbase. Most significant was the fresh look, with new, sharp-lined bodywork, triple projector-beam headlights, an LED taillight, all-new instrumentation with onboard computer, and a unique triple muffler, underseat exhaust.

Daytona 955i, 955i SE, Tiger

With the Daytona 955i close to the end of its model life, the only updates were a Graphite finish for the frame and swingarm, with a yellow color option for the 955 SE. Released in March 2004, the 2005 model

But for the black wheels and swingarm to match the frame, the Daytona SE was identical to the Daytona 955i. By 2005 Triumph's top-of-the-range sport bike had been left behind by the Japanese competition. *Triumph Motorcycles*

2005 Speed Triple and Sprint ST
(Differing from 2004)

Stroke	71.4mm
Capacity	1,050cc
Horsepower	128 at 9,100 rpm (Speed Triple); 123 at 9,250 rpm (Sprint ST)
Front Fork	45mm upside-down (Speed Triple)
Front Brakes	4-piston radial calipers (Speed Triple)
Rear Wheel	5.50x17-inch (Speed Triple)
Rear Tire	180/55ZR17 (Speed Triple)
Wheelbase	52.4 inches (1,457mm) (Sprint ST)
Dry Weight	462 pounds (210kg) (Sprint ST)
Colors	Neon Blue, Scorched Yellow (Speed Triple)

2005 Daytona 955i, 955i SE, Tiger
(Differing from 2004)

Wheels	Cast-alloy 14-spoke 19x2.5-inch, 17x3.5-inch (Tiger)
Colors	Yellow (SE); British Racing Green (Tiger)

year Tiger now featured revised front and rear suspension, new steering geometry, and cast-alloy wheels with tubeless tires. A 25.8-degree steering head angle, pulled back from 28 degrees, quickened the steering, and the suspension was shortened and stiffened. Even more than before, the Tiger was a large-scale adventure tourer more suited to the tarmac than off-road.

CHAPTER NINE

British Racing Green was a new color for the updated 2005 Tiger. Givi saddlebags were standard for the US. *Triumph Motorcycles*

Apart from the new fairing decal, there was little to distinguish the Daytona 650 from the previous Daytona 600. The additional capacity improved midrange power, but the Daytona was no longer eligible for Supersport racing. *Triumph Motorcycles*

Daytona 650, Speed Four

While the only specification change to the Speed Four was an additional color, Triumph followed Kawasaki's lead, with the Daytona 600 growing to 650cc and taking it beyond the traditional racing 600 category. Although factory rider Craig Jones won the final round of the 2004 British Supersport Championship on the Daytona 600, the 650 marked a shift in Triumph's strategy. More midrange street power was the next goal, a 3.2mm-longer stroke providing the extra capacity, making the Daytona 650 the first Triumph 650 in three decades. Other internal updates included a backlash eliminator gear, new seven-plate clutch (down from nine), and a new remote gear linkage. The engine received a new cylinder head and the Keihin EFI remapping. As the Daytona 600's chassis was never a cause for concern, it went unchanged. The red and yellow colors were also as for 2004, with silver dropped, and 2005 would be the final year for both of the smaller-displacement fours.

2005 Modern Classics
Thruxton 900, Bonneville, Bonneville SE, Bonneville T100

The Thruxton and Bonneville received the obligatory new colors, with the Bonneville T100 now powered

2005 Daytona 650, Speed Four
(Differing from 2004)

Stroke	44.5mm (Daytona 650)
Capacity	646cc (Daytona 650)
Horsepower	112 at 12,500 rpm (Daytona 650)
Compression Ratio	12.85:1 (Daytona 650)
Colors	Neon Blue (Speed Four)

The Bonneville SE was identical to the standard Bonneville but had a black-finished engine. *Triumph Motorcycles*

by a retuned version of the larger Thruxton engine, still with the 360-degree crankshaft. A Bonneville SE was also available, with a black engine finish instead of the standard polished cases.

2006 Model Year

Triumph continued their strong growth, with production increasing 18 percent to 37,400 units this year. Along with three new models, the Rocket III Classic, Scrambler, and Daytona 675 Triple, Triumph released their sixty8 accessories collection for the Bonneville. As part of a strategy for global expansion, Prince Andrew, Duke of York, officially opened a new factory (Factory Four) in Thailand to produce the Classic range.

2006 Cruisers
Rocket III, Rocket III Classic, Speedmaster, America

With the Rocket III successfully conquering the power-cruiser market, the Rocket III Classic was aimed at those requiring a more relaxed ride. Rider footboards, pull-back handlebars, and a more comfortable seat provided improved rider and passenger comfort, with hand-painted coachlines, chopped rear fender, chrome cam cover, black finished engine, and cone mufflers adding individual styling touches. The rest of the cruiser lineup continued unchanged.

2006 Urban Sports
Daytona 675

Replacing the four-cylinder Daytona for 2006 was the much-awaited Daytona 675 Triple. After several failed attempts at meeting the Japanese head-on, the 675 was the sportiest, most performance-oriented machine ever to emerge from Hinckley. The new engine was smaller than its predecessor, with a stacked gearbox minimizing engine length, and the four valves per cylinder were set at an included angle of 23 degrees, the engine fed by 44mm throttle bodies with Keihin closed-loop injection with multi-hole injectors. The cast-aluminum frame with two-piece cast-alloy swingarm was considerably more compact than the Daytona 650's, and the suspension was by an upside-down Kayaba fork and piggyback rear shock. Brakes were Nissin, radial four-piston on the front, with a radial master cylinder. Styled by factory technician-turned-stylist Chris Hennegan, with the chassis setup by ace tester David Lopez, the result was brilliant, with the Daytona 675 winning *Cycle World*'s middleweight comparison and the summary, "The 675 offers the best all-round performance, versatility, and visual flair in the middleweight sportbike class." The Daytona 675 also received the ultimate accolade, taking the 2006 International Bike of the Year award in a poll by 15 international motorcycle magazines. It was a banner year for Triumph, and for Triumph fans everywhere. The Urban Sports reigned supreme.

2005 Thruxton, Bonneville, Bonneville SE, Bonneville T100 (Differing from 2004)

Bore	90mm (Bonneville T100)
Capacity	865cc (Bonneville T100)
Horsepower	63 at 7,250 rpm (Bonneville T100)
Wheelbase	56.7 inches (1,490mm) (Thruxton); 59.1 inches (1,500mm) (Bonneville, SE and Bonneville T100)
Colors	Blue and silver, Yellow and silver (Thruxton); green, blue (Bonneville); white and Tangerine (Bonneville T100)

2006 Rocket III, Classic, Speedmaster, America (Differing from 2005)

Colors	Yellow (Rocket III); red and white, red and black (Classic); Jet Black and Sunset Red (Speedmaster); Goodwood Green and silver (America)

Speed Triple, Sprint ST, Daytona 955i, Tiger

The rest of the Urban Sports lineup was as much as for 2005, the Daytona 955i still with the earlier engine, but updated with the Speed Triple's high-pressure diecast crankcases, a new breather system, a primary drive backlash eliminator, and improved gear selector mechanism. But it was obvious that the Daytona 955i was marking time, and this would be its final year. The

New for 2006, the Rocket III Classic featured footboards and a more comfortable seat. Styling touches extended to a black-finished engine with chrome-plated cam cover, cone mufflers, and two-tone colors. *Triumph Motorcycles*

CHAPTER NINE

2006 Daytona 675

Type	Inline three-cylinder DOHC, 4 valves per cylinder liquid-cooled
Bore	74mm
Stroke	52.3mm
Capacity	675cc
Horsepower	123 at 12,500 rpm
Compression Ratio	12.65:1
Fuel Injection	Multipoint sequential electronic with forced air induction
Ignition	Digital inductive
Gearbox	Six-speed
Frame	Aluminum beam twin-spar
Swingarm	Braced aluminum twin-sided with adjustable pivot
Front Fork	41mm upside-down adjustable
Rear Suspension	Monoshock adjustable rising rate
Brakes	2x308mm disc 4-piston radial calipers front, 220mm disc rear
Wheels	17x3.5-inch front and 17x5.5-inch rear
Tires	120/70 ZR17 front and 180/55 ZR17 rear
Wheelbase	54.8 inches (1,392mm)
Dry Weight	363 pounds (165kg)
Colors	Graphite, Tornado Red, Scorched Yellow

RIGHT: Triumph's best sport effort yet, the three-cylinder Daytona 675 was considerably more compact than the four-cylinder Daytona and was voted International Bike of the Year. *Australian Motorcycle News*

BELOW: The Daytona 675's front end included a Kayaba upside-down fork with Nissin radial four-piston calipers. *Australian Motorcycle News*

2006 Speed Triple, Sprint ST, Daytona 955i, Tiger
(Differing from 2005)

Colors
Fusion White (Speed Triple); Sunset Red (Sprint ST); Jet Black (Tiger)

Tiger was also in its final incarnation as a 955, while the recently updated Speed Triple and Sprint ST continued unchanged.

2006 Modern Classics
Scrambler
Triumph's reputation in America was largely built on the success of its off-road TR6C and Desert Sled, and

2006 Scrambler
(Differing from the Bonneville)

Type	270-degree parallel twin DOHC, 4 valves per cylinder air-cooled
Bore	90mm
Capacity	865cc
Horsepower	54 at 7,000 rpm
Colors	Caspian Blue and white, Tornado Red and white

it was this that inspired the new Scrambler. The association with legends such as Steve McQueen didn't hurt, and while the new Scrambler was based on the Bonneville, it incorporated a few significant updates. As it was aimed squarely at the US market, the larger-capacity twin included the 270-degree crankshaft to provide more rumble in the exhaust, with the engine tuned for more midrange power. Chassis-wise, the rake was pulled in slightly, to 27.8 degrees, and the suspension raised to provide more ground clearance. Another product of John Mockett, rather than dual exhaust pipes exiting on the left as on the original TR6C, it had high pipes on the right. Not as elegant as the original, they did at least include effective heat shields. The long list of options extended to a set of number plates marked 278, Steve McQueen's old ISDT number. While the new Scrambler was definitely no stripped-down competition Desert Sled, it certainly conveyed a similar sense of purpose. Following the appearance of the Speed Triple in *Mission: Impossible 2*, Tom Cruise requested that Triumph supply a Scrambler for the next installment, the 2006 film *Mission: Impossible 3*. This model was a one-off created especially for the movie.

Thruxton 900, Bonneville, Bonneville T100

While the existing Modern Classic range of the Bonneville, T100, and Thruxton continued as before, new this year was a custom range for the Bonneville, the sixty8. These contemporary accessories moved away from tradition, with stylish laptop bags instead of panniers and iPod holders instead of tank bags. Appealing to a more style-oriented clientele, the sixty8 line seduced a number of celebrities, including George Clooney.

2007 Model Year

Now with more than 1,000 employees, Triumph built 41,125 motorcycles this year. The 2007 model year saw an all-new Tiger, with all the Classic engines now standardized to 865cc.

2003–2009

LEFT: With its high-rise exhaust system, there was no mistaking the Scrambler's TR6 heritage. But at more than 450 pounds dry, it was more about style than function. *Triumph Motorcycles*

BELOW LEFT: A special one-off Scrambler was built for the 2006 film *Mission: Impossible 3*. *Triumph Motorcycles America*

BELOW: One of the 2006 Classic "sixty8"-accessorized Bonnevilles. *Triumph Motorcycles*

2006 Thruxton, Bonneville, Bonneville T100 *(Differing from 2005)*

Bore	90mm (Bonneville T100)
Capacity	865cc (Bonneville T100)
Horsepower	66 at 7,200 rpm (Bonneville T100)
Colors	Silver (Bonneville); white and Tangerine (Bonneville T100)

2007 Cruisers
Rocket III, Rocket III Classic, Speedmaster, America

This year the America received the larger-displacement Speedmaster engine, and with both models so similar, the Speedmaster received a few additional styling touches and extra equipment. The Speedmaster retained a dual-disc front end and a tachometer in the gas tank panel, and while the America received chrome engine covers, the Speedmaster's engine was black. Other detail differences extended to the America's wider and more pulled-back handlebars, and traditional reverse cone mufflers rather than the

ABOVE: For 2007 the America received the larger Speedmaster engine, and cast-alloy wheels replaced the wire-spoke type. The front brake was still a single disc. *Australian Motorcycle News*

221

CHAPTER NINE

The Mulberry Red 2007 Speedmaster, now with new wheels and cutoff mufflers.
Australian Motorcycle News

2007 Rocket III, Classic, Speedmaster, America (Differing from 2006)

Bore	90mm (America)
Capacity	865cc (America)
Horsepower	54 at 6,750 rpm (America)
Wheels	Cast-alloy 12-spoke 18x2.5-inch, 15x3.5-inch (America)
Colors	Phantom Black, Mulberry Red (Rocket III); blue and white, black and red (Classic); Mulberry Red (Speedmaster); black and red, blue and white (America)

2007 Tiger (Differing from 2006)

Stroke	71.4mm
Capacity	1,050cc
Horsepower	114 at 9,400 rpm
Compression Ratio	12:1
Frame	Aluminum beam perimeter
Swingarm	Braced aluminum twin-sided with adjustable pivot
Front Fork	43mm upside-down adjustable
Brakes	2x320mm discs 4-piston radial calipers (front), 255mm disc (rear)
Wheels	3.50x17-inch and 5.50x17-inch
Tires	120/70ZR17 and 180/55ZR17
Wheelbase	59.4 inches (1,510mm)
Dry Weight	436 pounds (198kg)
Colors	Black, blue, yellow, white

Speedmaster's slash cut design. Both received new wheels, the Speedmaster a new five-spoke design, and the America also getting new cast-alloy wheels instead of the previous wire-spoked type. The Rocket III and Rocket III Classic continued as before, with new colors.

2007 Urban Sports
Tiger

Realizing the reality of a world with diminishing dirt roads, the Tiger finally made the transition to pure street bike. Work had begun back in August 2003, with the styling selected from drawings by Marbese Design in Milan. Within 18 months, in March 2005, the first prototype was running. The Tiger's frame design echoed the Speed Triple's, but with a Daytona 675–style swingarm, while the 1,050cc engine was retuned for more midrange power. The second-generation Keihin unit was also from the 675, as were the headlights and instruments. The entire package was considerably more compact, with a lower seat, shorter wheelbase, quicker steering, and less weight than before. But it was the shorter Showa upside-down front fork with Nissin radial brake calipers and sporty 17-inch wheels that announced the Tiger's days of beating around the bush were over.

2007 Speed Triple, Sprint ST (Differing from 2006)

Colors
Roulette Green (Speed Triple); Phantom Black (Sprint ST)

Daytona 675, Speed Triple, Sprint ST

Now the only outright sport model in the range, the award-winning Daytona 675 continued unchanged for 2007, and apart from a new color, so did the Speed Triple. To meet new Euro 3 emissions, the Speed Triple and Sprint ST received the Tiger's new Keihin ECU. The Speed Triple's silencer was revised, adding

2003–2009

2007 Scrambler, Thruxton, Bonneville, Bonneville Black, Bonneville T100
(Differing from 2006)

Bore
90mm (Bonneville)
Capacity
865cc (Bonneville)
Horsepower
66 at 7,200 rpm (Bonneville)
Colors
Green and silver (Scrambler); Tornado Red (Thruxton); Jet Black and white (Bonneville T100)

Resplendent in Tornado Red, the 2007 Thruxton had a black engine and color-matched fenders. This was the final year the classic range featured carburetors. *Australian Motorcycle News*

stainless steel heat shields, and with higher handlebars and a taller screen, the Sprint ST's focus was more on touring rather than sport, more like the pre-2005 ST.

2007 Modern Classics
Scrambler, Thruxton, Bonneville, Bonneville Black, Bonneville T100

Apart from the Bonneville T100, new features for the entire Classic range included black engine cases, with the Scrambler and Bonneville Black also receiving black engine outer covers. The Bonneville's engine was now the 865cc T100 unit, with polished outer covers, and new this year was an adjustable clutch lever. The Thruxton had a revised tank and seat cowl decal, along with color-matched fenders.

2008 Model Year

Now the fastest-growing motorcycle brand in the world, Triumph's production hit 48,929 units in 2008, finally eclipsing their 1969 record of 46,700. New models included the Street Triple and Rocket III Touring, along with a significantly updated Speed Triple. This year Triumph also announced their first official racing venture since reestablishing at Hinckley. Run from Bologna, Italy, by former Ducati team manager Stefano Caracchi, Team Triumph-SC fielded entries in both the World Supersport and Superstock 600 European Championships. This was also a year of celebration, and on March 24, the anniversary of Steve McQueen's 78th birthday, Triumph released an officially licensed vintage T-shirt. And to acknowledge the 50th anniversary of the Bonneville's release at the

TOP: With its sporty, 17-inch wheels and upside-down front fork, the new Tiger was much more street-oriented than its predecessor. The new 5.3-gallon steel gas tank was smaller than the previous plastic type. *Australian Motorcycle News*

ABOVE: The Sprint ST featured a higher screen and higher handlebars for 2007. Most were fitted with standard saddlebags. *Australian Motorcycle News*

223

CHAPTER NINE

RIGHT: With the Daytona 675's chassis, the Street Triple continued the Speed Triple's naked aggressive style. The Nissin front brakes were a lower-specification twin-piston design. *Australian Motorcycle News*

OPPOSITE TOP LEFT: A black-and-gold Daytona 675 Special Edition was available for 2008, but the specifications were unchanged from the standard version. *Triumph Motorcycles*

OPPOSITE TOP RIGHT: Glen Richards provided the Daytona 675 its first series victory, winning the 2008 British Supersport Championship. *Triumph Motorcycles*

1958 Earls Court Show, celebratory Belstaff and Ewan McGregor Bonnevilles were produced.

2008 Urban Sports
Street Triple

In the wake of the Daytona 675's extraordinary success, it was inevitable that it would spawn a naked version, and the Street Triple appeared for 2008. Rather than a fairing-less Daytona, the Street Triple was designed from the ground up with its own identity, the three-cylinder engine receiving lower lift camshafts to boost midrange power. The frame, swingarm, and wheels came from the Daytona, but with a shallower steering head angle of 24.3 degrees. The Nissin front brake calipers were also a lower-spec two-piston sliding pin type, but with the Daytona's headlights and instrument console, the Street Triple was lean, light, and agile—ready to attack the highly competitive naked middleweight category. As usual, a full range of accessories was available, including an Arrow exhaust that boosted power by 6 horsepower.

Daytona 675, 675 Special Edition

The Daytona 675 continued to set the sport middleweight benchmark, winning the international magazine Supertest Trophy for the third straight year. Ostensibly unchanged, the 675 this year received new fairing decals, a gold anodized top nut, and an improved headlight. Also available was a Special Edition, in black with gold wheels and decals. While success eluded the factory-supported World Supersport team, the veteran Australian rider Glen Richards won the British Supersport Championship on the MAP Embassy Daytona 675, winning 3 of the 11 races. Triumph was also successful

2008 Street Triple
(Differing from the Daytona 675)

Horsepower	107 at 11,700 rpm
Front Brakes	2-piston calipers
Wheelbase	54.9 inches (1,395mm)
Dry Weight	367 pounds (167kg)
Colors	Fusion White, Jet Black, Roulette Green

2008 Daytona 675, Daytona SE
(Differing from 2007)

Colors	Neon Blue, Jet Black (675); Phantom Black (SE)

in the German Supersport Championship, Arne Tode dominating on the G-LAB Racing Daytona.

Speed Triple, Tiger, Sprint ST

Now in its 14th year, the Speed Triple received a restrained restyle by Milan-based Marabese design. Apart from an updated gear change mechanism, the 1,050cc engine was much as before, with new features centering on the chassis. The front braking system included a Nissin 19mm radial master cylinder with four 34mm-piston Brembo radial calipers. Magura tapered aluminum handlebars provided the same riding position as the 2007 version, and the completely redesigned rear end included a 25mm-longer rear subframe, new bodywork, and silencer heat shields. The lighter, seven-spoke, alloy wheels were unique to the Speed Triple, and as usual a wide range of options was available. Apart from an ABS option and a Blazing Orange color option, the Tiger was unchanged this year, while the Sprint ST received an improved headlight and modified footrests.

2008 Cruisers
Rocket III Touring

Broadening the appeal of the Rocket III, the Rocket III Touring featured a retuned 2.3-liter engine, now with stump-pulling torque, new tubular steel twin-spine frame, and twin-sided steel swingarm. Shared with the Rocket III were only the front brake, mirrors, and taillight,

2008 Speed Triple, Tiger, Sprint ST
(Differing from 2007)

Dry Weight
443 pounds (201kg) (Tiger ABS); 469 pounds (213kg) (Sprint ST ABS)
Colors
Black, white, orange (Speed Triple); Blazing Orange (Tiger); Pacific Blue, Graphite (Sprint ST)

ABOVE: The Speed Triple received new wheels and Brembo front brakes for 2008. The restyled rear section allowed more room for a passenger. *Triumph Motorcycles*

LEFT: An accessorized 2008 Speed Triple with Arrow exhaust system and solo seat. *Triumph Motorcycles America*

CHAPTER NINE

A new model in the 2008 cruiser lineup was the Rocket III Touring. With new frame, fork, wheels, and lights, plus saddlebags and a screen, this was designed to provide a "cruiser-style" touring experience. This example is two-tone blue, Eclipse and Azure. *Australian Motorcycle News*

the Touring receiving 16-inch cast-aluminum wheels, a conventional Kayaba fork, instruments moved from the handlebars to the top of the tank, and a full-size quick-release screen, and a pair of 10-gallon panniers were standard. While confidence-inspiring to ride, even at a walking pace, the huge, near 800-pound Rocket III Touring wasn't for the short or the weak.

Rocket III, Rocket III Classic, Speedmaster, America

While the rest of the Rocket III lineup was as before, but with new colors, the Speedmaster and America received a larger, 5.1-gallon gas tank with chromed badges. During the model year, the Speedmaster and America also received Keihin electronic fuel injection with secondary air injection, which provided improved throttle action and a power increase.

2008 Rocket III Touring
(Differing from the Rocket III)

Horsepower	107 at 5,400 rpm
Front Suspension	43mm fork
Wheels	Alloy 25-spoke 16x3.5-inch front and 16x5.0-inch rear
Tires	150/80 R17 front and 180/70 R16 rear
Wheelbase	67.2 inches (1,705mm)
Dry Weight	798 pounds (362kg)
Colors	Black and silver, black and white, black and red, blue

2008 Rocket III, Classic
(Differing from 2007)

Fuel System	Multipoint sequential fuel injection (Speedmaster, America)
Horsepower	61 at 6,800 rpm (Speedmaster, America)
Colors	Claret (Rocket III); red and white, blue and silver (Classic); blue and silver (Speedmaster); blue and Aegean Blue (America)

2008 Modern Classics
Scrambler, Thruxton, Bonneville, Bonneville Black, T100

Initially the existing Classic range continued with only new colors for 2008, but during the year Keihin multipoint sequential electronic fuel injection replaced the carburetors. As the injection intakes featured carburetor-style bodies, visually the injection looked similar to before, and the power was increased slightly on all models except the Thruxton. This year also marked the 50th anniversary of the T120 Bonneville's release. Celebrating this launch, in May 2008 Triumph announced two special editions, one by Hollywood star Ewan McGregor and the other by high-end motorcycling fashion designer Belstaff. A renowned motorcycle enthusiast, McGregor incorporated a copper-plated gas tank in his design, with a Belstaff black waxed-cotton seat and side panels. McGregor explained, "I wanted to use a traditional waxed cotton material as it has such a resonance with the history of style in motorcycling and mix it with the tradition of coppering tanks. I am a huge fan of Steve McQueen and his films from the 1960s, and the font that I have chosen for the Triumph logo harks back to that golden era of biking." Ewan McGregor's specially designed Bonneville was auctioned for his chosen charity, UNICEF, for whom he is an ambassador. For their 50th Anniversary Bonneville, Belstaff designer Michele Mallenotti produced a black-and-gold color scheme synonymous with the fashion brand's identity. The design also included a gold seat cowl and a black-and-gold pannier bag.

Also celebrating the Bonneville's 50th anniversary, at the 2008 Bonneville Speed Week, Californian Triumph dealer Matt Capri set two new unfaired 1,000cc Blown Fuel land speed records on his turbocharged Bonneville. He recorded a two-way speed of 162.472 miles per hour in the kilometer run and 161.188 miles per hour

2008 Scrambler, Thruxton, Bonneville, Bonneville Black, T100
(Differing from 2007)

Fuel System
Keihin multipoint sequential fuel injection 36.5mm throttle bodies

Horsepower
59 at 6,800 rpm (Scrambler); 67 at 7,500 (Bonneville)

Colors
Silver and Tangerine (Scrambler); black and gold, silver and Red Stripe (Thruxton); white, Claret (Bonneville); Claret and silver, green and white (T100)

in the mile. And to promote the Thruxton, Triumph sponsored the Thruxton Cup Challenge, run in conjunction with the American Historic Racing Motorcycle Association's Historic Cup Roadrace Series events.

2009 Model Year

As it was no secret that Triumph was about to unleash possibly their most significant model yet, the new Thunderbird cruiser, for the 2010 model year, in the meantime the existing range was moderately honed. Along with an updated Daytona 675 came a higher-specification Street Triple R and a 15th Anniversary Speed Triple and 50th Anniversary Bonneville.

2009 Urban Sports
Street Triple, Street Triple R, Daytona 675, Daytona SE

In the wake of the Street Triple's success, with 7,531 built in 2008, Triumph released the higher-spec Street Triple R for 2009. As the engine was deemed powerful enough, with appropriate characteristics for a naked bike, updates for the R were limited to the chassis. This included a fully adjustable Kayaba upside-down front fork with Nissin four-piston radial calipers, an adjustable rear shock, and trick-looking Magura tapered aluminum handlebars.

Although the Daytona 675 was still winning awards and proving extremely successful, with 15,200 built since 2006, this highly competitive category was changing quickly, and Triumph updated the Daytona for 2009. Subtle restyling by Chris Hennegan resulted in a more aggressive and modern frontal aspect, with a new cockpit, headlight, and screen. Chassis updates included separate high- and low-speed compression

2009 Street Triple, Street Triple R, Daytona 675, Daytona SE
(Differing from 2008)

Horsepower
126 at 12,600 rpm (Daytona 675)
Front Fork
41mm upside-down adjustable (Street Triple R)
Rear Suspension
Adjustable monoshock (Street Triple R)
Front **Brakes**
4-piston radial calipers (Street Triple R)
Wheelbase
54.7 inches (1,390mm) Street Triple R
Dry Weight
356 pounds (162kg) Daytona 675
Colors
Graphite, orange (Street Triple R); white and blue (675 SE)

ABOVE: Belstaff's 50th Anniversary Bonneville was in black and gold, linked with their identity of producing traditional waxed-cotton motorcycle clothing. *Triumph Motorcycles America*

LEFT: Ewan McGregor with his interpretation of the 50th Anniversary Bonneville. Unique features included a copper-plated gas tank and waxed-cotton seat and side covers. *Triumph Motorcycles America*

BELOW: Matt Capri's turbocharged Bonneville set two new land speed records at Bonneville in 2008. *Triumph Motorcycles*

CHAPTER NINE

2009 Speed Triple, Tiger, Tiger SE, Sprint ST (Differing from 2008)

Colors
Matte Black (Speed Triple); Graphite and black (Tiger SE); Phantom Black (Sprint ST)

2009 Rocket III, Classic, Touring, Speedmaster, America (Differing from 2008)

Dry Weight
788 pounds (358kg) (Rocket III Touring)
Colors
Phantom Black and New England White (Speedmaster)

damping suspension adjustment and new Nissin four-piston Monobloc front brake calipers. A new cylinder head and revised exhaust system resulted in a small power increase, and with judicious lightening of various components the overall weight was slightly reduced. Already superb, it was simply a case of the best becoming even better. Triumph announced a 675 SE in May 2009, in Pearl White with a Sparkle Blue frame, and as usual a wide range of accessories was available, from powershifters to Arrow exhausts. Triumph again competed in the World Supersport Championship, the Italian-based PE1 team headed by former 500GP race winner Gary McCoy, but results this year were disappointing.

Speed Triple, Speed Triple 15th Anniversary Special Edition, Tiger, Tiger SE, Sprint ST

As the most popular Hinckley Triumph, with more than 35,000 built since 1994, a Speed Triple limited edition was offered for 2009. This carried a John Bloor signature on the gas tank, unique metallic Phantom Black colors, red-pinstriped wheels, and a color-matched belly pan and screen. Apart from the rear shock providing slightly less travel (134mm), the regular Speed Triple was unchanged, as were the Tiger and Sprint ST this year. A Tiger SE with standard ABS and custom paint was also available, while the 2009 Speed Triple made another significant movie appearance, ridden by Angelina Jolie in *Salt*.

2009 Cruisers
Rocket III, Rocket III Classic, Rocket III Touring, Speedmaster, America

With the bombshell Thunderbird imminent, the cruiser lineup continued unchanged for 2009. The only color change was for the Speedmaster, now also with color-coded side panels.

2009 Modern Classics
Scrambler, Thruxton, Bonneville, Bonneville SE, 50th Anniversary, T100

While the Scrambler, Thruxton, and T100 were ostensibly unchanged this year, the Bonneville and Bonneville SE received a pair of 17-inch cast-alloy wheels, styled similar to the original 1979 Meriden Bonneville Special T140D's cast-alloy Morris wheels. The Bonneville and SE also had shorter fenders and abrupt upswept mufflers. Although the 17-inch wheels may have looked a little incongruous on such a traditional retro model, they were considerably lighter and transformed the handling. For those still wanting the traditional look, the T100 was still available, along with a special Bonneville 50th Anniversary edition painted in the same Tangerine Dream colors of the first 1959 Bonneville. Echoing the engine capacity of the original, 650 numbered examples were built, each with a certificate of authenticity signed by John Bloor.

The Bonneville also celebrated its 50 years with four world speed records. British journalist Alan Cathcart rode Matt Capri's street-legal Thruxton Bonneville to a new two-way FIM world record for the

ABOVE: Celebrating 15 years of the iconic Speed Triple, this Anniversary Special Edition was the first model to include an authorized John Bloor autograph on the tank. *Triumph Motorcycles America*

OPPOSITE TOP: During 2009 this stunning white-and-blue Daytona 675 SE became available. *Triumph Motorcycles*

OPPOSITE BOTTOM: The Daytona 675 chassis was similar to before, but it shed 7 pounds for 2009, and the improved front brakes were Nissin Monoblock. *Triumph Motorcycles*

CHAPTER NINE

2009 Scrambler, Thruxton, Bonneville, Bonneville SE, Bonneville 50th Anniversary, T100
(Differing from 2008)

Wheels	Alloy 7-spoke 17x3.0-inch front and 17x3.5-inch rear (Bonneville, SE)
Tires	110/70 R17 front and 130/80 R17 rear (Bonneville, SE)
Wheelbase	57.2 inches (1,454mm) (Bonneville, SE)
Dry Weight	440 pounds (200kg) (Bonneville, SE)
Colors	Jet Black, Matte Khaki Green (Scrambler); black, blue and white (Bonneville SE); blue and orange (50th Anniversary)

TOP: The Bonneville and SE received cast-alloy wheels reminiscent of the Morris wheels on the ill-fated 1979 Bonneville T140D Special. This SE has an accessory Arrow two-into-one exhaust, also replicating the original. *Australian Motorcycle News*

ABOVE: With the Bonneville and SE featuring more modern wheels, the T100 was left as the traditional Bonneville. *Australian Motorcycle News*

RIGHT: Apart from the distinctive Matte Khaki Green paint, the Scrambler was much as before for 2009. The black engine cases first appeared in 2007. *Triumph Motorcycles*

flying mile at 152.678 miles per hour and another for the flying kilometer at 152.770 miles per hour, new records for the FIM's 1,000cc Normally-Aspirated Twin-Cylinder Unstreamlined category. On the turbocharged Bonneville Cathcart set two more new FIM world records for the 1,000cc Forced Induction Twin-Cylinder Unstreamlined class, with a two-way speed for the flying mile of 165.405 miles per hour and 165.672 miles per hour for the flying kilometer.

After the hiccup of the 2002 fire, Hinckley ended their second decade with a broader model lineup, with more emphasis on cruisers and classics than during the previous decade. Whereas specialist tourers and sport bikes initially drove the Triumph renaissance, by 2009, out of a 20-model lineup, only the Daytona 675 and Sprint ST represented these traditional categories. Production had increased to new levels, a factory had opened in Thailand, and Triumph was set to prosper even more. Not only was the factory rebuilt after the 2002 fire, so was the spirit of the company.

2003–2009

ABOVE: Five generations of the Speed Triple from the T509. More than any other model, the Speed Triple symbolized the Triumph renaissance as Hinckley approached the end of their second decade. *Triumph Motorcycles America*

LEFT: The Bonneville 50th Anniversary (left) with an original 1959 example (right). The 50th Anniversary also included a special seat with white piping and a gold logo, unique side cover logos, and a chromed cam cover. *Triumph Motorcycles*

231

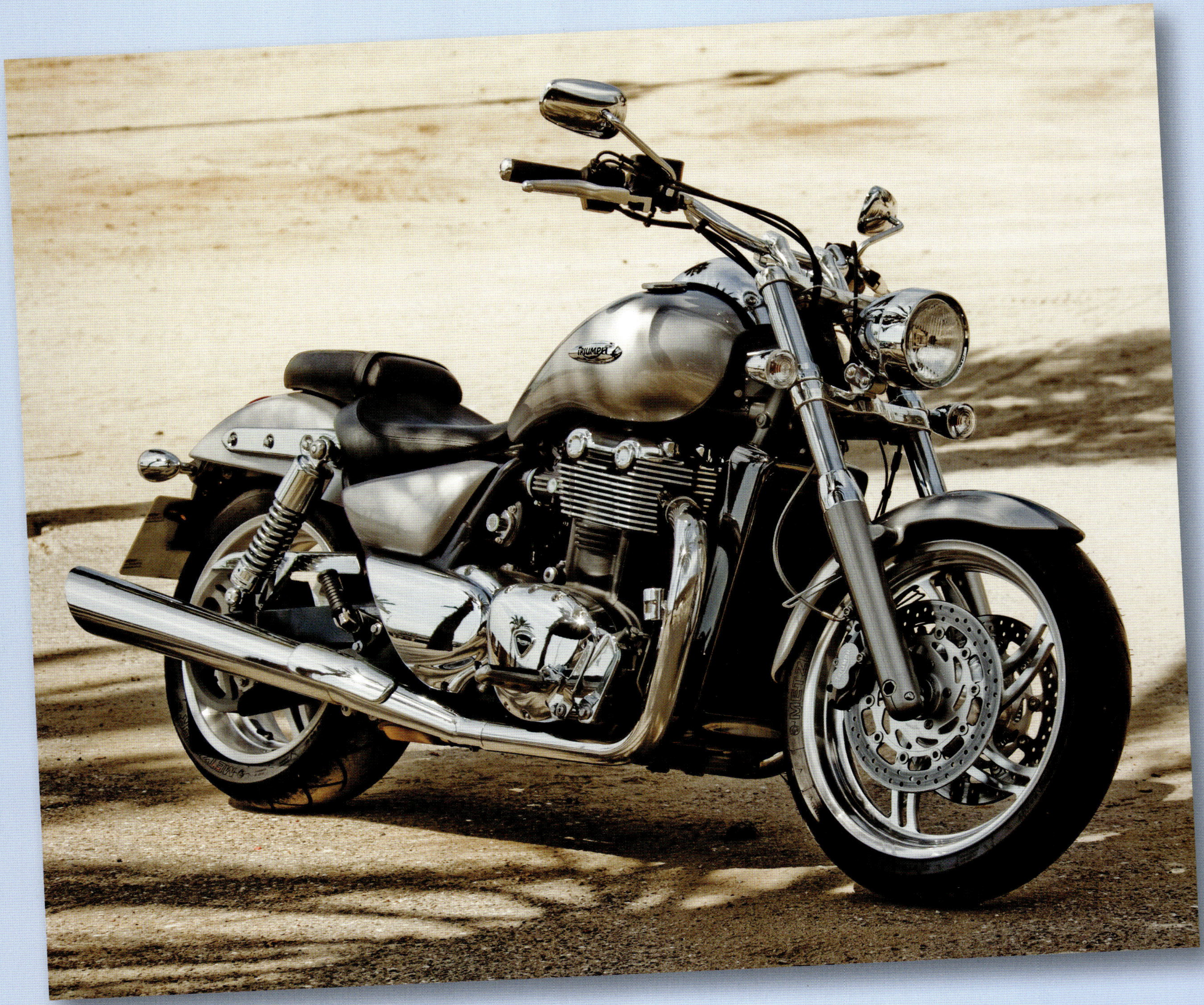

10 2010–2024

Building on Tradition

Larger Thunderbirds, Tigers, Trophys, Bonnevilles, Scramblers, and new Singles

Hinckley's third decade contrasted strikingly with the earlier era at Meriden. Meriden's third decade after World War II, the 1970s, was their most tumultuous, categorized by indecisive management, union disruption, dubious product development, and minimal reinvestment in plant and equipment. Under John Bloor the scenario was exactly the opposite, and with a strong emphasis on model development, quality, and niche products, Triumph was establishing new design standards. Edward Turner made the parallel twin a benchmark layout, copied and emulated, but eventually the inline four assumed that mantle. Triumph tried with four cylinders but found their niche of triples and parallel twins much more successful, and with strong roots in the past, these engine layouts would characterize Triumph's next decade.

2010 Model Year

Alongside the Thunderbird for 2010 were the Rocket III Roadster and two special editions, the Bonneville Sixty and the Speed Triple SE. Triumph also announced a partnership with Öhlins to supply accessory suspension units for their Urban Sport range, and a technical partnership with lubricant specialists Castrol. Group turnover for 2010 was £312.4 million and motorcycle production 45,501 units.

2010 Cruisers
Thunderbird

Between 2006 and 2008 US cruiser sales topped more than a million, with around as much again for the rest of the world, most sales larger-capacity V-twins. Faced with a huge displacement gap between the Rocket III and the Speedmaster/America, Triumph needed something new in their cruiser lineup. Although V-twins dominated this segment, Triumph was traditionally associated with the parallel twin,

ABOVE: With no sign of ugly and excessive chrome, the Thunderbird was tastefully presented, as only the British know how. *Triumph Motorcycles*

OPPOSITE: Just as the original Thunderbird established Triumph as the foremost manufacturer of large-capacity parallel twins, so did the new Thunderbird 60 years later. *Triumph Motorcycles*

CHAPTER TEN

RIGHT: Providing the primary balance of a 90-degree V-twin, the big parallel twin featured a 270-degree crankshaft and double overhead camshafts driven by a central chain. The all-new six-speed gearbox included quieter helical gears from second upwards. *Triumph Motorcycles*

BELOW: A wide range of accessories was available to customize the Thunderbird. *Triumph Motorcycles*

2004 Rocket III

Type	Liquid-cooled, DOHC, parallel twin
Bore	103.8mm
Stroke	94.3mm
Capacity	1,597cc
Horsepower	85 at 4,850 rpm
Compression Ratio	9.7:1
Fuel Injection	Multipoint sequential electronic fuel injection
Gearbox	Six-speed, helical 2nd to 6th gears
Final Drive	Belt
Frame	Tubular steel, twin spine
Swingarm	Twin-sided, steel
Front Fork	47mm Showa
Rear Suspension	Twin shocks with adjustable preload
Brakes	2x310mm disc 4-piston calipers front, 310mm disc rear
Wheels	Alloy 5-spoke 19x3.5-inch front and 17x6.0-inch rear
Tires	120/70 R19 front and 200/50 R17 rear
Wheelbase	63.5 inches (1,615mm)
Dry Weight	678 pounds (308kg)
Colors	Black, silver, blue

and that was the path they took. Just as the Rocket III ushered in Triumph's new era of the power cruiser, with the Thunderbird they entered the big-cruiser market with a similarly individual design.

The original Thunderbird had established Triumph's reputation in America, and the name came with strong connotations. Just as the original was also the largest parallel twin at the time, so was the new 1,600cc Thunderbird. With a 270-degree crank providing the sound and feel cruiser customers expected, twin balance shafts front and rear of the crank smoothed the pulses of the massive cylinders. With a cylinder bore larger than the Rocket III's, dual ignition was required, with Keihin closed-loop fuel injection. A must for a modern cruiser was belt final drive, the first for Triumph since 1922, and the frame consisted of steel tubes connected by a series of aluminum castings. As the Thunderbird was intended to compete directly with Harley-Davidson, it was no surprise that the styling was the work of Los Angeles–based designer Tim Prentice, also responsible for the Rocket III Touring. Prentice's design brief was to maintain a clean, classic look with muscular proportions and no fake parts, and with legendary Spanish tester David Lopez sorting out the chassis, the Thunderbird not only looked the part, but also handled as a Triumph should. Despite the huge dimensions and significant weight, unlike most other cruisers the Thunderbird was fun to ride on twisting roads. A huge range of custom accessories was available, from a 1,700cc big-bore kit to swingarm covers and radiator shrouds, and early in 2010 two hand-painted color options were available in limited numbers: Phantom Blue Haze and Phantom Red Haze. Launched mid-2009 as an early-release 2010 model, the Thunderbird impressed *Cycle World* enough to be named their Cruiser of the Year, a feat it repeated in 2010.

Rocket III Roadster, Rocket III Touring, Speedmaster, America

By 2010, 18,000 Rocket IIIs had been built, 34 percent of them destined for the US. The Rocket III had evolved through the Classic and Touring versions, and with the release of the Thunderbird, Triumph repositioned the Rocket III from a cruiser to performance naked bike, the Rocket III Roadster. A freer-flowing exhaust contributed to more power and a massive increase in torque, to 163 lbs/ft at only 2,750 rpm, and while the basic Rocket III chassis was unchanged, Tim Prentice's styling emphasized the street-fighter element, the seat lower and more forward, with a pair of trademark twin headlights and individual instruments. As the Rocket III Touring was a relatively new model, it continued largely unchanged, and the Rocket III Classic was deleted. New for the Speedmaster this year was a restyled speedometer, perched on the top triple clamp, with a tachometer in the gas tank nacelle,

234

2010–2024

2010 Rocket III Roadster, Touring, Speedmaster, America
(Differing from the Rocket III)

Horsepower
146 at 5,750 rpm (Roadster); 105 at 6,000 rpm (Touring); 60 at 6,800 rpm (Speedmaster, America)
Brakes
Including Nissin ABS (Roadster)
Wheelbase
66.7 inches (1,695mm) (Roadster)
Wet Weight
807 pounds (366kg) (Roadster); 869 pounds (395kg) (Touring); 550 pounds (250kg) (Speedmaster, America)
Colors
Phantom Black, Matte Black (Roadster); black and white (Touring)

but apart from this and a slightly lower claimed power, it and the America were unchanged.

2010 Urban Sports
Street Triple, Street Triple R, Daytona 675, 675 SE, Speed Triple, Speed Triple SE, Tiger, Tiger SE, Sprint ST

All Urban Sport models had a slightly lower claimed power, with the Street Triple, Street Triple R, and Daytona 675 also receiving small changes in wheelbase. The Daytona's graphics were new, and the 675 SE this year included white-striped wheels and carbon-fiber parts. For its final year in its current form, the Speed Triple SE received an updated suspension and a two-tone paint scheme with color-matched fly screen and seat cowl. The suspension included modified damping and a softer rear spring, features later introduced on the standard Speed Triple. The Tiger SE was a higher-specification Tiger with standard ABS, color-matched hand guards, and saddlebags.

Italian-based BE1 Racing ran Triumph's official entry in the World Supersport Championship, their most successful rider, Chaz Davies, finishing fourth overall. His best result was third, which he achieved three times during the season. Triumph also supported a one-make race series of BE1-prepared Street Triple Rs over seven rounds.

2010 Modern Classics
Scrambler, Bonneville, Bonneville SE, Bonneville Sixty, T100, Thruxton, Thruxton SE

An update to the Scrambler saw a small claimed power increase, while the power for all Bonnevilles and the Thruxton was slightly lower than before. All 2010 Modern Classics featured revised instrumentation with

2010 Street Triple, Street Triple R, Daytona 675, 675 SE, Speed Triple, Tiger, Tiger SE, Sprint ST
(Differing from 2009)

Horsepower	105 at 11,700 rpm (Street Triple, R); 124 at 12,600 rpm (675, SE); 128 at 9,250 rpm (Speed Triple); 111 at 9,400 rpm (Tiger); 123 at 9,100 rpm (Sprint ST)
Wheelbase	54.7 inches (1,390mm) (Street Triple); 54.5 inches (1,385mm) (Street Triple R); 55.7 inches (1,415mm) (675, SE); 56.1 inches (1,425mm) (Speed Triple)
Wet Weight	416 pounds (189kg) (Street Triple); 407 pounds (185kg) (675); 477 pounds (217kg) (Speed Triple); 502 pounds (228kg) (Tiger); 539 pounds (245kg) (Tiger SE); 530 pounds (241kg) (Sprint ST)
Colors	Red (Street Triple); black (Street Triple R); blue with gold wheels (675); red and white (Speed Triple SE); Matte Black and Graphite, orange (Tiger SE)

TOP: Now more street fighter than cruiser, and only available in two different shades of black, the Rocket III Roadster had more power and torque than before. *Triumph Motorcycles*

LEFT: The 2010 Daytona 675 SE had racing touches such as adjustable, machined alloy levers, similar to those found on the factory World Supersport racers, with a carbon-fiber heat shield, exhaust cap, cockpit infills, and rear hugger to complete the factory racer look. *Triumph Motorcycles*

CHAPTER TEN

ABOVE: Finishing fourth overall on the Team ParkinGO BE-1 Daytona 675, Welshman Chaz Davies was Triumph's most successful rider in the 2010 World Supersport Championship. *Triumph Motorcycles*

RIGHT: The Speed Triple SE for 2010 had a special two-tone paint scheme and improved suspension. *Triumph Motorcycles*

2010 Scrambler, Bonneville, Bonneville SE, Bonneville Sixty, T100, Thruxton, Thruxton SE
(Differing from 2009)

Horsepower
58 at 6,800 rpm (Scrambler); 66 at 7,500 rpm (Bonneville); 68 at 7,400 rpm (Thruxton)

Wheelbase
58.6 inches (1,490mm) (Bonneville)

Wet Weight
506 pounds (230kg) (Scrambler, T100, Thruxton); 495 pounds (225kg) (Bonneville)

Colors
Meriden Blue and Caspian Blue (Bonneville Sixty)

Extremely attractive in two-tone blue, the T100 Bonneville Sixty replicated the colors of the 1960 T120 Bonneville. *Triumph Motorcycles*

a digital odometer, and there was no change to the colors this year. A limited edition of 650 T100 Bonneville Sixtys, inspired by the 1960 T120 Bonneville, was also produced this year, each with a numbered plaque and certificate of authenticity, and during the model year a white Thruxton Special Edition with red frame was also available.

2011 Model Year

This year saw the most new model launches ever for Triumph in a single year. Heading the new models were the Tiger 800 and XC, joined by a significantly updated Speed Triple, the Supersport Daytona 675R, and a touring Sprint GT, along with variations of and updates to the existing Street Triple, Thunderbird, Speedmaster, and America. The 22-model range was now subdivided into six categories, and despite difficult global conditions, the introduction of many more models saw revenue rise 11 percent, to £345.3 million, and sales increase 7 percent, to 48,684. How many companies could say *that* so soon after the recession?

2011 Cruisers
Thunderbird Storm, Thunderbird, Rocket III Roadster, Rocket III Touring, America, and Speedmaster

Deciding that the Thunderbird platform was ripe for expansion, Triumph painted their big cruiser black, fitted the 1,700cc big-bore kit as standard, and gave it twin headlights and a flat drag-style handlebar on risers, and the brooding Thunderbird Storm muscle bike was born. While the Thunderbird cruiser continued unchanged but for new color options, a special-edition Thunderbird SE with factory-fitted touring accessories was also available. Apart from colors, the two Rocket III models were also unchanged this year.

Because they were still viewed as similar, the America and Speedmaster were significantly updated for 2011 as a means of differentiation. While mechanically unchanged, the classic cruiser America received new styling, with a wider, sweeping handlebar, larger headlight, a short, fat front wheel, deep fenders, and a plusher seat. Less conservative and traditional, the Speedmaster was the lean, stripped-down machine with a meaner attitude. The pulled-back bars were flatter and the headlight smaller, and skimpy fenders and a taller, skinny front wheel implied a sportier image. The front fork wasn't shrouded, and there was now only a single front disc. Aimed at making the smaller cruisers better suited to shorter riders, both the America and Speedmaster sat an inch lower, now with a 27.1-inch seat height. With newly distinct identities as entry-level cruisers, the America and Speedmaster opened the path to the larger Thunderbird and Rocket III.

2011 Adventure
Tiger 800, 800XC, 1050, 1050 SE

Triumph's continued expansion into new market segment continued with the release of the Tiger middleweight adventure tourer. Powered by an all-new 800cc 12-valve triple, this came in two versions: the tarmac-friendly Tiger 800 and the more dirt-focused Tiger XC ("Cross-Country") trail bike counterpart. Work on the new dual-purpose platform began in April 2007, with the focus designed to separate the new models between the more street-oriented Tiger 1050 and the touring Street Triple. To produce a broad spread of torque, the 799cc engine was ostensibly a long-stroked version of the 675, but with a slightly lower compression ratio and new gearbox and gear change mechanism. Both the 800 and 800XC shared a new steel trellis frame, with the XC's steering rake slightly steeper at 23.1 degrees (as opposed to 23.7 degrees). While the engine and rear wheel sizes were the same, the XC featured wire-spoked wheels and more suspension travel. ABS was an option and comfort and practicality key targets, with the seat height adjustable on each bike between two positions, along with a low-seat option. With moderate weight and

2011 Thunderbird Storm, Thunderbird, Rocket III Roadster, Touring, America, Speedmaster
(Differing from 2010)

Bore	107.1mm (Storm)
Capacity	1,699cc (Storm)
Horsepower	97 at 5,200 rpm (Storm)
Front Brake	Single 310mm disc (Speedmaster)
Front Wheel	Alloy 16x3.0-inch (America); 19x2.5-inch (Speedmaster)
Rear Wheel	Alloy 15x4.0-inch (America, Speedmaster)
Front Tire	130/90R16 (America); 110/90R19 (Speedmaster)
Rear Tire	170/80B15 (America, Speedmaster)
Wheelbase	63.3 inches (1,610mm) (America); 62.9 inches (1,600mm) (Speedmaster)
Wet Weight	746 pounds (339kg) (Storm)
Colors	Phantom Black, Matte Black (Storm); blue and white (Thunderbird); red, blue (Roadster); black and silver (Touring); Cranberry Red (Speedmaster)

TOP: Painted 95 percent black, the menacing 1700cc Thunderbird Storm was more muscle bike than cruiser. *Australian Motorcycle News*

ABOVE: Both the Speedmaster and America were restyled for 2011 to differentiate them. With its 19-inch front wheel, the Speedmaster (foreground) presented a sportier image, while the America (behind) was more traditional. Both now had a single front disc brake. *Australian Motorcycle News*

CHAPTER TEN

ABOVE: The evolution of the Tiger species resulted in the new 800 for 2011. This is the street-focused Tiger 800, with cast-alloy wheels. *Australian Motorcycle News* News

LEFT: With its longer travel suspension and wire-spoked wheels, the Tiger 800XC was more off-road oriented. The beak-like front fender was dirt-bike style. *Australian Motorcycle News*

dimensions, the Tiger 800 could lay claim to be the perfect all-around multipurpose motorcycle. But for a new color and a small power increase, the Tiger 1050 and 1050SE continued unchanged.

2011 Touring
Sprint GT, Sprint ST

The long-running Sprint ST sport-tourer evolved into the more touring-focused Sprint GT for 2011. Now with a low-slung exhaust system, the 1,050cc three-cylinder engine was retuned for slightly more power and torque.

With standard luggage and more passenger room, the Sprint GT was more touring-focused than the Sprint ST. *Triumph Motorcycles*

2011 Tiger 800, 800XC, 1050, 1050SE
(Differing from 2010)

Type	Inline three-cylinder DOHC, 4 valves per cylinder liquid-cooled (800, XC)
Bore	74mm (800, XC)
Stroke	61.9mm (800, XC)
Capacity	799cc (800, XC)
Horsepower	94 at 9,300 rpm (800, XC); 113 at 9,400 rpm (1050, SE)
Compression Ratio	11.1:1 (800, XC)
Fuel Injection	Multipoint sequential electronic 44mm throttle bodies (800, XC)
Gearbox	Six-speed (800, XC)
Frame	Tubular steel trellis (800, XC)
Swingarm	Twin-sided cast-aluminum (800, XC)
Front Fork	Showa 43mm upside-down adjustable (800); 45mm (800XC)
Rear Suspension	Showa Monoshock adjustable
Brakes	2x308mm disc 2-piston calipers front, 255mm disc rear
Wheels	Alloy 19x2.5-inch front and 17x4.25-inch rear (800); Spoked 21x2.5-inch front and 17x4.25-inch rear (XC)
Tires	110/80 ZR19 front and 150/70 ZR17 rear (800); 90/90 ZR21 front (XC)
Wheelbase	61.2 inches (1,555mm) (800); 61.7 inches (1,568mm) (XC)
Wet Weight	462 pounds (210kg) (800); 473 pounds (215kg) (XC)
Colors	Black, white, yellow (800); orange (XC); white (1050)

2011 Sprint GT, Sprint ST (Differing from 2010)

Horsepower
128 at 9,200 rpm (Sprint GT)
Wheelbase
60.5 inches (1,537mm) (Sprint GT)
Wet Weight
583 pounds (265kg) (Sprint GT)
Colors
Aluminum Silver, Pacific Blue (Sprint GT)

2010–2024

While the twin-spar aluminum frame was shared with the ST, just about everything else was new, including the Showa 43mm front fork, Showa rear shock with remote preload adjuster, 3-inch longer swingarm, lighter wheels, and improved brakes with standard ABS. Along with an updated cockpit design, the Sprint GT came standard with color-coordinated hard luggage, and it was one of the most effective high-performance tourers available. An updated Sprint ST was also available, but only for the United Kingdom, this retaining the underseat exhaust system and shorter swingarm but including the GT's new cockpit, wheels, and brakes.

2011 Supersports
Daytona 675R, Daytona 675

After five years as a standard-setting real-world sportbike, with production of more than 23,000, Triumph released the more track-focused Daytona 675R for 2011. Built in the minimum amount of 2,000 for World Supersport homologation, the 675R's development commenced in April 2009 with the intention of creating a serious contender for both Supersport and Superstock competition. While the engine and frame were shared with the Daytona 675, the 657R featured an Öhlins 43mm NIX30 front fork and a TTX36 rear shock absorber. Brembo supplied superbike-spec radially mounted Monobloc front brake calipers with an 18mm radial master cylinder. Other track-spec features included a standard quickshifter and a carbon-fiber hugger, silencer heat shield, and front fender.

A range of official Triumph accessories was also available, including an Arrow slip-on silencer, race-style CNC-machined levers, and a single seat cowl, along with a full factory race kit that included camshafts and valves, plus a programmable Keihin ECU.

Triumph initially re-signed with BE1 Racing to run the World Supersport team, but this partnership collapsed early in 2011, leaving Triumph without an entry. But in the British Supersport Championship, the Australian pair of Billy McConnell and Paul Young finished third and fourth respectively on Oxford Tag Daytona 675Rs. McConnell won two races and Young one.

2011 Daytona 675R
(Differing from the Daytona 675)

Front Fork
43mm Öhlins upside-down adjustable
Rear Suspension
Öhlins Twin Tube Monoshock adjustable rising rate
Front Brakes
4-piston Brembo Monobloc radial calipers
Wheelbase
54.9 inches (1,395mm)
Color
Crystal White

2011 Roadsters
Speed Triple, Street Triple, Street Triple R

The Speed Triple was Triumph's best-selling model, with sales of more than 65,000 since 1994, and designer Tim Prentice had the difficult task of creating a more modern version without destroying the soul of the original. With a pair of pentagonal twin headlights replacing the distinctive standalone round type, not only was the style new, so was the completed updated chassis underneath. The twin-spar aluminum frame was all new, with a lighter single-sided swingarm that was 0.73 inch longer. The engine was located and canted further forward, with the rider moved 1.73 inch closer to the front axle. With a steep steering head angle of 22.8 degrees, and with a longer wheelbase, the forward weight bias was now 50.9/49.2 percent (compared to the previous 48.6/51.4 percent). Other chassis updates included a return to the pre-2005 17x6.0-inch rear wheel. The new wheels saved over 6 pounds of unsprung weight, and while the 1,050cc three-cylinder engine was similar, the power was increased. ABS was also available

Bristling with quality components, the Crystal White Daytona 675R represented the pinnacle of Supersports performance. Along with Öhlins suspension and Brembo front brakes, the 675R had a red rear subframe and deeper engraving on the engine cases. *Triumph Motorcycles*

Restyled and lighter than before, the 2011 Speed Triple continued the familiar hooligan street-fighter image, but with improved weight distribution and handling. The familiar twin-headlight style was also updated. *Triumph Motorcycles*

239

2011 Speed Triple, Street Triple, Street Triple R (Differing from 2010)

Horsepower	133 at 9,400 rpm (Speed Triple)
Rear Brake	255mm disc (Speed Triple)
Rear Wheel	6.00x17-inch (Speed Triple)
Rear Tire	190/55ZR17 (Speed Triple)
Wheelbase	56.5-inch (1,435mm) (Speed Triple) 55.5-inch (1,410mm) (Street Triple, R)
Dry Weight	409 pounds (186kg) (Speed Triple)
Wet Weight	471 pounds (214kg) (Speed Triple)
Colors	Red (Speed Triple); blue, yellow (Street Triple)

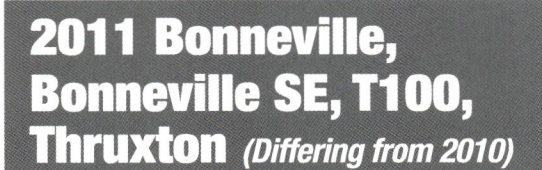

2011 Bonneville, Bonneville SE, T100, Thruxton (Differing from 2010)

Horsepower	67 at 7,500 rpm (Bonneville, SE, T100)
Colors	Orange and black, blue and white (Bonneville SE), cream and chocolate (T100)

2012 Tiger Explorer, 1050 SE (Differing from 2011)

Type	Inline three-cylinder DOHC, 4 valves per cylinder liquid-cooled (Explorer)
Bore	85mm (Explorer)
Stroke	71.4mm (Explorer)
Capacity	1,215cc (Explorer)
Horsepower	135 at 9,300 rpm (Explorer)
Compression Ratio	11.1 (Explorer)
Fuel Injection	Ride-by-wire Keihin 46mm throttle bodies (Explorer)
Gearbox	Six-speed (Explorer)
Frame	Tubular steel trellis (Explorer)
Swingarm	Single-sided cast-aluminum with shaft drive (Explorer)
Front Fork	Kayaba 46mm upside-dDown (Explorer)
Rear Suspension	Kayaba Monoshock adjustable (Explorer)
Brakes	2x305mm disc 4-piston calipers front, 282mm disc rear; switchable ABS (Explorer)
Wheels	Alloy 19x2.5-inch front and 17x4.0-inch rear (Explorer)
Tires	110/80 R19 front and 150/70 R17 rear (Explorer)
Wheelbase	60.2 inches (1,530mm) (Explorer)
Wet Weight	571 pounds (259kg) (Explorer)
Colors	Black, blue, Graphite (Explorer); red, white (1050 SE)

With the Tiger Explorer Triumph expanded beyond their traditional market into BMW GS territory, but with its cast-alloy wheels, it was more suited to road than off-road duties. *Triumph Motorcycles*

The 675cc Street Triple and Street Triple R didn't receive the same makeover as the Speed Triple, still retaining the twin round headlight design, but with new colors and decals.

2011 Classics
Scrambler, Thruxton, Bonneville, Bonneville SE, T100

It was very much business as usual for the range of Classics, with the Scrambler and Thruxton unchanged, while the Bonneville SE and T100 had a slightly higher claimed power ratings and new colors.

2012 Model Year
Following on from the model proliferation of 2011, this year Triumph initially launched three new motorcycles, the Tiger 1200 Explorer, the Speed Triple R, and the Steve McQueen Special. As Triumph was celebrating 110 years, later in the model year a 110th Anniversary Bonneville was released.

2012 Adventure
Tiger Explorer, 1050, 1050 SE, 800, 800XC

Triumph's most significant new model for 2012 was the Tiger Explorer, an unashamed attempt to muscle in on BMW's highly acclaimed and successful R1200GS maxi-enduro market. The Explorer project began back in July 2006, and with a target of around 130 horsepower envisaged to be competitive, rather than adapting the existing 1,050cc triple, Triumph's engineers designed a completely new 1,200cc engine. Durability was essential, so the service intervals were stretched to 10,000 miles, final drive was by shaft, and firsts for Triumph included a ride-by-wire throttle and cruise and traction control. New features included a high-capacity 950-watt alternator positioned behind the cylinders.

Triumph's familiar dual-tube steel twin-spar frame was also new, as was the single-side aluminum swingarm housing the shaft drive. This included a torsional damper to reduce loading and a parallelogram

as an option. The result was a bike that maintained the in-your-face hooligan attitude of the original, but with improved handling and rideability. Triumph was adhering to the philosophy of if it ain't broke, don't fix it, and they succeeded.

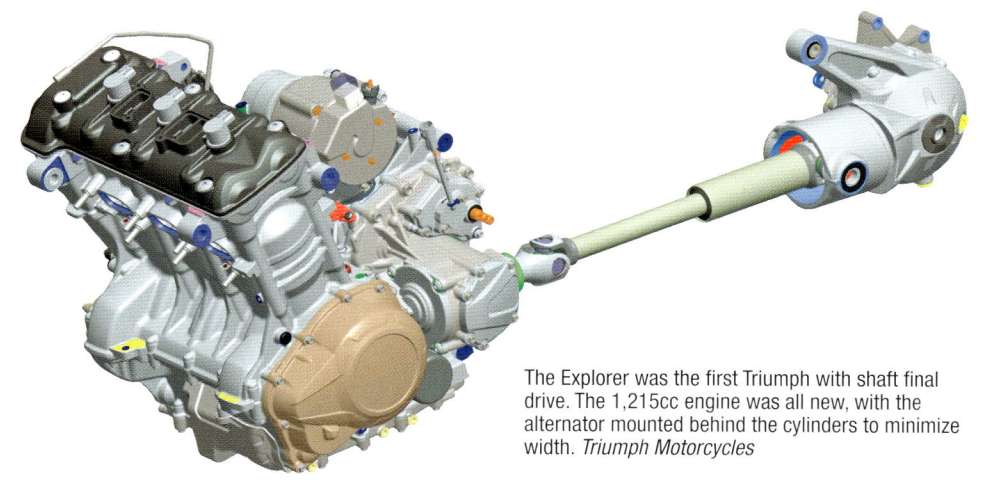

The Explorer was the first Triumph with shaft final drive. The 1,215cc engine was all new, with the alternator mounted behind the cylinders to minimize width. Triumph Motorcycles

With the introduction of the Explorer, the Tiger 1050 SE's profile was more street-focused for 2012. The SE still came standard with ABS and luggage. Triumph Motorcycles

linkage. Suspension was Kayaba, with Nissin brakes, but despite the four-piston calipers the Explorer's considerable weight taxed braking. With a near 34-inch seat height, the Explorer was a large and intimidating motorcycle, more suited to the tarmac than off-road. But Triumph was hoping this wouldn't deter the 6,000 potential buyers in this popular segment, particularly in the United Kingdom and Italy.

But for a Graphite-colored frame, the Tiger 800 and 800XC continued unchanged, but the Tiger 1050 and 1050 SE were updated for 2012 with revised ergonomics, improved suspension, and numerous detail changes. These included lower, black anodized, tapered aluminum handlebars, redesigned suspension, new graphics, and a black finish for many components.

2012 Roadsters
Speed Triple, Speed Triple R, Street Triple, Street Triple R

Although it had gotten the green light back in 2007, with R&D concentrating on other projects, the Speed Triple R didn't appear until 2012, despite the Speed Triple being one of their most successful models. As with the Daytona 675R's development, the Speed Triple R was essentially a stock Speed Triple with Öhlins suspension, lighter wheels, and Brembo Monobloc front brake calipers. The 1,050cc triple's gearbox was redesigned with 10 new gears, new shafts, and selector drum, while the 43mm front fork was an NIX30 with a TTX36 twin-tube rear shock absorber. Equally important in contributing to the improved handling were a pair of forged aluminum PVM wheels, 3.75 pounds lighter than those on the standard Speed Triple. The four-piston Monobloc Brembo brake calipers increased braking performance by 5 percent, and a switchable electronic ABS was optional. Subtle differences in the Speed Triple included R-branded tapered black alloy bars, carbon-fiber radiator, mudguard and tank covers, unique decals, and a red-colored subframe.

This year the Street Triple and Street Triple R featured updated styling, with a pair of headlights similar in style to the Speed Triple, but were otherwise very similar to 2011.

2012 Speed Triple R, Speed Triple, Street Triple R, Street Triple
(Differing from 2011)

Front Fork
43mm Öhlins adjustable (Speed Triple R)
Rear Suspension
Öhlins monoshock adjustable (Speed Triple R)
Front Brake
Brembo 4-piston Monobloc (Speed Triple R)
Wet Weight
466 pounds (212kg) (Speed Triple R)
Colors
Red, white (Speed Triple R); white, red, (Street Triple R); purple (Street Triple)

2012 Classics
Scrambler, Thruxton, Bonneville, Bonneville SE, T100, Steve McQueen Edition, 110th Anniversary Limited Edition

As this was the year of limited-edition Bonneville T100s, the Scrambler, Thruxton, Bonneville, Bonneville SE, and T100 continued much as before. The limited-edition T100s kicked off with the Steve McQueen version, the fascination with legendary actor Steve

ABOVE: A new headlight design and Imperial Purple color distinguished the 2012 Street Triple. Triumph Motorcycles

LEFT: Although it took five years to appear, the Street Triple R was worth the wait. Lighter wheels and Öhlins suspension significantly improved the handling, while the front Brembo brakes were state-of-the-art. Triumph Motorcycles

CHAPTER TEN

2012 Scrambler, Thruxton, Bonneville, Bonneville SE, T100 (Differing from 2011)

Colors Matte Black (Scrambler); gold (Bonneville); red and white, Graphite and black (T100); Matte Khaki Green (McQueen); Brooklands Green and Aluminum Silver (110th)

McQueen living on more than three decades after his death. Triumph joined forces with McQueen's estate to create an officially licensed version of the classic Bonneville T100, paying tribute to the late actor's love of British bikes. Inspired by the Trophy TR6 of the famous stunt scene in *The Great Escape*, it included military-style Matte Khaki Green paint, a stencil-style Triumph decal on the tank, and a McQueen signature on the side covers. In addition to a solo seat and black luggage rack, other features were a skid plate, a small black-bodied headlamp, and black wheel rims, hubs, handlebars, rear springs, and front fender supports. Only 1,100 were produced, each with an individually numbered plaque and certificate of authenticity.

To celebrate their 110th anniversary, Triumph launched a special-edition Bonneville T100 inspired by the 1902 No.1 model. This featured an anniversary crest, directly inspired by the early Triumph tank badges, its nostalgic appeal further enhanced by retro-inspired colors of Brooklands Green and Aluminum Silver, complemented by black fenders as on the original. The Brooklands Green commemorated Triumph's first-ever motorcycle race win at the Surrey-based circuit in 1908, while the tank crest paid tribute to the Queen in her Jubilee year, with three dots added to represent the locations of Triumph factories over the last 110 years: Coventry, Meriden, and Hinckley. Only 1,000 examples were manufactured, each with a numbered plaque and certificate, with 250 allocated to the United States.

2012 Touring and Supersports
Sprint GT, Daytona 675, 675R

This year there was no Sprint ST, and as the Sprint GT was a new 2011 model year release, it continued unchanged for 2012. The Daytona 675 received new graphics, a black belly pan, and embossed clutch and alternator covers. Triumph continued to develop the Daytona 675 for Supersport racing, but in the World Supersport Championship the Italian Suriano Power Team Triumphs were disappointing, their best result a third-place finish by Alex Baldoloni. However, in the British Supersport series the Triumph Daytonas were totally dominant, winning 15 of the 24 races, with Glen Richards repeating his 2008 result by taking the championship on the Smiths Triumph. In the United States Triumph also made a roadracing return with entries in the AMA SportBike Championship. It was a moderately successful season, Jason DiSalvo coming second by the narrowest of margins (0.048 seconds)

BELOW: In Matte Khaki Green with black highlighting, the officially licensed Steve McQueen Limited Edition Bonneville T100 had a utilitarian look about it. The period-style single seat was more generously padded than stock. *Triumph Motorcycles America*

RIGHT: The 110th Anniversary Limited Edition T100's special colors and a crest were designed to accentuate the brand's nostalgic appeal. Only 1,000 were produced. *Triumph Motorcycles*

FAR RIGHT: DiSalvo came within 0.048 second of clinching a repeat Daytona 200 victory for Triumph. *Triumph Motorcycles America*

2012 Sprint GT, Daytona 675, 675R (Differing from 2011)

Colors
Black (Sprint GT); yellow (675)

2010–2024

LEFT: At Bonneville in 2012 Jason DiSalvo set a new world record of 174.276 miles per hour on the Carpenter Racing Rocket III. *Triumph Motorcycles America*

BELOW: Powered by a pair of turbocharged Rocket III engines, Triumph's Castrol Rocket is planned to pass the 400-mile-per-hour barrier. *Triumph Motorcycles America*

in the Daytona 200 on the Team Latus Motors Racing Daytona, and winning at Infineon to end the season fifth overall. At Daytona DiSalvo's 675 carried the same #9 and paint scheme as Gary Nixon's 1967 race winner.

2012 Cruisers
Thunderbird, Thunderbird Storm, Rocket III Roadster, Rocket III Touring, America, Speedmaster

The cruiser Thunderbird, Rocket III, America, and Speedmaster ranges continued largely unchanged. This year also saw the Rocket III Cruiser surprisingly set a world speed record in the Modified (Normally-Aspirated) Fuel Class up to 3,000cc at Bonneville. Prepared by Carpenter Racing in New Jersey and piloted by racer Jason DiSalvo, the 240-horsepower Rocket III achieved an average speed of 174.276 miles per hour.

2012 Thunderbird, Thunderbird Storm, Rocket III Roadster, Rocket III Touring, America, Speedmaster
(Differing from 2011)

Wheelbase
63.6 inches (1,617mm) (America); 63.3 inches (1,606mm) (Speedmaster)

Colors
Red and black (Thunderbird)

2013

As Triumph and speed were synonymous, in 2013 Triumph joined forces with Castrol to relive the period from 1955 to 1970 when, but for a brief 33-day hiatus, Triumph could claim to be "The World's Fastest Motorcycle." Triumph aimed to restore that title by beating Rocky Robinson's 2010 motorcycle land speed record of 376.156 miles per hour. A joint project between aerodynamic engineer Matt Markstaller and engine builder Bob Carpenter, piloted by Jason DiSalvo, the Castrol Rocket chassis was a carbon-Kevlar monocoque, with power provided by two turbocharged Rocket III engines producing more than 1,000 horsepower at 9,000 rpm. Testing was undertaken in 2013 and 2014 for an attempt at 400 miles per hour sometime in the future.

243

CHAPTER TEN

ABOVE: The Explorer XC was similar to the Explorer, but it now had wire-spoked wheels with tubeless tires, foglights, and hand guards. *Triumph Motorcycles*

RIGHT: Triumph gave the venerable Tiger 1050 a new lease on life in the Tiger Sport. Only available in the United Kingdom and Australia, it was designed for those requiring a less expensive, physically smaller, and less technologically intimidating Adventure model. *Triumph Motorcycles*

2013 Adventure
Tiger Explorer, Explorer XC, 1050, Sport, 800, 800XC

Triumph expanded their adventure touring models by fitting the Explorer with wire-spoked wheels and aluminum rims, hand guards, engine guards, and foglights to create the Explorer XC. A wire-wheeled version of the standard Explorer was also available for Europe and Australia, and unlike previous wire-spoked wheels these now accepted tubeless tires. With the Tiger now Hinckley's longest-running model (with 46,182 produced since 1993), an all-new Tiger Sport was built for the UK and Australian markets. The Tiger Sport included 120 new parts, and the engine also produced more power, thanks to a freer-flowing airbox, stainless steel exhaust, and new injection mapping. While the cast-aluminum frame was unchanged, the swingarm was 1.2 inches longer, and a new rear subframe provided a lower seat height and a new handlebar with improved ergonomics. The bodywork was more angular than before, with four reflector headlights replacing the older-style projector type. With its earlier-style instrument panel, apart from switchable ABS, the Tiger Sport was an older style of motorcycle, devoid of modern electronic aids. The other Tiger models were unchanged for 2013, the regular Tiger 1050 in its final year and available only for the United Kingdom.

2013 Supersports
Daytona 675, 675R

When Triumph released the Daytona 675 back in 2006 it was unique, and as the first 675cc three-cylinder Supersport it set a new class benchmark. But seven years on and 29,406 examples later, and despite its 2009 update, the Daytona 675 was no longer the Supersport yardstick. It was a sign of Triumph's maturity as a motorcycle manufacturer that they could create a virtually all-new motorcycle in a little more than three years. While the most obvious difference was the new low-mounted exhaust system with single muffler, almost the entire motorcycle was new, the more powerful engine with a shorter stroke, a new separate closed-deck cylinder block, higher-compression forged pistons, titanium inlet valves, larger (35mm) big-end bearings, lighter crank and alternator, and redesigned gearbox. With 44mm throttle bodies, twin multi-spray injectors per cylinder were a first for Triumph, and rather surprisingly the electronic systems weren't as sophisticated as those on the Adventure bikes, with a conventional cable throttle and no provision for alternative engine maps or traction control.

Beneath the familiar bodywork was an all-new cast-aluminum chassis. Lighter than before, this featured fewer sections (8 rather than 11) and a steeper, 22.9-degree steering head angle. As before, suspension was KYB (formerly Kayaba) on the 675 and Öhlins on the 675R, and a shorter cast-aluminum swingarm provided a reduction in wheelbase. The 675 received new versions of the Nissin radial Monobloc calipers, with the 675R now with Brembo brakes all-round, including the rear single-piston caliper and slightly larger (310mm) and thicker (4.5mm) front discs. Revised ergonomics, with a half-inch lower seat height,

---

2013 Tiger Explorer XC, Sport (Differing from 2012)

Horsepower
123 at 9,400 rpm (Sport)

Wheels
Wire-spoked 19x2.5-inch front and 17x4.0-inch rear (Explorer XC)

Wheelbase
60.6 inches (1,540mm) (Sport); 60.2 inches (1,530mm) (800)

Wet Weight
518 pounds (235kg) (Sport); 586 pounds (267kg) (Explorer XC)

Colors
White, red (Sport); Khaki Green (Explorer XC, 800XC); blue (800)

2010–2024

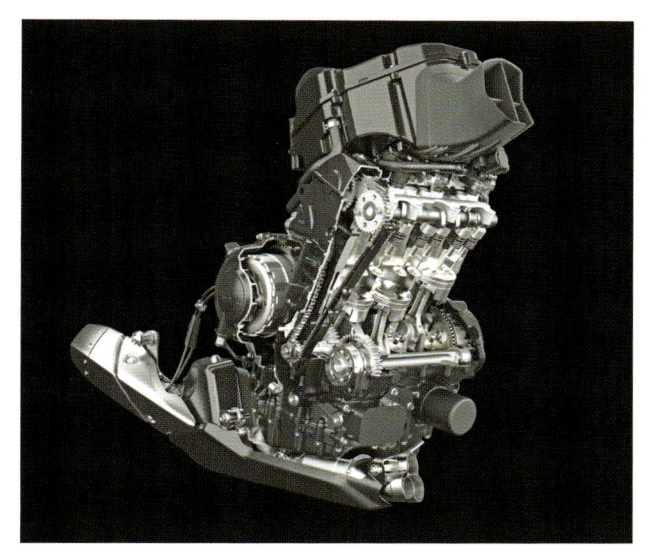

ABOVE: Still with a gear-driven counterbalancer, the Daytona's new short-stroke engine now produced 126 horsepower. New intake ducts provided a 39 percent airflow increase to the larger airbox. *Triumph Motorcycles America*

RIGHT: With a new engine and frame, the Daytona 675 was all new for 2013. *Triumph Motorcycles America*

and lighter wheels aided agility, and again the Daytona 675 set the standard for middleweight sportbikes.

Despite the all-new Daytona, race success was disappointing. Daytonas only won three races in the British Supersport Championship, Billy McConnell finishing third overall on the Smiths entry, while in AMA Pro SportBike Triumph was completely swamped by Yamaha.

2013 Roadsters
Speed Triple, Speed Triple R, Speed Triple SE, Street Triple, Street Triple R

As the Speed Triple and Speed Triple R were new models, they continued unchanged, along with an additional Speed Triple SE. This featured a blue frame, tank stripes, color-matched fly screen, belly pan, and seat cowl, and extensive carbon fiber.

Triumph's most popular model, the Street Triple and Street Triple R, received a significant makeover for 2013. While the 675cc three-cylinder engine was carried over, with new throttle bodies and ECU, the chassis was completely updated, significantly lighter than before and with improved weight distribution. The twin-spar eight-part aluminum frame was new, as was the two-piece high-pressure diecast rear subframe, with a 4mm-lower and adjustable swingarm pivot, lighter dual-sided swingarm, and a steeper steering head angle of 24.1 degrees. Although the seat was taller, at 32.3 inches, it was narrower to accommodate shorter riders. One-pound-lighter wheels contributed to a noticeable decrease in unsprung weight, and with sharper styling and a lighter low-slung exhaust, the new Street Triple set a new standard for middleweight naked sportbikes. The Street Triple R offered a higher-specification adjustable KYB fork with radial

2013 Daytona 675, 675R
(Differing from 2012)

Bore	76mm
Stroke	49.58mm
Horsepower	126 at 12,500 rpm
Compression Ratio	13.1:1
Front Brakes	2x310mm discs
Wheelbase	54.1 inches (1,375mm)
Wet Weight	405 pounds (184kg)
Colors	Black, white, red (675); white (675R)

2013 Speed Triple R, Speed Triple, Speed Triple SE, Street Triple R, Street Triple
(Differing from 2012)

Wet Weight	403 pounds (183kg) (Street Triple, Street Triple R)
Colors	Yellow (Speed Triple); Graphite (Speed Triple SE, Street Triple R); blue (Street Triple)

Lighter and with improved weight distribution, the 2013 Street Triple R was also one of the most impressive all-round middleweight motorcycles available. *Triumph Motorcycles*

245

CHAPTER TEN

Nissin four-piston brake calipers, stiffer springs, a red subframe, and color-coded radiator cowl. Smoother, softer, and more refined than before, and offering an impeccable balance between power and weight, the new Street Triple was simply without equal in this class.

2013 Touring
Trophy, Trophy SE, Sprint GT

While the Sprint GT continued unchanged for the United Kingdom and Australia only, after an absence of ten years Triumph extended into a new sector, the full-dress luxury tourer, with a new Trophy. Based on the Tiger Explorer engine and drivetrain, the Trophy provided a comfortable riding position, standard lockable luggage, and a host of luxury equipment. For their first venture into this category Triumph was amazingly successful. With a redesigned exhaust and airbox, the three-cylinder engine was retuned for a wider spread of torque, while the cylinders were offset 6mm to improve both torque and fuel economy. Weighing only 25 pounds, the twin-spar aluminum frame was Triumph's first employing the engine as a stressed member, while the cast-aluminum single-sided swingarm containing the shaft drive was half an inch longer than the Explorer's. Another first for Triumph was WP suspension, chosen for the electronic control available on the SE. The linked Nissin braking system included non-switchable ABS, while the Trophy's bodywork consisted of 34 separate panels and an electrically adjustable windscreen. The equipment list was vast, from a heated seat to a USB audio system link to a locker containing a credit-card clip for toll roads. The Trophy SE included not only the electronically adjustable suspension, but also a tire pressure monitoring system, and while the Trophy was wide and large, it was also extremely fast and efficient: a mile cruncher par excellence.

Ten years after the demise of the previous four-cylinder Trophy 1200, Triumph released a new full-dress luxury tourer, the three-cylinder shaft-drive Trophy SE. This came with full equipment, including electronically adjustable WP suspension and a quality audio system. Triumph Motorcycles

2013 Cruisers
Rocket III Roadster, Rocket III Touring, America, Speedmaster, Thunderbird, Thunderbird Storm

Although the power remained unchanged, the removal of the torque limiter in the first three gears provided the Rocket III Roadster with even more astonishing acceleration. Many black components replaced chrome to provide a more menacing appearance, including the radiator cowls, fender rails, airbox cover, fork protectors, and headlight bezels. Other new cosmetic touches saw a pair of center stripes on the fenders and gas tank and new seat vinyl and stitching. Largely unchanged since its release in 2007, the Rocket II Touring included a number of accessories as standard equipment, as well as hand-painted coach lines by a former Meriden

2013 Trophy, Trophy SE, Sprint GT
(Differing from 2012)

Type	Inline three-cylinder DOHC, 4 valves per cylinder liquid-cooled (Trophy)
Bore	85mm (Trophy)
Stroke	71.4mm (Trophy)
Capacity	1,215cc (Trophy)
Horsepower	132 at 8,900 rpm (Trophy)
Compression Ratio	11.1 (Trophy)
Fuel Injection	Ride-by-wire Keihin 46mm throttle bodies (Trophy)
Gearbox	Six-speed (Trophy)
Frame	Aluminum twin-spar (Trophy)
Swingarm	Single-sided cast-aluminum with shaft drive (Trophy)
Front Fork	WP 43mm upside-down (Trophy)
Rear Suspension	WP Monoshock adjustable (Trophy)
Brakes	2x320mm disc 4-piston calipers front, 282mm disc rear; linked brakes ABS (Trophy)
Wheels	Alloy 17x3.5-inch front and 17x6.0-inch rear (Trophy)
Tires	120/70ZR17 front and 190/55ZR17 rear (Trophy)
Wheelbase	60.7 inches (1,542mm) (Trophy, SE)
Wet Weight	653 pounds (297kg) (Trophy); 662 pounds (301kg) (Trophy SE)
Colors	Silver, blue (Trophy)

2013 Rocket III Roadster, Rocket III Touring, America, Speedmaster, Thunderbird, Thunderbird Storm (Differing from 2012)

Colors

Black and red or white stripes (Rocket III Roadster); black and red (Rocket III Touring); Caribbean Blue and white (America); blue and black (Speedmaster); blue and black (Thunderbird); Graphite (Thunderbird Storm)

2010–2024

The Rocket III Roadster received more low-down power and a number of black components instead of chrome. These included the mirrors, horn cover, and ABS pulse rings. The new tank badge was chrome with black text, and the paint included twin stripes on the tank and fenders. *Triumph Motorcycles*

2013 Scrambler, Thruxton, Bonneville, T100
(Differing from 2012)

Colors
Graphite (Scrambler); purple and white (Bonneville); red and white (T100); green (Thruxton)

LEFT: Apart from a new color of Matte Graphite, the Scrambler continued unchanged for 2013. *Triumph Motorcycles America*

BELOW: New colors for the T100 of Cranberry Red and Fusion White further emphasized tradition. Factory accessories here include the King and Queen seat, sissy bar, leather panniers, and screen. *Triumph Motorcycles*

craftsman. Along with new colors, Triumph's entry-level cruisers, the Speedmaster and America, both received a more aesthetically pleasing engine-cooling fin finish and more discreet oil-cooler pipes.

2013 Classics
Scrambler, Bonneville, T100, Thruxton
With the regular Bonneville SE dropped, apart from a black-and-red Bonneville SE for Australia, the Classic range was unchanged for 2013. The Australia-only Bonneville SE boasted a red frame and a number of individual features, while the T100 Black was a budget T100 in Jet Black.

2014–2019
By 2014 Triumph had expanded to include nine subsidiaries, with six factories: a CKD (completely knocked down) facility in Brazil alongside the two manufacturing facilities in Hinckley and three in Chonburi, Thailand. Employees now exceeded 1,600 worldwide, and the five-category range included 27 models. Triumph had come a long way since the establishment of the T1 factory in Hinckley, and although the Meriden era was

CHAPTER TEN

2014 Thunderbird, Commander, LT, America, America LT, Speedmaster
(Differing from 2013)

Bore	107.1mm (Commander, LT)
Stroke	94.3mm (Commander, LT)
Capacity	1,699cc (Commander, LT)
Horsepower	93 at 5,400 rpm (Commander, LT)
Wheels	Alloy 5-spoke 17x3.5-inch front and 17x6.0-inch rear (Commander); wire-spoked 16x3.5-inch front and 16x5.5-inch rear (LT)
Tires	140/75 R17 front and 200/50 R17 rear (Commander); 150/80R16 front and 180/70R16 rear (LT)
Wheelbase	65.5 inches (1,665mm) (Commander, LT); 63.3 inches (1,610mm) (America, LT); 62.9 inches (1,600mm) (Speedmaster)
Dry Weight	766 pounds (348kg) (Commander); 836 pounds (380kg) (LT); 594 pounds (270kg) (America LT)
Colors	Red and silver (Thunderbird); red and crimson, black and gray (Commander); blue and white, red and black (LT); red and red (America); Graphite and black (Speedmaster)

still relatively recent, it seemed like another age. Each year Triumph concentrated on expanding and updating one specific category, and for 2014 it was the extremely popular and lucrative cruiser lineup. This year saw two new Thunderbirds, the range now in two capacities with two differing frames, resulting in four distinctly different models.

Triumph was also more aggressive in supporting racing during 2014, both flat-track and road racing. While Rick Martin's AMA Pro Flat-Track season on a Bonneville was disappointing, in roadracing the Triumph Daytona 675 was incredibly successful, and victories at Daytona and the Isle of Man continued Triumph's earlier traditions.

2014 Cruisers
Thunderbird Commander, Thunderbird LT, Thunderbird Storm, Thunderbird, Rocket III Roadster, Touring, America, America LT, Speedmaster

Although the existing Thunderbird and Thunderbird Storm were already targeting the US cruiser market, the line expanded with two new Thunderbird models, the Commander and LT ("Light Touring"). Both shared the Thunderbird Storm's 1,700cc engine, but with slightly less power and a new frame, airbox, and swingarm. The twin backbone open-cradle tubular steel was redesigned with a 1.2-inch-lower seat pan, with new steering geometry (a 29.9-degree steering head angle) to accommodate the wider tires. Beautifully finished, with artist-signed, hand-painted pinstripes, the Commander LT came with 16-inch wire-spoked wheels with whitewall radial tires specially commissioned from Avon. Standard features included a shrouded front fork, quickly detachable windshield, leather saddlebags, and deep fenders. Designed to cover long distances in luxurious comfort, the rider and pillion seats were made from soft, pliant, dual-layer, dual-density foam. The Commander was effectively a naked version of the LT with an additional headlight, round exhaust tailpipe, smaller fenders, and 17-inch alloy wheels. Both were intended to provide a serious

ABOVE: Now with black highlighting, the minimalist hot-rod-style Speedmaster was even more aggressive. *Triumph Motorcycles*

RIGHT: Two new Thunderbirds were released for 2014, both 1,700cc with a new frame. The Thunderbird Commander (left) was a stripped version, with alloy wheels and Triumph's distinctive twin headlights, while the LT (right) provided a classic touring style with a screen, luggage, ocean deep chrome, and special whitewall tires. *Triumph Motorcycles*

248

and better-value alternative to Harley-Davidson in the large-capacity cruiser market, complementing the existing Thunderbird and Thunderbird Storm (which continued unchanged for 2014). The existing Rocket III Roadster and Rocket III Touring were also unchanged this year.

Both of the entry-level cruisers, the America and Speedmaster, were updated for 2014, with an additional America LT. New features for both the America and Speedmaster included machined engine block and cooling fins and new mufflers. The America now included more chrome, plus rider footboards (with heel-and-toe shift), while the Speedmaster highlighting was black (headlight, handlebar, wheels, pedals, and springs). While maintaining the America's chrome finish and polished detail, the America LT incorporated factory-fitted touring accessories, including saddlebags, sissy bar, and windshield.

2014 Supersports
Daytona 675, 675R

As they were both new for 2013, the Daytona 675 and 675R continued unchanged, but the new Daytona finally lived up to its promise in 2014 with major track success. This began in March, with Danny Eslick winning the only endurance race on the AMA Pro Road Racing schedule, the Daytona 200. Providing Triumph their first Daytona victory since Gary Nixon's 1967 win, Eslick took pole position and fought race-long with the Yamahas of Jake Gagne and Jake Lewis, ultimately winning by only 10.975 seconds. Bobby Fong on another Triumph Daytona was fourth. Jason DiSalvo continued the Daytona 675's success with a victory in the AMA Pro SportBike race at Mid-Ohio.

History-making victories weren't confined to the United States. At the Isle of Man in June, Triumph won their first Supersport TT since Bruce Anstey's victory back in 2003. In an extremely tight race, Smiths Triumph rider Gary Johnson took the first Supersport TT at 124.526 miles per hour, defeating Anstey's Honda by only 1.5 seconds after four laps of racing in difficult conditions. Unfortunately a crash in the Superstock race saw Johnson injured and unable to compete in the second Supersport TT.

Championship victories were also more forthcoming during 2014, with Kenny Riedmann winning the Canadian Pro Sport Bike Championship, while in the British Superbike Championship the two Smiths Triumphs of Billy McConnell and Graeme Gowland fought season long for the title, McConnell ultimately coming out the victor.

2014 Adventure
Tiger Explorer, Explorer Wire-Wheel, Explorer XC, Sport, 800, 800XC, 800XC SE

Apart from the introduction of a Tiger 800XC Special Edition, the Tiger range was unchanged for 2014. In

ABOVE: Building on the standard America's more conservative and traditional style, the America LT added a screen, sissy bar, and saddlebags. *Triumph Motorcycles*

LEFT: Danny Eslick's Riders Discount Racing Daytona 200 mile race-winning 675R. Triumph later released a commemorative Daytona 675R Eslick Edition. *Triumph Motorcycles America*

BELOW: After taking pole position for the 73rd running of the Daytona 200, Eslick won a close-fought race by nearly 11 seconds. *Triumph Motorcycles America–Tim White*

CHAPTER TEN

Kenny Riedmann won the Canadian Pro Sport Bike Championship on the Daytona 675R. *Triumph Motorcycles America–John Walker*

With 11 years elapsing since Triumph's last TT win, Gary Johnson on the Smiths Triumph was victorious in the 2014 Supersport TT. *Triumph Motorcycles*

2014 Tiger Explorer, Explorer Wire-Wheel, Explorer XC, Sport, 800, 800XC, 800XC SE
(Differing from 2013)

Colors
White (Explorer, XC); red (Explorer Wire-Wheel); Khaki Green (800XC); black and red (800XC SE)

the United Kingdom the Tiger 800 was the second-best-selling bike above 500cc, so the formula of all-round useable street-oriented adventure bike was very successful. The 2014 800XC Special Edition was ostensibly a cosmetic rendition, finished in Volcanic Black, with a red frame and black highlighting.

2014 Touring
Trophy, Trophy SE, Sprint GT SE

New for Europe only for 2014 was the Sprint GT SE. With standard heated handgrips, touring screen, gel seat, and top box, it offered a lighter, sportier alternative to the full-specification Trophy and Trophy SE, which continued unchanged.

2014 Roadsters
Speed Triple, Speed Triple R, Street Triple, Street Triple R

With more than 70,000 examples produced, by 2014 the Speed Triple was well established as the benchmark street fighter, and to keep it at the forefront it received a few cosmetic updates this year. These included a color-matched fly screen and belly pan, a rubber tank pad, and black exhaust heat shields. The standard Speed Triple included gold decals and pinstriped wheels, with the Speed Triple R receiving red pinstripes and an additional seat cowl. As expected, both the Street Triple and Street Triple R continued unchanged, with Australia receiving a learner-approved 660cc version.

250

2010–2024

2014 Speed Triple, Speed Triple R, Street Triple
(Differing from 2013)

Wet Weight
471 pounds (214kg) (Speed Triple R)
Colors
Blue (2014 Speed Triple); green (2014 Street Triple)

2014 Trophy, Trophy SE, Sprint GT SE
(Differing from 2013)

Colors
Champagne, red (2014 Sprint GT SE)

2014 Classics
Scrambler, Thruxton, Bonneville, T100, T100 Special Edition

The successful formula of building on Triumph's heritage continued during 2014 with two new limited-edition Bonneville T100s and a Lunar Silver and Diablo Red Scrambler replicating Steve McQueen's ISDT bike of 50 years earlier. All Classics received machined cylinder head cooling fins for 2014, the Scrambler with black highlighting of the handlebars, wheel rims and hubs, and rear brake master cylinder. This year the Thruxton had a color-matched fly screen, and it, the Bonneville, and the T100 received black oil-cooler lines and new mufflers, these with a more stirring exhaust note.

ABOVE LEFT: With standard ABS, the Volcanic Black and red Tiger 800XC SE featured a number of black highlights, the black paint incorporating red metallic highlights. *Triumph Motorcycles*

ABOVE RIGHT: While maintaining the successful street-fighter profile, the 2014 Speed Triple now had a standard fly screen and belly pan and came with gold decals and gold wheel highlights. *Triumph Motorcycles*

LEFT: For 2014 the Sprint GT SE with touring screen and top box was available in Europe only. This is the Cranberry Red version. *Triumph Motorcycles*

BELOW: Painted to replicate one of the last Meriden Bonnevilles, the 1982 eight-valve TSS, the Jet Black and Lunar Silver T100 Special Edition included hand-painted gold pinstripes and a brush finish on the engine covers. *Triumph Motorcycles*

251

CHAPTER TEN

Still epitomizing the factory café racer, the 2014 Thruxton now included a standard fly screen, new mufflers, chrome chain guard, and black oil cooler lines. Triumph Motorcycles

Celebrating 50 years since the 1964 ISDT, only 50 examples of this Special Edition T100 were built. This is with Roy Peplow's gold medal–winning factory T100C ISDT Trophy, a 1962 T100C fitted with a 350cc engine. Triumph Motorcycles

2014 Scrambler, Thruxton, Thruxton Ace, Bonneville, T100, T100 Special Edition
(Differing from 2013)

Colors
Blue, silver and red (Scrambler); white and black (Thruxton Ace); silver, blue and white (Bonneville); white and gold, black and red (T100); black and silver (T100 Special Edition)

With a seemingly insatiable nostalgic market, special-edition Bonnevilles had been a successful addition to Triumph's Classic range for many years, so it was no surprise to see another two for 2014. Replicating the black and silver colors of the short-lived 1982 eight-valve TSS, the 2014 T100 Special Edition came with hand-painted gold coach lines, a chrome grab rail, and a brushed finish for the clutch, sprocket, alternator, and cam covers. Another special edition of only 50 examples was produced in celebration of 50 years of the 1964 ISDT. In this event, the same that Steve McQueen (#278) competed in, four British Triumph riders won gold medals: John Giles (#277) and Ken Heanes (#279) on TR6s, and Roy Peplow (#210) and Ray Sayer (#253) on T100s. On sale for one day only, July 12, 2014, at the biennial Triumph Live in Gaydon, Warwickshire, in keeping with the nostalgic theme, this special edition's tank included the signatures of the four riders and, for the first time in 30 years, the original Triumph gas tank logo. Other classic adaptations included a cross-brace handlebar, engine sump plate, center stand, and bespoke seat with ISDT-style integrated toolbag.

The London truckers' Ace Café became a favorite haunt of Britain's bikers in the early 1960s, and to celebrate the Triumph's history with this café icon, a new special edition, the Thruxton Ace, was revealed as part of the café's annual Brighton Burn-Up celebrations. During September 2014, more than 25,000 café racer fans descended on the Brighton seafront to share and celebrate their passion for dropped handlebars, roaring exhausts, and old-style rock 'n' roll. Cobranded with Ace Café logos, the engine was that of the Thruxton, and the Bonneville-based chassis included a 41mm KYB fork, Nissin disc brakes, and wire-spoked wheels with aluminum rims. This Special Edition Thruxton became available late in 2014.

2015

While introducing two new cruisers and three limited-edition Bonnevilles for 2015, Triumph mainly concentrated on developing the Tiger 800 range of adventure bikes. The increasingly wide range of models over six different families resulted in a 4.5 percent increase in sales worldwide, to 54,432 units. Significant marketing developments included the establishment of Triumph Motorcycles India, and although only 1,300 motorcycles were sold in the first year, Triumph saw these emerging markets as pivotal to future growth.

Triumph Motorcycles America's racing presence during 2015 spanned AMA Pro Flat Track, the MotoAmerica AMA/FIM North American Road Racing Series, and the Canadian Superbike Championship. After finishing eighth in AMA Pro Flat Track during

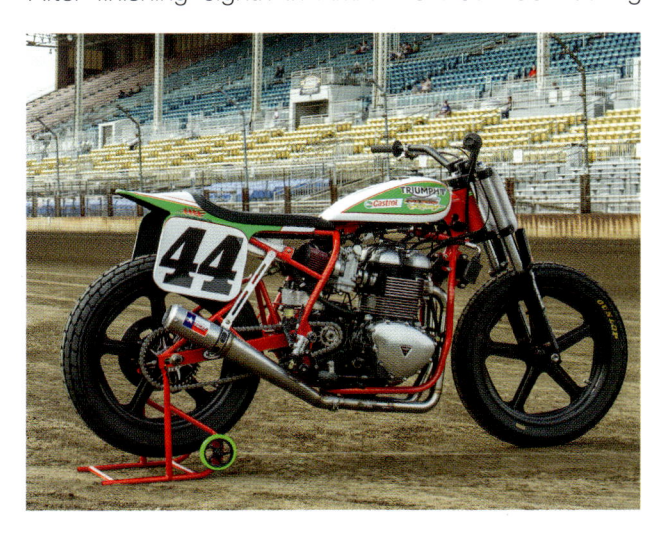

The Team Latus Motors Bonneville flat-track racer was moderately successful in the 2015 AMA Grand National Championship. Brandon Robinson finished sixth overall. Triumph Motorcycles

252

2014, 24-year-old Brandon Robinson joined the Latus Motors Racing team for 2015. Jake Shoemaker rode the Bonneville Performance bike. Robinson rewarded Triumph with its most successful season yet in the highly competitive series, finishing sixth overall in GNC1 (Shoemaker ended finishing 15th overall).

Unfortunately, just as the Bonneville was becoming competitive, Latus withdrew from the Grand National Championship to concentrate on Superstock.

2015 Cruisers
Thunderbird Nightstorm, Thunderbird Commander, Thunderbird LT, Thunderbird Storm, Thunderbird, Rocket III Roadster, Touring, America, America LT, Speedmaster

The existing cruiser range of Thunderbird, Rocket III, America, and Speedmaster continued unchanged, with two new limited-edition models added for 2015. The Thunderbird Nightstorm was a special-edition Thunderbird Storm, featuring custom paint, black exhaust, and a deluxe seat. The basic Thunderbird Storm architecture was unchanged. This included the 98-horsepower 1,699cc 270-degree parallel twin, 47-millimeter Showa front fork and Showa shocks, and twin front-disc brakes with four-piston calipers. The seat height remained a low 700 millimeters. Stripped of chrome, the Nightstorm included special styling touches to emphasize a moodier presence. The custom hand-painted gas tank included ghost flames in a two-tone combination of Phantom Black and Silver Frost. Along with the black painted exhaust were black handlebars, mirrors, and levers. A ribbed, stitched, and embossed seat also set the Nightstorm apart.

Since its launch in 2004, the Rocket III had been continually developed and garnered a significant cult following. Still the largest capacity motorcycle in production, the Rocket III's 10 years were celebrated with the Rocket X Limited Edition. Built as a numbered edition of 500 units, each example had numbered side panels and a billet aluminum badge. Premium custom paint specialists 8-Ball applied the paint scheme, with "grinded" polished metal stripes offset by several black components.

One of two special-edition cruisers released for 2015, the Thunderbird Nightstorm promoted a menacing image. *Triumph Motorcycles*

Celebrating 10 years of the Rocket III, The X Limited Edition featured quality custom paint. *Triumph Motorcycles*

2015 Thunderbird, (Storm, Nightstorm, Commander, LT), Rocket III (Roadster, Touring, X) America (LT), Speedmaster
(Differing from 2014)

Colors
Black/silver (Nightstorm); black, red (America); black/gold, red (Rocket III Touring); Gloss Jet Black (X Limited Edition)

CHAPTER TEN

2015 Daytona 675, 675R
(Differing from 2014)

Colors
Black/graphite, white/blue, red/black (675); white/black, black/silver (675R)

2015 Trophy SE, Sprint GT SE
(Differing from 2014)

Colors
Pacific Blue (Trophy SE); Caspian Blue, Cranberry Red (Sprint GT SE)

2015 Speed 94, Speed 94R, Street Triple Rx Special Edition Speed Triple (R), Street Triple (R)
(Differing from 2014)

Colors
Yellow, black (94, 94R); red (Speed Triple); white, black (Speed Triple R); Aluminum Silver (Street Triple Rx); red (Street Triple); Matte Black (Street Triple R)

2015 Supersports
Daytona 675, 675R

For 2015 Bobby Fong returned to Latus Motors Racing, piloting a Daytona 675R in the Supersport class in the MotoAmerica AMA/FIM Road Racing Series. After an early-season injury, Fong managed several podium finishes to end sixth in the point standings.

The Street Triple Rx Special Edition included styling cues from the Daytona 675R. Triumph Motorcycles

Another celebratory model was the Speed 94R, this time celebrating 21 years of the Speed Triple. Triumph Motorcycles

Kenny Riedmann successfully defended his 2014 Canadian Pro Sport Bike championship on the 675R, winning the final five races. The production Daytona was unchanged but for colors, the 675R continuing with high-spec Öhlins TTX rear suspension and NIX30 upside-down fork, and Brembo front-brake calipers.

2015 Touring
Trophy SE, Sprint GT SE

The Touring range was reduced to only two models for 2015, with the standard Trophy discontinued. As before, the Trophy SE continued with the 1,215cc three-cylinder 134-horsepower engine and the Sprint GT SE with the smaller 1,050-cc 130-horsepower unit.

2015 Roadsters
Speed 94, Speed 94R, Street Triple Rx Special Edition, Speed Triple, Speed Triple R, Street Triple, Street Triple R

The standard Street Triple, Street Triple R, Speed Triple, and Speed Triple R received refreshed decals and logos but were otherwise unchanged. To celebrate 21 years of the Speed Triple, first released in 1994, Triumph released two special editions: the Speed 94 and Speed 94R. Colors and styling cues echoed the original. Enhancements included black exhausts; silencer wraps; color-matched fly screen, seat cowl, and belly pan; Speed Triple logos in the original 1994 typeface; and a commemorative Speed Triple 94 tank-mounted plaque. The higher-spec Speed 94R added black-anodized, fully-adjustable Öhlins NIX30 43mm front forks, Öhlins TTX36 rear shock, and Brembo monobloc radial-brake calipers. With the existing Speed Triple about to be replaced by the Speed 94, it was a fitting swan song.

Also available for 2015 was the Street Triple Rx Special Edition that borrowed cues from the TT- and Daytona 200–winning Daytona 675R Supersports, notably the angular seat unit complemented by a seat cowl, belly pan, and fly screen. The basic engine and chassis were shared with the Street Triple R, with the same 105-horsepower 675cc three-cylinder engine; eight-piece, aluminum twin-spar frame with two-piece die cast rear subframe; 41-millimeter KYB adjustable inverted front fork; and Nissin radial four-piston calipers with switchable ABS and twin 310-millimeter floating discs.

2015 Classics
Bonneville Newchurch, Bonneville Spirit, Bonneville T214, Thruxton Ace, Scrambler, Thruxton, Bonneville, T100, T100 Black

The year 2015 was the end of the line for the 865cc air-cooled engine. To keep it alive one more year Triumph released four special-edition models: three Bonnevilles (the Newchurch, T214 Land Speed Record, and Spirit) and the Thruxton Ace Limited Edition. The Bonneville, T100 Black, and Thruxton were unchanged, with the T100 and Scrambler available with new color options.

The special-edition Newchurch was created to celebrate a special annual Triumph motorcycle event. Every year the small town of Neukirchen in the Austrian mountains played host to the world's biggest Triumph party. The town's name was changed to Newchurch for the weeklong party. The Newchurch model featured a number of blacked-out features and a custom lower seat. Continuing the spirit of customization that had always typified the Bonneville were the T100-based Bonneville Spirit and the T100 Black-based T214. Further celebrating Johnny Allen's 1956 two-way average speed of 214.4 miles per hour, the T214 was limited to 1,000 individually numbered bikes.

The fourth special edition, the Thruxton Ace, celebrated Triumph's historical association with the Ace Café on London's North Circular. The bike was co-branded with Ace Café logos on the side panels, tank, and tail section. While the engine and chassis were shared with the standard Thruxton, the Ace received a sculpted Oxblood custom seat with detachable seat hump, authentic bar-end mirrors, and a handlebar plaque.

2015 Newchurch, Spirit, T214, Thruxton Ace, Scrambler, Thruxton, Bonneville, T100
(Differing from 2014)

Colors
Red/white, blue/white (Newchurch); blue/white (Spirit, T214); white/black (Thruxton Ace); red/silver (Scrambler); silver/black, blue/black (T100)

2015 Adventure
Tiger 800XRt, 800XCA, 800XRx, 800XCx, 800XR, 800XC, Sport, Explorer, Explorer XC

The Tiger 800 expanded to an almost bewildering six-model range for 2015, with three models in two

The T100-based Bonneville Spirit Special Edition featured custom colors, fenders, wheel rims and hubs, and a minimalistic Scrambler black headlight. *Triumph Motorcycles*

BELOW: The Bonneville T214 Special Edition included details echoing Johnny Allen's 1956 *Texas Ceegar*. *Triumph Motorcycles*

BOTTOM: The Thruxton Ace combined the evocative silhouette of a classic Triumph racer with a one-off Pure White and Jet Black livery. *Triumph Motorcycles*

CHAPTER TEN

2015–2017 Tiger 800XRt, 800XCA, 800XRx, 800XCx, 800XR, 800XC, Sport, Explorer, Explorer XC

(Differing from 2014)

Horsepower	94 at 9,250rpm (800)
Fuel Injection	Keihin multipoint sequential electronic 42mm throttle bodies (800)
Front fork	White Power 43mm upside-down adjustable (800XC)
Rear suspension	White Power monoshock adjustable (800XC)
Wheelbase	60.2 inches (1,530mm) XR; 60.8 inches (1544mm) XC
Wet weight	468 pounds (213kg) XR; 475 pounds (216kg) XRx; 480 pounds (218kg) XC; 486 pounds (221kg) XCx
Colors	Orange (XRt); Matte Khaki Green (XCA); white, black, blue (XRx, XCx); white, black (XR, XC); red/black, white/black (Sport); white, Graphite, red (Explorer); white, Graphite, Khaki Green (Explorer XC)

In the more street-oriented Tiger 800 XR series, the XRx was the middle-spec model. All Tigers had an engine protection plate. *Triumph Motorcycles*

The Tiger 800XCA was the highest specification model in the cross-country Tiger 800XC series. Standard equipment included LED fog lights and a rack for optional luggage. *Triumph Motorcycles*

series: the road-biased XR and the off-road XC. In the XR series the higher-spec XRt headed the slightly less equipped 800XRx and more basic XR, while in the XC (cross-country) series, the 800XCA sat above the 800XCx and the more basic XC. The larger Tiger Sport (1,050cc) and Explorer and Explorer XC (1,215cc) were unchanged.

To become Euro4-compliant, the second-generation 800cc three-cylinder engine included a ride-by-wire throttle that also provided traction control. The XRx and XCx included three riding modes and cruise control and all 800cc engines were 17 percent more fuel-efficient. The potential range for the 5-gallon (19-liter) gas tank was now 272 miles. Gearshift components from the Daytona 675 resulted in a more precise gear-change action, allowing the rider to use the engine to its 10,000rpm limit. ABS was standard on all models, but the XRt and XCA received a larger-capacity 650-watt alternator to power the standard heated seats and grips, LED fog lights, and many available accessories.

Chassis updates were mainly confined to the XC series, these now with an adjustable White Power upside-down 43-millimeter front fork and shock absorber. This higher-quality suspension provided 220 millimeters of front-wheel and 215 millimeters of rear-wheel travel. The XR continued with cast-alloy 19- and 17-inch wheels with the XC still carrying a 21- and 17-inch wire-spoked type. The steering geometry also differed between the XR and XC, the XR with a steeper steering-head angle and less trail. Stylistically, all Tiger 800s received a new radiator shroud and fuel tank side panels, while rider comfort was improved, with all versions featuring adjustable seats and handlebars and a low-seat option. The high-spec XRt and XCA came fitted with standard pannier rails for an optional aluminum luggage system.

After five years, the updated Tiger 800cc range provided noticeably improved adventure performance. This was particularly evident with the XC models with White Power suspension, but as it was more suitable for pavement, the XR was arguably the consummate all-around middleweight tourer.

2016

While Triumph originally planned an attempt on the motorcycle land-speed record in August 2015, TT legend Guy Martin's crash at the Ulster Grand Prix early in August saw this postponed to 2016. Developed by Triumph, Hot-Rod Conspiracy, and Carpenter Racing, and powered by two turbocharged Rocket III engines running on methanol, the purpose-built Streamliner produced around 1,000 horsepower. At the Bonneville

Salt Flats on August 8, Martin set the fastest ever speed for a Triumph, at 245.667 miles per hour.

Motorcycle sales increased slightly, to 56,253 for the financial year ending June 30, 2016. This year also saw the best sales ever for Hinckley Triumph in North America—sales reminiscent of the Meriden era and eclipsing 13,000 units for the first time.

2016 Adventure and Touring
Tiger Explorer XRt, XCa, XRx, XCx, XR, XC, Sport, 800XRt, 800XCA, 800XRx, 800XCx, 800XR, 800XC, Sprint GT, Trophy SE

Following on from the 2015 expansion of the Tiger 800 lineup was a new six-model Tiger Explorer range and a new Tiger Sport. The Tiger Explorer range now followed the same format as the Tiger 800, with three models in two series: the road-biased XR, XRx, and XRt, plus the off-road XC, XCx, and XCa. As with the 800, the XRt and XCa were the highest specification versions. With a new exhaust system, the 1,215cc three-cylinder engine produced slightly more power and torque, but the most significant updates included cornering-optimized Traction Control and ABS and TSAS semi-active White Power suspension system on the XCx, XCa, XRx, and XRt. Other new features were an electrically adjustable screen, a Hill Hold Control system, and five Rider Modes. WP adjustable suspension, with ABS and traction control, featured on the XR and XC. Wider wheels allowed for larger section tires. Low-seat XRx and XCx variants were also available.

The Tiger Sport was also updated for 2016. With 104 changes, the 1050cc Triple featured new cylinder heads with a revised combustion chamber, inlet ports and pistons, and a new ride-by-wire throttle system. Along with a 38 percent freer-flowing exhaust, power torque and fuel economy were improved. Rider modes, traction control, cruise control, and ABS were now standard, and slip assist reduced clutch effort by

2016–2017 Tiger Explorer XRt, XCa, XRx, XCx, XR, XC, Sport (Differing from 2015)

Horsepower	137 at 9,300rpm (Explorer); 124 at 9,450rpm (Sport)
Compression ratio	12.25:1 (Sport)
Front fork	White Power 48mm upside-down adjustable (XR, XC)
Rear suspension	White Power monoshock adjustable (XR, XC)
Wheels	Alloy 19×3-inch front and 17×4.5-inch rear (XR) spoked (XC)
Tires	120/70 R19 front and 170/60 R17 rear (XR, XC)
Wheelbase	59.8 inches (1,520mm) XR, XC
Wet weight	538 pounds (244kg) XR; 543 pounds (246kg) XRx, XC; 560 pounds (254kg) XRt; 558 pounds (253kg) XCx; 569 pounds (258kg) XCa; 480 pounds (218kg) Sport
Colors	Matte Black, Aluminum Silver (Sport)

48 percent. New styling included revised engine covers, adventure hand guards, screen aero diffusers, and an adjustable tinted windshield. The basic chassis, with a 43-millimeter Showa fork and monoshock, 320-millimeter Nissin front brakes, and 17-inch wheels, remained. While not exactly a lithe roadster, the Tiger Sport was designed for the motorcyclist who required a single bike capable of multiple duties. The recently introduced six-model Tiger 800 range continued unchanged this year, as did the Sprint GT and Trophy SE, but by 2018 the touring models were discontinued.

The Tiger Explorer was also expanded to include six models for 2016. A high-spec street XRt (left) and basic street XR (right) flank the high spec off-road XCA (center). Triumph Motorcycles

Also updated for 2016 was the Tiger Sport, which received new styling and a more advanced electronics package. Triumph Motorcycles

CHAPTER TEN

One of the most significant new model releases for 2016 was the T120 Bonneville (left). The T120 Black is on the right. *Triumph Motorcycles*

2016–2020 Street Twin, Bonneville T120, T120 Black, Thruxton, Thruxton R

Type	Liquid cooled, 8 valves, SOHC, 270°parallel twin
Bore	84.6mm (Street Twin); 97.6mm (T120, Thruxton)
Stroke	80mm
Capacity	899cc (Street Twin); 1197cc (T120, Thruxton)
Horsepower	54 at 5,900rpm (Street Twin); 79 at 6,550rpm (T120); 96 at 6,750rpm (Thruxton)
Compression ratio	10.55:1 (Street Twin); 10.0:1 (T120); 11.0:1 (Thruxton)
Fuel system	Keihin multipoint sequential electronic fuel injection 38mm throttle
Gearbox	Five-speed (Street Twin); six-speed (T120, Thruxton)
Frame	Tubular steel cradle
Swingarm	Twin-sided tubular steel (Street Twin); Twin-sided box-section steel (T120); Twin-sided aluminum (Thruxton)
Front fork	41mm Kayaba telescopic, 43mm Showa upside-down (Thruxton R)
Rear suspension	Twin Kayaba shock absorbers, Twin Öhlins (Thruxton R)
Brakes	310mm disc two-piston caliper front, 255mm disc rear (Street Twin); twin 310mm disc front (T120, Thruxton)
Wheels	18×2.75-inch front and 17×4.25-inch rear (Street Twin, T120); 17×3.5-inch and 17×5-inch (Thruxton)
Tires	100/90-18 front and 150/70R17 rear (Street Twin, T120); 120/70 ZR17 and 160/60 ZR17 (Thruxton)
Wheelbase	56.7 inches (1,439mm) Street Twin; 56.9 inches (1,445mm) T120; 55.7 inches (1,415mm) Thruxton
Dry weight	436 pounds (198kg) Street Twin; 493 pounds (224kg) T120; 453 pounds (206kg) Thruxton; 447 pounds (203kg) Thruxton R
Colors	Cranberry Red, Aluminum Silver, Matte Black, Jet Black, and Phantom Black (Street Twin); Cranberry Red/silver, black/white, black/red (T120); black, Matte Graphite (T120 Black); black, white, green (Thruxton); red, silver (Thruxton R)

2016 Classics
Street Twin, Bonneville T120, T120 Black, Thruxton, Thruxton R

New from the ground up, the 2016 Bonneville initially overlapped the existing range that still included the Bonneville, T100, 214, Newchurch, Spirit, Scrambler, and Thruxton. Introduced to celebrate the 60th anniversary of Johnny Allen's 1956 world land-speed record, the five new Bonnevilles shared the same parallel-twin platform with virtually nothing carried over from the previous version. Designed and developed in the United Kingdom over a four-year period, all were manufactured at the company's three factories south of Bangkok. They combined new levels of sophistication, power, and reliability with a retro feel in a modern classic.

The first new Bonneville was the entry-level Street Twin, replacing the previous air-cooled T100. Joining soon afterward were the T120, 120 Black, Thruxton, and Thruxton R. The new Bonnevilles accounted for one-third of Triumph's sales in 2016. The Bonneville T120 was awarded *Rider* Magazine's 2016 Motorcycle of the Year.

Now liquid-cooled to meet Euro 4 regulations, the new parallel-twin featured crankpins spaced at 270 degrees to replicate the sound and perfect primary balance of a 90-degree V-twin. Twin balance shafts were positioned fore and aft of the crank, and a single overhead camshaft operated the four valves per cylinder rather than twin overhead camshafts as on the previous engine. The design emphasis was on high torque rather than peak power. The engine was 20 millimeters narrower than before, and while the Street Twin made do with a five-speed transmission, the T120 and Thruxton had a six-speed. On the T120 the two 38-millimeter Keihin throttle bodies were styled to replicate the look of original Amal monobloc carburetors. The T120 exhaust system with twin peashooter

The Thruxton (right) and Thruxton R (left) epitomized the factory café racer style. *Triumph Motorcycles*

mufflers was also visually authentic, but was cleverly designed with twin skins, the outer hiding the inner exhaust that went to a catalytic converter underneath the engine. As the throttle was ride-by-wire, the T120 included alternative throttle maps and two new riding modes as well as traction control and ABS.

With nonadjustable Kayaba suspension and Nissin brakes, the Street Twin and T120 chassis spec was similar to that of previous versions. The Street Twin had a single front disc, the T120 twin discs. While the Street Twin made do with cast alloy wheels, the T120 rolled on traditional wire spokes with chrome rims. The Street Twin was also available with several optional "inspiration" kits to allow personalization.

The second-generation Thruxton and Thruxton R were also eagerly awaited. While celebrating the T120R's success at the Hampshire racing circuit between 1962 and 1969, the new Thruxton was considerably more potent than the earlier edition. While sharing the T120's 1200cc liquid-cooled, eight-valve SOHC parallel-twin in Thruxton guise, it was labeled "High Power" rather than the T120's "High Torque." The performance increase came through a lighter 270-degree crankshaft, higher compression ratio, a larger airbox, revised mapping for the Keihin ECU, and a new exhaust system. This included a pair of twin upswept exhausts with reverse-megaphone cones (chrome on the Thruxton, brushed stainless steel on the Thruxton R). Three riding modes were now available.

Both Thruxtons shared the tubular steel main frame with the T120, but the shorter aluminum swingarm reduced the wheelbase 1.18 inches (30 millimeters). The front wheel was reduced in diameter to 17 inches, and the steering was significantly quicker. Both wire-spoked wheels featured light aluminum rims. While the Thruxton shared the T120's conventional Kayaba suspension, the Thruxton R received a fully adjustable 43-millimeter Showa upside-down big-piston front fork and a pair of remote-reservoir Öhlins shock absorbers. Completing the superior Thruxton R spec was a pair of Brembo monobloc four-piston radial brake calipers. Triumph offered more than 160 accessories for the Thruxton, including a range of Vance & Hines exhausts and three inspiration kits.

Harking back to a time when British twins set the performance standard, the second generation Thruxton and Thruxton R provided the appeal of a 1960s café racer without the heartache, combining stunning retro styling with modern technology, performance, and handling.

2016 Cruisers
Thunderbird, Thunderbird Storm, Thunderbird Nightstorm, Thunderbird Commander, Thunderbird LT, Rocket III Roadster, Touring, America, America LT, Speedmaster

The cruiser range was unchanged for 2016, and as Triumph began to concentrate on Bonneville-based classics, the entire cruiser line dwindled. By 2017, the only cruisers offered were the two Rocket III models, the Thunderbird Storm and Thunderbird LT.

2016 Roadsters and Supersports
Speed Triple S, Speed Triple R, Street Triple, Street Triple R, Street Triple Rx, Daytona 675, 675R

Last revamped for 2011, the Speed Triple had become considered too mild stylistically and was significantly updated for 2016. The three-cylinder engine now incorporated 104 developments, including new pistons and crankshaft; revised combustion chamber, cylinder head, and inlet ports; ride-by-wire throttle (a first for the Speed Triple); 70 percent freer-flowing exhaust; new airbox; and a smaller, more efficient radiator. Power and torque were increased, and five ride

CHAPTER TEN

Restyled for 2016, the Speed Triple received new styling with a lower and more aggressive front aspect. The Speed Triple R here had higher-specification Öhlins suspension. Triumph Motorcycles

2016–2017 Speed Triple S, Speed Triple R
(Differing from 2015)

Horsepower
138 at 9,500rpm
Dry weight
422 pounds (192kg)
Colors
Red, black (S), white, Graphite (R)

modes and three ABS settings were available. The Keihin ECU was the same as the Daytona 675.

While both the Speed Triple S and Speed Triple R now featured Brembo radial monobloc brake calipers, the R retained the higher-spec Öhlins NIX30 upside-down front fork and TTX36 RSU shock absorber. The Speed Triple R also featured a machined handlebar clamp and risers, along with machined swingarm spindle and rear wheel covers. New aggressive styling extended to a narrower seat, a more sculpted tank, and a pair of low DRL headlights. The front end was noticeably lower, with the highest point now the gas cap. According to chief engineer Stuart Wood, "A lot of effort went into combining elegance with the Speed Triple's trademark aggressive look." Lighter, more powerful, and more refined, the new Speed Triple continued as the class-leading streetfighter with attitude.

The Street Triple, Street Triple R, Street Triple Rx, and Daytona 675 and 675R were unchanged for 2016. To use up 2015 stock, Triumph Motorcycles America offered a Triumph Racing Partners Purchase Program and Performance Bonus for 2015 Daytonas. This was later extended to include 2016 versions.

2017

For the financial year ending June 30, 2017, Triumph sold 63,404 motorcycles, an 11.1 percent increase year to year. Although key components such as crankshafts and camshafts were still produced at Hinckley, 80 percent of all Triumphs were now built in the three factories in Thailand. Located in Chonburi, 50 miles southeast of Bangkok, these plants produced around 1,800 motorcycles a week. In August 2017, Triumph announced a nonequity global partnership with India's third largest manufacturer, Bajaj Auto. The intent was to develop a range of mid-capacity motorcycles focusing on the collective strengths of both companies and to allow Triumph entry into the growing Indian market.

2017 Classics
Bonneville Bobber

Five new models joined the Bonneville lineup for 2017, and this now consisted of ten models on two distinct platforms. The most significant was the Bonneville Bobber. Bobbers appeared after World War II and were ostensibly stripped custom motorcycles created by returning GIs. Eschewing superfluous details, many were based on Triumph Tigers and Speed Twins, customizers generally adding a tiny tank, single seat with rigid rear end, slash-cut exhausts, and fatter tires for dirt tracks. Triumph's Bobber took these style considerations to heart, cleverly adding modern features. While the Bobber's spirit harked back 70 years, in terms of function it was totally up-to-date.

The T120's High Torque parallel twin was retuned to produce slightly less power and torque, but the most significant feature was a new frame and linkage monoshock rear suspension styled to replicate a rigid rear end. While the rear brake was the usual disc, the rear wheel included a drum brake–inspired hub. The rear wheel was a fat 16 inches and the wheelbase stretched to nearly 60 inches. A unique feature was an adjustable riding position with a floating single seat adjustable fore and aft on the rear of the beam frame. The Bobber theme extended to a wide flat handlebar and minimal bodywork, lighting, and instrumentation. Sales were two-and-half-times better than projected and Triumph initially struggled to meet demand. As the fastest selling model in Triumph's history, the Bobber was a significant contributor to the 15 percent increase in 2017 American sales. It continued through 2019 unchanged.

2017–2020 Bonneville Bobber (Differing from the T120)

Horsepower
76 at 6,100rpm
Swingarm
Twin-sided tubular steel
Rear suspension
Kayaba monoshock
Front brake
Single 310mm disc Nissin 2 piston caliper
Wheels
19×2.50-inch front and 16×3.50-inch rear
Tires
100/90-19 front and 150/80R16 rear
Wheelbase
59.4 inches (1,510mm)
Dry weight
502 pounds (228kg)
Colors
Ironstone, red, green/silver, black

2017–2018 Street Scrambler (Differing from the Street Twin)

Front wheel
19×2.5-inch
Tires
100/90×19
Wheelbase
56.9 inches (1446mm)
Dry weight
469 pounds (213kg)
Colors
Black, Matte Khaki Green, red/silver

Street Scrambler

Built on a legend that began with the TR5 Trophy and culminated with the TR6C Trophy and T120C, the Scrambler was one of Triumph's more important nostalgic models. Since its introduction in 2006, sales numbered more than 17,000. But as the earlier 865cc engine was discontinued and the Scrambler was largely unchanged, by 2017 it was time for an update.

Based on the Street Twin platform (same basic frame, Kayaba 41-millimeter front fork, and Nissin brakes), but for a high-rise stacked brushed stainless-steel crossover exhaust on the right, the 900cc five-speed Street Scrambler engine was unchanged. The uprated Kayaba rear shock absorbers were longer and the wire front wheel 19 inches. Compared to the outgoing Scrambler the new Street Scrambler had a shorter wheelbase, sharper steering, and a lower seat. The high-rise handlebar, single round instrument, and

ABOVE: Capturing the style of an earlier era, the Bobber was extremely successful. *Triumph Motorcycles*

BELOW: Although more suited to pavement, the Street Scrambler was still capable off-road. *Triumph Motorcycles*

CHAPTER TEN

large headlight all evoked the 1960s. While off-road prowess was never a priority, the Street Scrambler's improved suspension was surprisingly capable off-road. But its natural habitat was city streets. The Street Scrambler was unchanged before a new model replaced it for 2019.

T100, T100 Black, Street Cup

Based on the 900cc Street Twin, the T100 and T100 Black were ostensibly downsized versions of the T120 and T120 Black. The T100 shared the T120's classic 1960s Bonneville styling with a traditional peashooter exhaust, wire wheels with chrome rims, and a retro-styled gas tank with classic rubber kneepads. The frame was similar to the Street Twin, but with slightly slower steering and a longer wheelbase. Also similar were the Kayaba suspension and Nissin brakes. As with the T120, the handlebars were wider and the instrument layout included twin analog dials with LCD displays. The T100 Black featured black wheel rims, a twin skin peashooter exhaust in a matte black finish, and blacked out engine covers. Compared to the previous T100, the new model was smoother and the 18-inch front wheel and upgraded suspension provided better handling.

ABOVE: With its café racer style the Street Cup was billed as a contemporary urban custom street racer. *Triumph Motorcycles*

RIGHT: While the T100's style was similar to the T120, the 900cc engine and chassis were based on the Street Twin. The front brake was a single disc. *Triumph Motorcycles*

> ## 2017–2020 T100, T100 Black, Street Cup
> ### (Differing from Street Twin)
>
> **Wheelbase**
> 57.1 inches (1450mm) T100; 56.5 inches (1435mm) Street Cup
> **Dry weight**
> 469 pounds (213kg) T100; (200kg) Street Cup
> **Colors**
> Blue/orange and white, black (T100); Jet/Matte Black (T100 Black); yellow/silver, black/silver (Street Cup)
> **Colors**
> Ironstone, red, green/silver, black

Also based on the Street Twin, the Street Cup was worlds apart from the T100 in style and application. Blending components from several examples of the modern classic family, it delivered café racer function to match the styling. The 900cc engine with black cases was shared with the T100 Black, but the Street Cup received a period brushed aluminum exhaust system. To quicken steering, the Street Cup included longer rear shock absorbers, and the café racer style dictated Thruxton-style rear-sets and Ace-style handlebars. The cast-alloy 18- and 17-inch wheels were a lighter split-spoke type, and the single front disc brake included a Nissin twin-piston sliding axial front caliper and floating 310-millimeter disc. An Alcantara seat with color-matched seat cowl and small fly-screen completed the sporting style. All three T100s continued through 2019 unchanged.

2017 Roadsters
Street Triple S, Street Triple R, Street Triple RS

Since its release for 2008, the stripped-down Street Triple had become Triumph's biggest selling model, with over 50,000 sales. Much hinged on the new version's success.

The project commenced in 2014 when chief engineer Stuart Wood and his team decided to bore and stroke the existing engine to 78×53.4 millimeters, the largest internal dimensions possible with the existing castings. With only 10 percent carryover of parts from the 675cc triple, the new engine employed 80 new components, including crankshaft, pistons, Nikasil plated aluminum barrels, and revised balancer shaft. Also new was a lighter freer-flowing exhaust, revised air box, gearbox (with shorter first and second ratios), and slipper clutch (on the R and RS). A new ride-by-wire throttle allowed for five riding modes. As the power characteristics differed for each version, they also had bespoke camshafts. For the basic Street Triple S, the power increased 6.6 percent, for the Street Triple RS, 16 percent.

While the frame was carried over from the previous model, a new gullwing rear swingarm, with increased longitudinal torsional stiffness combined with an engineered reduction in lateral stiffness, improved high-speed stability. This was accompanied by a general higher-level specification of suspension, brakes, and tires. The Street Triple S had a Showa upside-down front fork and preload-adjustable piggyback reservoir rear monoshock while the R included a separate-function front fork and damping-adjustable Showa rear shock. The top-of-the-line RS had the highest spec Showa big-piston front fork and a premium Öhlins STX40 piggyback reservoir monoshock. The braking specification also varied between the three version, the S with Nissin two-piston sliding calipers on the front and a Brembo single-piston sliding caliper on the rear. Replacing the Nissin calipers on the R were Brembo M4.32 four-piston radial monobloc calipers and on the RS range-topping Brembo M50 four-piston radial monobloc calipers. The switchable ABS package was from Continental.

The "nose-down" attitude of the latest generation Speed Triples inspired the styling—the bodywork also included a new fly screen, radiator cowls, and integrated air intake. The R and RS featured a full-color 5-inch TFT dash and new switchgear incorporating a five-way joystick; the RS included a color-coded belly pan and pillion seat cover. The Street Triple R was also available as a lower-ride-height version (Street Triple R LRH), with unique suspension and seating setup.

Weighing an impressive 4.4 pounds (2kg) less than its predecessor, the new Street Triple was not only more powerful, but more competent and user-friendly. It entered a hotly contested 750–900cc market, but less weight and useful power kept the Street Triple at the head of the pack. The engine growing to 765cc also hastened the demise of the Daytona Supersports, which did not receive the larger engine.

During 2017 Triumph Motorcycles signed a three-year contract as the exclusive engine supplier to the FIM Moto2 World Championship from the 2019 season onward. The race-tuned 765cc engine was based on the powerplant for the 2017 Street Triple RS, but with a modified cylinder head with a 14:1 compression ratio, revised inlet and exhaust ports, and titanium valves and stiffer valve springs. A lighter, low-output race-kit alternator reduced inertia, and other modifications included a taller first-gear ratio, lower second gear, and a tunable slipper clutch. Magneti Marelli supplied the race ECU. Narrower engine covers reduced width, and a different cast-alloy sump allowed a more tucked-in run for the Arrow 3-1 titanium exhaust

2017–2019 Street Triple S, Street Triple R, Street Triple RS

Horsepower	137 at 9,300rpm (Explorer); 124 at 9,450rpm (Sport)
Type	Inline three-cylinder DOHC, four valves per cylinder, liquid-cooled
Bore	78mm
Stroke	53.4mm
Capacity	765cc
Horsepower	111 at 11,250rpm (S); 116 at 12,000rpm (R); 121 at 11,700rpm (RS)
Compression ratio	12.65:1
Fuel system	Multipoint sequential electronic fuel injection with SAI; electronic throttle control
Gearbox	Six-speed
Frame	Aluminum-beam twin spar with two-piece high-pressure diecast rear section
Swingarm	Twin-sided, cast aluminum
Front fork	Showa 41mm upside-down adjustable
Rear suspension	Showa monoshock adjustable (Öhlins RS)
Brakes	2×310mm disc four-piston radial calipers front; 220mm Brembo disc rear
Wheels	17×3.5-inch front and 17×5.5-inch rear
Tires	120/70 ZR17 front and 180/55 ZR17 rear
Wheelbase	55.5 inches (1,410mm)
Dry weight	365 pounds (166kg)
Colors	Red, black (S); black, silver, white (R); silver, black (RS)

During 2017, Triumph announced it would supply 765cc three-cylinder engines for the Moto2 World Championship. *Triumph Motorcycles*

The Street Triple grew to 765cc for 2017 with the high-spec RS now at the top of the range. *Triumph Motorcycles*

CHAPTER TEN

Triumph's Moto2 prototype was based on a production Daytona, with off-the-shelf K-tech suspension. The minimum weight for 2019 was reduced to 341.7 pounds (155kg). *Triumph Motorcycles*

With its fat front tire and beefier fork, the Bobber Black presented a more muscular image than the standard Bobber. The front brake was a Brembo twin disc. *Triumph Motorcycles*

headers. While retaining the offset cam chain drive, the rev ceiling was raised to 14,000rpm. The Moto2 engine produced over 135 horsepower at 14,000rpm, and the contract specified each sealed engine was allowed a 3 percent variation from 130 horsepower. With three-cylinder racers absent from the MotoGP grid for nearly 50 years, fans eagerly anticipated the prospect of 30 howling triples on the track.

2018

By 2018, Triumph's lineup numbered more than 30 models. With the decline of the Cruiser, Touring, and Supersports lines, the emphasis shifted to expanding and consolidating the more popular Classic, Adventure, and Roadster ranges. The 900cc and 1200cc SOHC parallel twin now powered the entire Classic range, in 2018 growing to include the Bobber Black and Speedmaster. The T120 became available with an "Escape" inspiration kit that included black waxed cotton and leather panniers, windshield, comfort "King & Queen" seat, chrome engine bars, pillion backrest, and polished detailing.

2018 Classics
Bonneville Bobber Black

Building on the Bobber's success, the Bonneville Bobber Black not only featured more muscular and minimalist styling but higher specification components. Along with the black finish and detailing, a fat 16-inch front wheel, chunkier 47-millimeter Showa cartridge-type front fork, and twin Brembo front disc brakes set the Bobber Black apart from the Bobber. Other higher-spec features included an LED headlight and single-button cruise control. Also setting the Bobber Black apart was a pair of black painted slash-cut exhausts. Like the Bobber, the Bobber Black was designed with customization in mind; more than 120 custom accessories were available.

2018–2020 Bonneville Bobber Black
(Differing from the Bobber)

Front fork
47mm Showa telescopic
Front brake
Twin 310mm discs, Brembo two-piston caliper
Front wheel
16×2.50-inch
Front tire
130/90B16
Dry weight
524 pounds (237.5kg)
Colors
Black, Matte Black

Bonneville Speedmaster

Largely unchanged since 2011, the Speedmaster returned to the Bonneville classic range for 2018. Derived from the Bobber Black, the Speedmaster shared the higher-torque 1,200cc 76-horsepower engine, twin Brembo front disc brakes, and hardtail-looking cage-type swingarm. The front fork was a 41-millimeter Kayaba and the steering geometry slightly quicker. The Speedmaster's style also combined T120 features, notably the 12-liter sculpted Bonneville gas tank, polished engine covers, and chrome-plated exhausts. A classic nacelle headlight, swept handlebar, forward pegs, fixed rear fender, and 16-inch wire wheels created a specific custom look with a laidback riding style. Modern features extended to a full LED headlight, two riding modes (Road and Rain), latest generation ABS, switchable traction control, a torque-assist clutch, and single-button cruise control. Creating a niche in the Bonneville range, the Speedmaster offered classic cruiser style with modern performance and reliability.

2018–2020 Bonneville Speedmaster
(Differing from the Bobber Black)

Front fork
41mm Kayaba telescopic
Dry weight
541 pounds (245.5kg)
Colors
Black, red, white/black

2018 Adventure
Tiger 1200, 800 XR, XC

Increased competition in the continually evolving adventure/touring market had seen the Tiger Explorer and 800 eclipsed by new offerings from BMW and KTM. Although recently updated (the Tiger Explorer in 2016, the Tiger 800 in 2015), Triumph introduced a heavily updated Adventure range for 2018. The 1,050cc Tiger Sport continued unchanged. The Tiger range still consisted of multiple variants across two platforms, but the range was simplified slightly to include three XR (XR, XRx, XRt) and two XC (XCx, XCA) versions, in both 1,200 and 800cc. The XR was still road-based, with cast-alloy wheels, and the off-road XC variants were fitted with wire-spoked wheels and off-road suspension.

With more than 26,800 sales over its six-year production run, it made sense for Triumph to spend four years updating their top-end triple. A name change saw "1200" replacing the Explorer nomenclature and the third-generation Tiger 1200 received 100 updates. This resulted in 22 pounds (10kg) of weight savings across the engine, chassis, and exhaust, while the increased power was now available in a lower rev range. With a lighter flywheel, the crank assembly alone was now 6.6 pounds (3kg) lighter. Along with sharper looking bodywork and new graphics, wheels, and colors was a new-generation electronics package, including a six-axis inertial measurement unit (IMU), Continental integrated braking system, optimized cornering ABS and traction control, hill hold, ride-by-wire throttle, shift assist, and up to six riding modes. Other high-spec components included improved Brembo monobloc four-piston brake calipers and adjustable WP Semi-Active suspension. A Low Ride Height XRx model variant provided an 815- to 835-millimeter seat (20 millimeters lower than standard).

As the most powerful shaft-driven adventure motorcycle on the market, with its more advanced electronics package the Tiger 1200 now covered all bases. Regardless of the interpretation of adventure, the Tiger 1200 XR and XC could do it all comfortably and efficiently.

Joining the updated Tiger 1200 was a new Tiger 800. This included over 200 upgrades, the engine

2018–2021 Tiger 1200 XR, XC, 800 XR, XC
(Differing from the 2017)

Horsepower
139 at 9,350rpm (1200)
Wet weight
536 pounds (243kg) 1200 XR; 548 pounds (248kg) 1200 XC; 445 pounds (202kg) 800XR; 459 pounds (208kg) 800XC
Colors
Red, white, Marine (1200); white, Khaki Green (800)

Combining features of the Bobber and the T120, the Speedmaster built on Triumph's classic custom heritage. Triumph Motorcycles

benefiting from closer tolerances and reduced gear backlash. The exhaust system was lighter and a lower first gear aided off-road riding. Other updates included color TFT instrumentation, Brembo twin-piston sliding calipers (except the XR), and new bodywork. Ergonomic development saw the handlebars moved back 10 millimeters and a new five-position adjustable windscreen with aero diffusers. As the weight was significantly reduced over the previous version, the new Tiger 800 was more capable, especially off-road.

2018 Roadsters
Speed Triple S, Speed Triple RS

Given the 765cc Street Triple's all-round performance, it arguably threatened the existence of the 1,050cc Speed Triple. But with a timeline extending nearly 25 years, Triumph's large-displacement Speed Triple was historically important. Thus, only two years since its last major update, another significantly revised

CHAPTER TEN

ABOVE: Updated significantly for 2018, the Tiger 1200 received more power and new styling. The 1200 XCA (left) and 1200 XRt (right) represented the highest specification of the respective cross-country and street ranges. *Triumph Motorcycles*

RIGHT: The Tiger 800 XCA was the most suited in the Adventure range to off-road duties. *Triumph Motorcycles*

FAR RIGHT: The Speed Triple RS was the highest specification and most powerful Speed Triple in the 24-year history of the model. *Triumph Motorcycles*

2018–2020 Speed Triple S, Speed Triple RS
(Differing from 2017)

Horsepower
148 at 10,500rpm
Compression ratio
12.92:1
Wheelbase
56.9 inches (1,445mm)
Dry weight
422 pounds (188kg) RS
Colors
White, black

Speed Triple appeared. Still offered in two versions, the highest specification Speed Triple yet was the RS, replacing the previous R.

Engine development saw 105 new components, updates including a lighter crank gear, Nikasil aluminum cylinder liners, and a smaller starter motor and alternator. New pistons and cylinder head provided a higher compression ratio, while a new sump lowered the engine oil level, reducing drag. The rerouted oil system now ran internally, allowing the removal of external oil pipes. The result was a 1,000rpm increase in redline with the power up 7 percent. While the chassis was carried over from the previous generation, a new electronics package included 5-inch color TFT instrumentation, five riding modes, and cruise control. Along with Öhlins suspension, the RS featured cornering ABS and traction control, Arrow sports mufflers, a carbon-fiber front fender, and color-coded belly pan. The RS was slightly lighter than before, and an adjustable front master cylinder lever ratio now made it more suitable for track use. With more power and equipment than ever, the Speed Triple maintained its tag as the ultimate "hooligan" bike.

2019

After offering a bewildering number of models, Triumph's range was significantly rationalized for 2019. The five-model Roadster and Adventure line-ups were unchanged from 2018, and the only cruiser was the Rocket III Roadster, primarily for the US market. In some other markets, notably Australia, the Thunderbird (Storm, Commander, LT) was still available, but by now this line was ostensibly discontinued.

The Bonneville platform now formed the backbone of Triumph's sales strategy and encompassed 17 models (including the carryover superseded Street Twin and Street Scrambler). New models included the Speed Twin, a pair of 1,200cc Scramblers, updated Street Twin and Street Scramblers, two limited-edition Bonneville T120s, and the first of the new Triumph Factory Custom line. With Triumph set to power the Moto2 grid, they announced two-time World Superbike Champion James Toseland as racing ambassador.

Speed Twin

Heading the 2019 range of modern classics was another resurrection of one of Triumph's great names from the past. Ostensibly a roadster based on the higher performance Thruxton, a new magnesium cam cover and lighter engine covers reduced engine weight by 5.5 pounds (2.5kg). The frame and twin-sided aluminum swingarm was also derived from the Thruxton R, while the suspension featured a Thruxton-style 41-millimeter Kayaba cartridge fork and preload adjustable shocks. Twin Brembo four-piston axial calipers gripped twin front discs, and the wheels were 17-inch cast aluminum. Modern rider aids included ABS, switchable traction control, and three riding modes. The modern stripped style extended to classic twin instruments. Premium detailing encompassed brushed-aluminum fenders, anodized forged-aluminum headlight brackets, and Monza-style locking fuel cap. While obviously not as groundbreaking as its iconic twentieth-century namesake, combining T120 ergonomics with the Thruxton's power and torque resulted in an extremely appealing modern roadster.

Scrambler 1200XC, 1200XE

Hoping to emulate the success of their mid-1960s desert racers, Triumph released two Bonneville-based Scramblers for 2019: the base 1200XC and the XE. Powering both was a new version of the Bonneville 1,200cc twin, tuned midway between the T120 and Thruxton/Speed Twin. This engine bolted into the usual tubular steel frame but with a new aluminum swingarm and surprisingly serious suspension and brakes. A pair of Öhlins piggyback rear shocks combined with an upside-down Showa front fork and Brembo M50 monobloc brakes.

The XE received a 47-millimeter fork with 250 millimeters of travel, while the XC made do with a 45-millimeter fork providing 200 millimeters of travel. The XE swingarm was also 32 millimeters longer but both versions featured side-laced aluminum-rimmed wire-spoked wheels with tubeless tires. The steering geometry also differed slightly—the XE with more rake and trail to suit off-road use. Both models had multiple riding modes but the XE received an Off-Road Pro mode that disabled the ABS and traction control. The XE also gained cornering ABS and cornering traction control while both received a new TFT instrument pack, keyless ignition, and cruise control. Dozens of accessories included two inspiration kits.

So convinced was Triumph in the Scrambler's ability as a serious off-roader, they entered desert racer and stunt rider Ernie Vigil in the 2018 Baja 1000. Minor mods included a race-tuned engine, sump guard, and engine bars, but the suspension was standard XE spec. Unfortunately, Virgil sustained a serious ankle injury in training, forcing Triumph to withdraw from the event.

2019–2020 Speed Twin
(Differing from the Thruxton)

Wheelbase
56.3 inches (1,430mm)
Dry weight
432 pounds (196kg)
Colors
Silver/gray, red/gray, black

Resurrecting another great name from the past, the Speed Twin combined the Street Twin's style with Thruxton power. *Triumph Motorcycles*

Inspired by Californian desert racers of the 1960s, the 1200XC was a base model but still received high-quality suspension and brakes. *Triumph Motorcycles*

The higher spec 1200XE had two-tone paint, gold colored front forks, and hand guards. *Triumph Motorcycles*

CHAPTER TEN

2019–2023 Scrambler 1200 XC, XE
(Differing from the T120)

Horsepower	89 at 7,400rpm
Compression ratio	11.0:1
Swingarm	Twin-sided aluminum
Front fork	45mm Showa upside-down (XC); 47mm (XE)
Rear suspension	Twin Öhlins
Front brakes	Twin 320mm discs
Wheels	21×2.15-inch front and 17×4.25-inch rear
Tires	90/90-21 front and 150/70R17 rear
Wheelbase	60.2 inches (1,530mm) XC; 61.8 inches (1,570mm) XE
Dry weight	452 pounds (205kg) XC; 456 pounds (207kg) XE
Colors	Black, green (XC); white/green, blue/black (XE)

After recovering, Vigil returned to off-road endurance racing with a Scrambler 1200XE at the NORRA Mexican 1000 in April 2019.

Street Twin, Street Scrambler

Since its launch three years earlier, the entry-level Street Twin had notched nearly 18,000 sales, becoming Triumph's single best-selling model. The stripped-down look also made it a suitable starting point for custom builds. For 2019 the Street Twin was boosted with a significant power increase, higher equipment specification, and more technology.

As the previous version struggled to maintain a decent power output with Euro 4 compliance, Triumph's engineers improved the power, increasing it 18 percent and nearly matching the previous Euro 3 865cc Bonneville. Lighter engine internals and revised cam timing also provided a broader powerband with a 500rpm higher redline than before. Road and Rain riding modes were standard.

Chassis updates included a higher specification 41-millimeter Kayaba cartridge front fork. Rather than add a second front disc, Triumph improved braking by retiring the Nissin twin-piston caliper in favor of a Brembo four-piston. New ten-spoke cast-alloy wheels, a more comfortable seat, and improved ergonomics completed the package.

Although barely two years old, a new Street Twin and Street Scrambler was released for 2019. Triumph Motorcycles

2019–2020 Street Twin, Street Scrambler
(Differing from 2018)

Horsepower
64 at 7,500rpm
Compression ratio
11.0:1
Front brake
Brembo four-piston caliper
Wheelbase
55.7 inches (1,415mm) Street Twin; 56.9 inches (1,445mm) Street Scrambler
Dry weight
437 pounds (198kg)
Colors
Ironstone, red, black (Street Twin); white, red, Khaki Green (Street Scrambler)

A garage full of custom inspired parts included Vance & Hines silencers, Fox rear shocks, bench seats, fly screens, rear fender removal kits, and compact bullet LED turn signals. Two inspiration kits included the stripped-back Urban Ride kit and the contemporary urban custom style Café Custom kit. Introduced alongside the Street Twin for 2019 was an updated Street Scrambler. Replacing the 2017 version, this shared the improvements in power, spec, and technology with the Street Twin. Other new features included new instruments, side cover graphics, and a wider-spaced front fork. Alongside the wide range of accessories was a custom-cool, stripped-back Urban Tracker inspiration kit, including panniers, a Vance & Hines muffler, and high front fender.

Bonneville T120 Ace, Bonneville T120 Diamond Edition

Two one-off limited-edition Bonnevilles were also available for 2019: the T120 Ace and T120 Diamond Edition. Based on the T120 Black, the T120 Ace followed the earlier Thruxton Ace by celebrating the café racer culture and London's legendary Ace Café, the spiritual home of classic British motorcycling. Commemorating the Bonneville's 60th anniversary, the T120 Diamond Edition was based on the T120.

Both were ostensibly cosmetic renditions, the T120 Ace providing a blacked-out street racer theme,

2019 Bonneville T120 Ace, T120 Diamond Edition
(Differing from the T120 and T120 Black)

Colors
Gray (T120 Ace); white/silver (T120 Diamond)

with the chrome-adorned T120 Diamond Edition a more classic focus. 1,400 T120 Aces were available worldwide, each receiving a numbered certificate signed by Triumph CEO Nick Bloor and Ace Café managing director Mark Wilsmore. With its unique paint and premium detailing the Diamond Edition also came with a signed certificate but was limited to 900 units.

Thruxton TFC (Triumph Factory Custom)

Expanding the emphasis on custom models was the first of an exclusive new high-end series, the Thruxton Triumph Factory Custom. Embodying the pinnacle of British custom design, it offered enhanced performance, technology, engineering, specification, and finish. The genesis came from the 2014 Triumph TFC Bobber and TFC Scrambler. Alongside the Thruxton TFC was a concept Rocket TFC.

A host of engine upgrades saw the power increasing 9 horsepower over standard. These changes included lighter engine components, higher compression pistons, and revised ports and cam profiles. The exhaust system featured a Vance & Hines titanium silencer with carbon fiber endcaps. Higher-specification equipment included a fully adjustable Öhlins upside-down NIX30 front fork and Öhlins piggyback-reservoir rear shocks with machined aluminum adjusters. The premium brake setup saw Brembo four-piston M4.34 radial monobloc calipers, twin floating 310-millimeter front discs, and a radial master cylinder. Also unique to the Thruxton TFC were three riding modes (Rain, Road, and Sport). All the lighting was LED, including the 7-inch headlight with daytime running light. An aluminum engine cradle, carbon fiber bodywork, and the removal of the rear fender resulted in an 11-pound (5kg) weight savings. Gold detailing offset the blacked-out overall effect. Individually numbered, and limited to 750 worldwide, the Thruxton TFC was Triumph's most exclusive modern classic, setting a new standard as the definitive factory custom.

Rocket 3 TFC, Rocket 3 R, and Rocket 3 GT

The Rocket 3 TFC was released in July 2019. The 2,500cc TFC engine was the most powerful Triumph to date and included lightweight titanium inlet valves and unique Arrow silencers with carbon-fiber endcaps. The aluminum frame and single-sided swingarm, Showa suspension, and Brembo Stylema brakes were the same as those used for the new Rocket 3 R and GT, which followed shortly afterward. High-quality equipment included an interchangeable leather single or dual seat, carbon-fiber bodywork, and an individual numbered plaque and custom-build look. Only 750 were built.

Following the Rocket 3 TFC, the Rocket 3 R and Rocket 3 GT also included the larger capacity

Two new limited edition Bonnevilles were also released for 2019. The T120 Diamond Edition (left) celebrated 60 years of the Bonneville while the T120 Ace (right) paid homage to the legendary Ace Café. Triumph Motorcycles

The Thruxton Triumph Factory Custom offered a premium of finish and equipment in an exclusive package. Triumph Motorcycles

2019 Thruxton TFC
(Differing from the Thruxton R)

Horsepower
106 at 8,000rpm
Front fork
Öhlins 43mm NIX30 upside down
Dry weight
437 pounds (198kg)
Colors
Carbon Black

CHAPTER TEN

2019–2023 Rocket 3 TFC, Rocket 3 R, Rocket 3 GT

Type
In-line three-cylinder DOHC, 4 valves per cylinder, liquid cooled
Bore
110.2mm
Stroke
85.9mm
Capacity
2458cc
Horsepower
165 at 6,000 rpm (182 TFC)
Compression ratio
10.8:1
Fuel injection
Ride-by-wire electronic fuel injection
Gearbox
Six-speed
Final drive
Shaft
Frame
Aluminum
Swingarm
Single-sided, cast aluminum
Front fork
Showa 47mm upside-down adjustable
Rear suspension
Showa Monoshock adjustable
Brakes
2x320mm disc, 4 piston calipers front, 316mm disc rear
Wheels
Alloy 17x3.5-inch front, 16x7.5-inch rear
Tires
150/80 R17 front, 240/50 R16 rear
Wheelbase
64.7 inches (1677mm)
Dry weight
638 pounds (290kg) TFC, 640 pounds (291kg), 649 pounds (294kg) GT
Colors
Red, Black (R), Silver/Gray, Black (GT)

engine, with 11 percent more power. Additionally, the new Rocket 3 incorporated modern technology, less weight, and premium brakes and suspension. Two versions were offered: the Rocket 3 R and touring Rocket 3 GT. With a larger bore and shorter stroke, the new three-cylinder engine was the largest displacement, with the highest torque, of any production engine. The maximum torque of 221Nm was produced at only 4,000rpm. The engine featured a new crankcase assembly, new dry sump lubrication system with integral oil tank, and new balancer shafts, saving 18kg over the earlier design. A stronger six-speed helical gearbox was designed to handle the increased torque, while a distinctive three-header exhaust delivered a unique soundtrack. Both Rocket 3s featured a new aluminum frame, as well as a single-sided swingarm with fully adjustable offset Showa Monoshock rear suspension unit with piggyback reservoir. The front suspension was a 47mm adjustable Showa front fork with high specification Brembo M4.30 Stylema Monobloc front disc brakes. The new Rocket 3 was 40kg lighter than the previous generation. A feature of both new Rocket 3s was the range of rider-adjustable ergonomics, and the GT included a wider touring handlebar. Electronic updates included second-generation TFT instrumentation, optimized cornering ABS and traction control, and four riding modes. In February 2020, a new Rocket 3 set a new Triumph production motorcycle 0–60-mile-per-hour acceleration record of 2.73 seconds.

2020

Although 2020 would ultimately be disrupted by the COVID-19 pandemic, in January of that year Triumph announced a long-term, non-equity partnership with Bajaj Auto India to build a new range of mid-capacity motorcycles. Triumph also entered the new decade with several updated models. This season was the second to feature Triumph-powered Moto2 machines, but a technical freeze early in the season resulted in a pause in chassis, swingarm, and aerodynamic developments until the end of 2021. With four riders theoretically able to win the world championship, Enea Bastianini ultimately prevailed on his Kalex in a final race showdown. Alongside the official Moto2 World Championship, Triumph launched a Triumph Triple Trophy. This award recognized standout performances in top speed, pole position, and fastest laps over the entire season; the winner was Marco Bezzecchi.

Marco Bezzecchi won the 2020 Triumph Triple Trophy, and with it a Street Triple RS. *Triumph Motorcycles*

Daytona Moto2 765 Limited Edition

In the wake of the first season, where Triumph provided three-cylinder 765cc engines for the Moto2 World Championship, the existing 675cc Daytona evolved into the Moto2 765 Limited Edition. The three-cylinder

2020 Daytona Moto2 765 Limited Edition

Type
In-line three-cylinder DOHC 4 valves per cylinder, liquid cooled
Bore
78mm
Stroke
53.38mm
Capacity
765cc
Horsepower
130 at 12,250 rpm
Compression ratio
12.9:1
Fuel system
Multipoint sequential electronic fuel injection with SAI; electronic throttle control
Gearbox
Six-speed
Frame
Aluminum beam twin spar with two-piece high-pressure die-cast rear
Swingarm
Twin-sided, cast aluminum
Front fork
Öhlins 43mm upside-down NIX30 fork
Rear suspension
Öhlins TTX36 twin tube monoshock
Brakes
Twin 310mm with Brembo Stylema 4-piston radial Monobloc calipers; Brembo single-piston caliper 220mm disc
Wheels
17x3.5-inch front, 17x5.5-inch rear
Tires
120/70 ZR17 front and 180/55 ZR17 rear
Wheelbase
54.3 inches (1,379mm)
Dry weight
363 pounds (165kg)
Colors
Black, Gray/silver

engine featured several components and performance upgrades derived directly from the Moto2 engine development program. These included titanium inlet valves, stronger pistons, DLC-coated gudgeon pins, new cam profiles, new intake trumpets, modified con rods, intake ports, cranks and barrels, and an higher compression ratio. The engine revved higher than the Street Triple RS engine, with redline now 13,250 rpm. The TIG-welded, high-performance Arrow titanium exhaust system contributed to the highest power output yet for a production 765cc Triumph. The new gearbox also received Moto2 track-optimized gear ratios.

The new Daytona shared the same lightweight chassis as the Moto2 engine-development prototype and Isle of Man–winning Daytona R. Further weight savings came through full carbon-fiber single-seat bodywork, and the lightest five-spoke, 17-inch cast-aluminum alloys wheels in its class. Other high-performance features included Brembo Stylema brakes, with matching span- and ratio-adjustable brake lever, and MCS radial master cylinder. The suspension was effected with race-specification Öhlins front and rear: a fully adjustable 43mm NIX30 front fork and TTX36 rear shock absorber. Technological features included full-color TFT instrumentation and five riding modes: Rain, Road, Rider Configurable, Sport, and Track. Production was limited to a series of individually numbered motorcycles, with only 765 for the US and Canada, and 765 for Europe, Asia, and the rest of the world. Each bike featured a billet-machined, laser-etched Moto2-branded aluminum top yoke displaying a unique limited-edition number.

Street Triple RS, Street Triple R, Street Triple S

While the basic formula was unchanged, the range-leading Street Triple RS received a significant performance Euro 5 upgrade courtesy of the Moto2 program. Engine updates to the 765cc triple included a new exhaust cam to optimize midrange performance and higher precision machining on the crankshaft, clutch, and balancer. This reduced engine mass resulted in a 7 percent reduction in rotational inertia.

The Daytona Moto2 765 Limited Edition featured the Union Jack–style livery of the Moto2 engine development bike. The solo seat bodywork was carbon fiber and included official Moto2 branding. Triumph Motorcycles

CHAPTER TEN

2020–2022 Street Triple RS, Street Triple R, Street Triple S (Differing from 2019)

Bore
76mm (S)
Stroke
48.5mm (S)
Capacity
660cc (S)
Horsepower
123 at 11,750rpm (RS); 118 at 12,000rpm (R); 47.6 at 9,000rpm (S); 95.2 at 11,250rpm *derestricted* (S)
Compression ratio
12.54:1 (RS); 12.5:1 (R); 12.1:1 (S)
Rear suspension
Öhlins STX40 (RS)
Brakes
2x310mm disc Brembo 4-piston radial calipers front (RS, R); Nissin twin piston calipers (S); 220mm Brembo disc rear (RS, R, S)
Wheelbase
55.3 inches (1,405mm) (RS, R); 55.5 inches (1,410mm) (S)
Dry weight
365 pounds (166kg) (RS); 370 pounds (168kg) (R, S);
Colors
Black, Silver (RS); Black, Matte Silver (R); White (S)

The three-model Street Triple range was significantly updated for 2020. The entry-level Street Triple S had lower-spec Nissin front brakes. *Triumph Motorcycles*

A new, freer-flowing silencer with carbon-fiber endcap and new intake duct contributed to a 9 percent increase in power and torque. The gearbox received shorter first- and second-gear ratios, while higher precision machining enabled removal of the anti-backlash gears. The chassis equipment was extremely high specification, including a Showa 41mm upside-down big-piston front fork, Öhlins STX40 shock absorber, and front Brembo M50 4-piston radial Monobloc calipers. More aggressive styling accompanied the increased engine performance, including more sporting, angular bodywork with twin LED headlights. Technological updates extended to enhanced TFT instrumentation, MyTriumph connectivity, and five improved riding modes.

Both the Street Triple R and Street Triple S were also updated, each sharing many features with the Street Triple RS. Not all of the engines were Euro 5 compliant. The 765cc R, similar to the RS, featured lighter internal engine components to reduce inertia, lower gear ratios, and a slip and assist clutch. The S was only available with a 660cc engine in two versions: restricted for A2 licenses and unrestricted. The R received high-specification Brembo M4.32 4-piston radial Monobloc front calipers and Showa suspension, while the S included twin sliding-piston Nissin calipers on the front. The R and S also shared the updated, aggressive, styling of the RS, with a more pronounced flyscreen and air intake. Both the Street Triple R and S were available in Low Ride Height (LRH) versions, with a 45mm lower seat height at 780mm. Three riding modes were available on the R and two on the S. All three Street Triple versions built on a successful formula, combining useable power in a lightweight package to create an arguably perfect all-round, useable motorcycle.

Bobber TFC

The success of the first two Triumph Factory Custom models led to a third in 2020, the Bobber TFC. Along with a premium custom finish and detailing, the 1,200cc engine received significant upgrades and many lighter engine components. This included a low-inertia crankshaft, clutch, balance shafts, rare-earth alternator, magnesium cam cover, and thin-walled engine covers and header pipes. The 39 percent lower engine inertia resulted in a 500rpm increase in revs, with a 13 percent power increase over the standard Bobber. The bespoke exhaust system included Arrow silencers with carbon fiber endcaps. Higher-specification equipment included a fully adjustable Öhlins upside-down front fork, adjustable Öhlins rear suspension unit, and Brembo M50 4-piston radial Monobloc calipers clamping twin floating Brembo front discs. The master cylinder was a Brembo radial unit with MCS span- and ratio-adjustable brake lever. Technological updates included the latest-generation ride-by-wire throttle with

2020 Bobber TFC
(Differing from the Bobber)

Horsepower
86 at 6,250rpm
Front suspension
Öhlins 43mm, USD, fully adjustable, NIX 30
Rear suspension
Öhlins adjustable
Front brake
Dual Brembo 310mm discs, M50 4-piston radial Monobloc calipers

three riding modes (Rain, Road, and Sport), while all the lighting was LED. Premium components included carbon-fiber bodywork, billet-machined aluminum-fork triple clamps and oil filler cap, leather seat, and gold detailing. Classic Bobber features were retained, such as the heritage-inspired battery cover and "drum brake"-look inspired rear hub. Only 750 Bobber TFCs were sold worldwide, making it one of Triumph's most exclusive and desirable Modern Classics. Each featured an individually numbered plaque and personalized custom build book.

Thruxton RS

The café racer Thruxton evolved into a higher-performance RS version for 2020. Engine upgrades for the 1,200cc twin included higher compression pistons, revised port and cam profile, and a secondary air system. As on the Street Triple RS, lighter engine components included a low-inertia crankshaft, clutch, balance shafts, rare-earth alternator, magnesium cam cover, and thin-walled engine covers. These modifications delivered a 20 percent reduction in inertia, allowing the engine to rev 500rpm higher. The upswept silencers included a Euro 5–spec catalytic converter. Power was up 8 horsepower over the Thruxton R. Weight savings saw the Thruxton RS scale 6 kilograms lighter than the Thruxton R.

Higher-specification equipment included a race-bred 43mm Showa big-piston upside-down front fork with Brembo 4-piston M50 radial Monobloc calipers and twin floating Brembo front discs. A pair of fully adjustable Öhlins piggyback-reservoir rear shock absorbers controlled the rear. The RS's fully blacked-out style extended to black anodized wheels and engine covers. With the RS, the Thruxton cemented its position as the quintessential factory café racer.

The Bobber TFC was one of Triumph's most exclusive models. Along with premium equipment, a silver foil Union Jack was incorporated in the tank kneepads. Triumph Motorcycles

A higher-spec Thruxton RS was released in 2020. This has a solo bullet-type seat as standard. Triumph Motorcycles

From 2020 Thruxton RS
(Differing from the Thruxton R)

Horsepower
103 at 7,500rpm
Compression ratio
12.066:1
Brakes
Twin 310mm disc front Brembo M50 radial Monobloc
Dry weight
43 pounds (197kg)
Colors
Black, Gray/Silver

CHAPTER TEN

The Bud Ekins Bonneville T120 and T100 Special Editions celebrated the 1950s and 1960s racer and stuntman Bud Ekins. *Triumph Motorcycles*

Five versions of the Tiger 900 were available for 2020. The range was headed by the 900 GT Pro (*left*) and 900 Rally Pro (*right*). *Triumph Motorcycles*

Bud Ekins Bonneville T120 and T100 Special Editions

As discussed in Chapters 2 and 3, Bud Ekins was one of Triumph's most successful off-road racers during the 1950s and 1960s before pivoting to a career as a motion-picture stuntman. Ekins also owned a successful Triumph dealership in Sherman Oaks, California. Through this business, he forged close friendships with many Hollywood stars, including Paul Newman, Clint Eastwood, and Steve McQueen. To celebrate Ekins's legend, Triumph released two Bonneville Special Editions in 2020. These were based on the T120 and T100, but featured a bespoke two-color paint scheme with hand-painted coach lining and a heritage Triumph logo. Along with specific black engine badges, the tank and front fender received a unique California "flying globe" Bud Ekins logo. Each example was supplied with a certificate of authenticity signed by Nick Bloor and Ekins's daughters, Susan and Donna.

Tiger 900, 900 Rally, 900 Rally Pro, 900 GT, 900 GT Pro

For 2020, the Tiger 900 replaced the successful Tiger 800. The new engine was more powerful, with higher torque, and the updated model was significantly lighter. The Tiger 900 range comprised five models: the base road-focused Tiger 900; the Tiger 900 GT and GT Pro for long-distance travel and urban adventure; and the Tiger 900 Rally and Rally Pro, emphasizing off-road adventure with road capability.

The 888cc engine was now Euro 5 compliant and featured a new 1,3,2 firing order. This provided improved character and tractability, with the sound and feel of a twin lower in the rev range while maintaining the midrange torque and top-end power of a triple. Peak torque increased more than 10 percent over the previous Tiger 800, with a 9 percent increase in peak power. A new twin radiator arrangement enabled the engine to be positioned 40mm forward and 20mm lower for a lower center of gravity. Sump design improvements ensured the revised engine location didn't compromise ground clearance.

The new steel trellis frame was lighter, and its modular construction included a bolt-on aluminum rear subframe. The suspension was tailored specifically for each model: the Tiger 900 and 900 GT received Marzocchi components (adjustable on the GT), while the Rally employed Showa. The GT Pro featured electronically adjustable rear suspension. All Tiger 900s models included front Brembo Stylema 4-piston Monobloc calipers and a radial master cylinder. Long-distance travel was improved with a larger 20-liter fuel tank, an adjustable windscreen, and a seat adjustable for height by 20mm. State-of-the-art riding technology included TFT instrumentation, up to six riding modes, and optimized cornering ABS.

2020–2023 Tiger 900, Rally, Rally Pro, GT, GT Pro *(Differing from the Tiger 800)*

Bore
78mm
Capacity
888cc
Horsepower
93.9 at 8,750rpm
Compression ratio
11.27:1
Front fork
Marzocchi 45mm upside-down fork (900, GT, GT Pro); Showa 45mm upside-down fork (Rally, Rally Pro)
Rear suspension
Marzocchi Monoshock (900, GT, GT Pro); Showa (Rally, Rally Pro)
Brakes
2x320mm disc 4 piston Brembo calipers front, 255mm Brembo disc rear
Wheels
Alloy 19x2.5in front and 17x4.25in rear (900, GT, GT Pro) Spoked 21x2.15in front and 17x4.25in rear (Rally, Rally Pro)
Tires
100/90x19 front and 150/70R17 rear (900, GT, GT Pro), 90/90x21 front (Rally, Rally Pro)
Wheelbase
61.2 inches (1,556mm) 900, GT, GT Pro; 61.1in (1,551mm) Rally, Rally Pro
Wet weight
422 pounds (192kg) 900; 427 pounds (194kg) GT; 436 pounds (198kg) GT Pro; 431 pounds (196kg) Rally; 442 pounds (201kg) Rally Pro
Colors
White (900), Red, Black, White (GT, GT Pro); Khaki, Black, White (Rally, Rally Pro)

From 2020 Tiger 1200 Desert and Alpine Special Editions *(Differing from the XCx and XRx)*

Horsepower
139 at 9,350rpm
Wet weight
541 pounds (246kg) desert; 534 pounds (242kg) alpine
Colors
Sandstorm (Desert); White (Alpine)

New special editions of the Tiger 1200 were also introduced for 2020. The Tiger 1200 Desert Edition (*left*) and Tiger 1200 Alpine Edition (*right*). *Triumph Motorcycles*

Tiger 1200 Desert and Alpine Special Editions

Two specific Tiger 1200 adventure editions, the Desert and Alpine, were introduced for 2020. Inspired by epic adventures undertaken in some of the most inhospitable deserts of the world—from the Sahara to the Kalahari—the Desert Special Edition was based on the mid-spec, off-road-focused XCx. The Alpine Special Edition was inspired by epic alpine adventures over breathtaking mountain ranges and based on the mid-spec, road-biased XRx.

Engine developments included a lighter flywheel, lighter crankshaft, and a magnesium cam cover. A lightweight Arrow titanium silencer contributed to a slight power increase. Premium chassis equipment included high-spec Brembo Monobloc brakes, adjustable White Power suspension controlled by Triumph Semi-Active Suspension (TSAS), an adjustable seat height (835–855mm), adjustable full-color TFT instrumentation, optimized-cornering ABS and traction control, up to five riding modes, and an electrically adjustable windscreen.

CHAPTER TEN

Another special edition was the Scrambler 1200 Bond Edition. This was ostensibly a custom Scrambler 1200 XE. *Triumph Motorcycles*

Scrambler 1200 Bond Edition

In December 2019, Triumph announced a partnership with the twenty-fifth official James Bond film, *No Time to Die*, to provide Tiger 900s and Scrambler 1200s for the picture's action sequences. The Scrambler 1200 Bond Edition was a result of this partnership: it was the first official motorcycle directly linked to the Bond franchise. The Scrambler 1200 Bond Edition was based on the Scrambler and 1200 XE, and was limited to 250 examples. It featured an exhaust number board with 007 branding, premium leather seat, a specific paint scheme, and a number of blacked-out components. The Arrow silencer included carbon-fiber endcaps, and topping the specification was a pair of Brembo M50 radial Monobloc front brake calipers.

2021

After the interruptions of the previous year, Triumph sold a record number of 81,541 motorcycles in 2021, representing a growth over 2020 of 29 percent. Phase 2 of the electric project TE-1 was also completed this year, with new models for 2021 centered on the entry-level Trident 660 and Tiger 850 Sport. Triumph expanded their competition program to include the British Supersport Championship in partnership with Simon Buckmaster's Performance Technical Racing. Kyle Smith and Brandon Paasch rode the Dynavolt Street Triple 765 RS. Smith and Paasch finished seventh and eighth in the championship. This was a prelude to an official entry in the Supersport World Championship in 2022. The 2021 Moto2 Championship was closely fought: Remy Gardner won the championship, narrowly beating his Red Bull KTM teammate Raul Fernandez. Fernandez had the consolation of winning the Triumph Triple Trophy. In 2021 Triumph renewed their contract as the exclusive engine supplier to the FIM Moto2 World Championship for three more years. They announced a forthcoming factory entry in the

OPPOSITE: Remy Gardner narrowly won the 2021 Moto2 World Championship on a Triumph-powered KTM. *Triumph Motorcycles*

276

Motocross and Enduro world championships. Triumph reached a modern-era milestone in December with their millionth motorcycle built since the brand's return in 1990. The millionth Triumph was a one-off, custom-painted Tiger 900 Rally Pro.

Trident 660

Triumph entered the middleweight roadster category with the all-new the Trident 660. Although an entry-level model, the Trident 660 offered competitive engine performance and technology in a light, minimalistic package. The three-cylinder engine was developed specifically for the Trident and incorporated sixty-seven new components. The engine tune emphasized a linear power and torque delivery, with 90 percent of the maximum torque available across most of the rev range. A restricted A2-compliant kit was also available. While a budget model, its technological features still included a TFT display, LED lighting, rain and riding modes, and switchable traction control. The suspension was Showa and the brakes Nissin, and the all-new tubular-steel chassis provided excellent handling with comfortable ergonomics.

The Trident 660 was an all-new entry-level model for 2021. The style was minimalist and the equipment modest. *Triumph Motorcycles*

From 2021 Trident 660
(Differing from the Street Triple S)

Horsepower
80 at 10,250rpm; 46 at 8,750rpm *restricted*
Compression ratio
11.95:1
Rear brake
255mm Nissin disc
Wheelbase
55.2 inches (1,401mm)
Wet weight
417 pounds (189kg)
Colors
Silver/Red, Black/Silver, White, Black

From 2021 Tiger 850 Sport *(Differing from the Tiger 900)*

Horsepower
84 at 8,500rpm
Colors
Graphite/Red, Graphite/Blue

Tiger 850 Sport

The Tiger 900 platform expanded with the Tiger 850 Sport in 2021. Building on the base Tiger 900, the Tiger 850 Sport's T-plane crank engine was retuned for a more manageable power delivery. In Europe, a dealer-fit A2 license compliance kit was also available. The Marzocchi suspension and Brembo Stylema brakes were shared with the Tiger 900, and over sixty accessories were offered for individual customization.

The Tiger 850 Sport was the most affordable model in the Tiger range and available with optional Givi luggage. *Triumph Motorcycles*

CHAPTER TEN

For 2021 the legendary Speed Triple received a 1200cc engine and was the most powerful version yet. *Triumph Motorcycles*

From 2021 Speed Triple 1200 RS
(Differing from 2020)

Bore
90mm
Stroke
60.8mm
Capacity
1,160cc
Horsepower
177.5 at 10,750 rpm
Compression ratio
13.2:1
Rear brake
Single 220mm disc Brembo twin piston caliper
Wet weight
437 pounds (198kg)
Colors
Black, Matte Silver

Speed Triple 1200 RS

The range-leading Speed Triple 1200 RS was significantly updated for 2021. Previously 1,050cc, it now received a 1,160cc engine developed with input from the Moto2 racing program. Every component was reengineered to increase performance and reduce mass, resulting in weight savings of 7 kilograms. Despite delivering 30 more horsepower, the engine was more compact than before. And with a new freer-flowing intake and exhaust, plus 12 percent less powertrain inertia, the engine revved 650rpm higher. Along with a lighter slip-and-assist clutch, the new stacked 6-speed gearbox provided smoother gear changes than before. The mass-centralized cast-aluminum frame was 17 percent lighter; it saved 10 kilograms over the previous version, delivering the highest power-to-weight ratio of any Speed Triple, over 25 percent greater than before and nearly double that of the original 1994 Speed Triple. Premium equipment included twin Brembo Stylema radial Monobloc front brake calipers and Metzeler Racetec RR or optional Pirelli Diablo Supercorsa SC2 track-approved tires. New 5-inch TFT instrumentation, optimized cornering ABS, cornering traction control, and five riding modes were part of the electronics package. The new Speed Triple 1200 RS was the most capable and technologically advanced Speed Triple yet.

Bonneville T120, T120 Black, T100, Street Twin, Street Twin Gold Line; Speedmaster, Bobber, Speed Twin

The complete Bonneville range was updated for 2021. The T120's engine received a lighter crankshaft, along with a new clutch and balance shafts to reduce inertia. Lighter aluminum wheel rims and a higher-spec Brembo front brake contributed to a weight savings of 7 kilograms. The power of the T100 engine was increased by 10 horsepower while a low inertia crankshaft, lighter balancer shafts, a thin-walled clutch cover, and a magnesium cam cover contributed to a 4-kilogram weight savings. The T100 received a higher-spec Brembo twin-piston front brake and a new cartridge front fork. Other updates included new cast wheels, new bodywork, and a more comfortable seat. The Street Twin now included new cast wheels with machined detailing, and a more comfortable seat was joined by a Gold Line Limited Edition; these latter featured Matte Sapphire black paint with hand-painted gold pinstripes and were limited to one thousand examples. The Bonneville Speedmaster received a lower-inertia 1,200cc high-torque engine, while chassis updates included a larger, 47mm Showa cartridge fork and higher-spec Brembo brakes. An even more comfortable twin-seat setup was fitted to the 2021 update. For improved comfort, the rider's seat now had a separate lumbar support and a sculpted deep foam construction. The Bobber received the low inertia engine as well, while chassis updates included a 16-inch front wheel (as on the previous Bobber Black) and a 47mm Showa fork. The 12-liter fuel tank was larger and the style was more blacked out and chunkier.

Since its release for the 2019 model year, the Speed Twin had set a benchmark for Modern Classic performance and was significantly updated for 2021. A lightweight crankshaft and alternator contributed to a 17 percent reduction in inertia, allowing for a

From 2021 Bonneville T120, T120 Black, T100, Street Twin, Street Twin Gold Line; Speedmaster, Bobber, Speed Twin
(Differing from 2020)

Horsepower
64 at 7,400rpm (T100); 98.6 at 7,250 (Speed Twin)
Compression ratio
11.0:1 (T100); 12.1:1 (Speed Twin)
Front fork
47mm Showa telescopic (Speedmaster, Bobber); 43mm USD Marzocchi fork (Speed Twin)
Front brake
Twin 320mm discs, Brembo M50 4-piston radial Monobloc calipers (Speed Twin)
Wheelbase
57.1 inches (1,450mm) T120, T120 Black, Street Twin; 59.1 inches (1,500mm) Speedmaster; 59.1 inches (1,500mm) Bobber; 55.6 inches (1,413mm) Speed Twin
Wet weight
493 pounds (224kg) T120; 503 pounds (228 kg) T100; 476 pounds (216kg) Street Twin; 580 pounds (263kg) Speedmaster; 553 pounds (251kg) Bobber; 476 pounds (216kg) Speed Twin
Colors
Black, Red/Silver, Blue/Silver (T120); Black, Matte Black/Matte Graphite (T120 Black); Blue/ White, Black, Red/ White (T100); Blue, Matte Ironstone, Black (Street Twin); Black, White/ Black, Hopper (Speedmaster); Gray/Ironstone, Red, Black (Bobber); Red, Gray, Black (Speed Twin)

500rpm redline increase. Higher-compression pistons, a new camshaft, and new brushed stainless-steel megaphone upswept silencers resulted in a small power increase. A higher-spec, upside-down 43mm Marzocchi fork was fitted, and braking improved through a pair of Brembo M50 radial Monobloc brake calipers with 320mm discs. The 17-inch cast-aluminum wheels were a lightweight 12-spoke design. Premium finishes extended brushed aluminum fenders, a classic Monza fuel cap, and a clear anodized aluminum swingarm.

TOP: The 2021 Bonneville lineup. *From left:* T120, T120 Black, T100, Street Twin, Bobber, Speedmaster. *Triumph Motorcycles*

BOTTOM: The class-leading Speed Twin received a significant update for 2021. The front fork and brakes were higher specification and the engine was more powerful. *Triumph Motorcycles*

CHAPTER TEN

The Rocket 3 GT Triple Black Limited Edition featured a distinctive three-shade black paint scheme. Triumph Motorcycles

Only 1,000 examples of the Scrambler 1200 Steve McQueen Edition were built. These were the highest-spec Scrambler yet. Triumph Motorcycles

From 2021 Scrambler 1200 XC, XE, Steve McQueen Edition
(Differing from 2020)

Colors
Cobalt Blue, Matte Khaki Green, Sapphire Black (Scrambler 1200); Green (Steve McQueen)

Rocket 3 R Black and Rocket 3 GT Triple Black Limited Editions

With the limited editions proving extremely successful, two Rocket 3 Limited Editions were offered early in 2021. The Rocket 3 R Black and Rocket 3 GT Triple Black were ostensibly mild cosmetic custom versions of the basic Rocket 3 R and Rocket 3 GT. The Rocket 3 R Black featured an all-black color scheme, while the Rocket 3 GT Triple Black sported three shades of paint. Along with a carbon-fiber fender, the black finish extended to the exhaust headers, intakes, and numerous components. These included badges, levers, footrests, and pedals. Each model was limited to one thousand examples worldwide and came with a certificate of authenticity.

Scrambler 1200 XC and XE; Scrambler 1200 Steve McQueen Edition

The Scrambler 1200 XC and XE shared the T120 Euro 5 engine updates for 2021. The engine received a low-inertia crank and high-compression cylinder head; it was specifically tuned to produce a wide torque spread. As it was a relatively new model, the basic chassis was unchanged. The Scrambler 1200 Steve McQueen Edition was based on the higher-spec Scrambler 1200 XE. Unique features included a Competition Green tank with brushed-foil kneepads, hand-painted gold lining, gold heritage Triumph logos, dedicated Steve McQueen tank graphic, brushed-aluminum Monza cap, a brushed stainless-steel tank strap, painted aluminum fenders, and stitched brown bench seat. The high specification was complemented by a selection of premium Scrambler accessories fitted as standard. A numbered limited edition, only one thousand were available worldwide.

2010–2024

Street Scrambler, Street Scrambler Sandstorm Limited Edition

As it was based on the Street Twin, the Street Scrambler was updated for 2021. The high-torque 900cc twin-cylinder engine received the same Euro 5 updates. While the basic Street Scrambler chassis was unchanged, new details featured a side panel with aluminum number board and a new leather-and-textile–inspired seat covering. The Street Scrambler Sandstorm Limited Edition appeared in this year, including many standard Scrambler accessories and a custom Sandstorm paint scheme. This incorporated Matte Storm Gray and Ironstone accents in a tritone style on the tank. Worldwide, 775 Sandstorm Editions were made available.

From 2021 Street Scrambler Sandstorm Limited Edition
(Differing from 2020)

Wet weight
492 pounds (223kg)
Colors
Black, Gray, Matte Khaki/Ironstone (Street Scrambler); Gray (Sandstorm)

Another limited edition for 2021 was the Street Scrambler Sandstorm. This also included a number of premium Street Scrambler accessories as standard. *Triumph Motorcycles*

The display at Silverstone in August 2022 celebrating Triumph's 120 years was headed by the original 1901 prototype, which had recently been discovered. *Triumph Motorcycles*

2022

Motorcycle sales increased in 2022 to 83,389 through eight hundred dealers worldwide. Triumph celebrated their 120-year anniversary, which included a celebratory parade lap at Silverstone prior to the Moto2 Race at the British Grand Prix on August 7. The event highlighted some of Triumph's most significant and historic models, such as the recently discovered original 1901 prototype, a 1915 Model H "Trusty," 1949 6T Thunderbird, Ken Heans's 1961 ISDT TR6, Steve McQueen's *The Great Escape* 1961 TR6 Trophy, 1973 X75 Hurricane, 1982 TR65T Tiger Trail, 1994 Speed Triple, 2004 Daytona 675, 2017 Moto2 Prototype, 2019 Scrambler 1200 XE Bond Bike (*No Time to Die*), 2021 Rocket 3R 221 Edition, 2021 Speed Triple 1200 RS, 2022 Speed Triple 1200 RR, and the 2022 Triumph Triple Trophy Street Triple 765 RS.

281

CHAPTER TEN

After a season racing in the UK, Brandon Paasch returned to America in 2022 and won the Daytona 200 on the Street Triple RS. *Triumph Motorcycles*

Stefano Manzi was Triumph's most successful rider in the 2022 World Supersport Championship. *Triumph Motorcycles*

Augusto Fernandez won the 2022 Moto2 World Championship on the Triumph-powered KTM. *Triumph Motorcycles*

2010–2024

With a change in the FIM World Supersport Next Generation regulations, Triumph's racing program expanded to include the Supersport World Championship and saw a factory return to the Daytona 200. On March 12, Brandon Paasch rode the TOBC Triumph Racing Team Street Triple RS to a narrow victory in the eightieth Daytona 200. His teammate Danny Eslick finished 6th. In the Supersport World Championship, the Dynavolt Triumph team engaged experienced Moto2 rider Stefano Manzi and the twenty-three-year-old Estonian Hannes Soomer. Manzi won Race 1 in Portugal and finished 6th overall. In the Triumph-powered Moto2 championship, Augusto Fernandez, on the Kalex-framed Red Bull KTM Ajo entry, comfortably won the World Championship. Jeremy Alcoba won the revised Triumph Triple Trophy, which was rewarded to riders who made the best race progression from start to finish.

While Triumph was still actively involved in road racing, their next area of competition participation was Motocross and Enduro. For this they engaged Motocross legend Ricky Carmichael and five-time Enduro World Champion Iván Cervantes in testing and development. In July, Cervantes won the Trail category in the Baja Aragón on a Tiger 900 Rally Pro. Cervantes dominated the 450km race, winning by 1 hour and 6 minutes. He followed this in October with a win in the Hard Trail category of the 1000 Dunas Raid in Morocco.

Champion Enduro rider Iván Cervantes won the Trail category in the Baja Aragón on a Tiger 900 Rally Pro. *Triumph Motorcycles*

283

CHAPTER TEN

The TE-1 electric development project was completed during 2022. The overall style was similar to the existing Street Triple, but the performance was on a par with the Speed Triple. Triumph Motorcycles

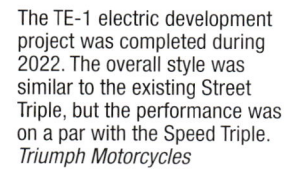

ABOVE, RIGHT: *The café racer–style Speed Triple 1200 RR included a half fairing and semi-active electronic Öhlins suspension.* Triumph Motorcycles

TE-1

The collaboration stage for Phases 3 and 4 of the TE-1 electric development project was completed in 2022. The final prototype test results exceeded project objectives and saw a 161-kilometer/100-mile range, 175 horsepower peak power, 0–60 mile-per-hour acceleration in 3.6 seconds and 0–100 miles per hour in 6.2 seconds, and a 20-minute charge time (0–80 percent). The latter acceleration was faster than the Speed Triple 1200 RS. The 220-kilogram weight was up to 25 percent lighter than comparable available electric motorcycles. To maintain Triumph's signature style and presence, the ergonomics, geometry, and weight distribution were developed to be similar to the Speed Triple, with the scale and visual impact of a Street Triple, while the throttle and torque delivery map were equivalent to a Speed Triple 1200 RS. Regenerative braking was successfully implemented, with scope for further optimization, and motor generator and transmission efficiencies were added to improve the range further in the future. Daytona 200 champion racer Brandon Paasch participated in the final testing phase. The success of the TE-1 project provided a platform for the future development of electric motorcycles.

Speed Triple 1200 RR

The first new model release for 2022 was the Speed Triple 1200 RR. This was a variation on the existing Speed Triple 1200 RS and a modern take on the traditional café racer. The 1200 RR was notable primarily for a small cockpit fairing with a classic single round headlight. While the engine and basic chassis were shared with the 1200 RS, the 1200 RR featured Öhlins S-EC 2.0 OBTi system electronically adjustable

From 2022 Speed Triple 1200 RR
(Differing from the Speed Triple 1200 RS)

Wheelbase
56.6 inches (1,439mm)
Wet weight
438 pounds (199kg)
Colors
Red/Gray, White/Gray

semi-active suspension, carbon-fiber detailing, and a color-coordinated belly pan and seat cowl. The clip-on handlebars and rear-set footpegs complemented the café racer style.

Tiger Sport 660

The naked middleweight Trident 660 evolved into the Adventure Tiger Sport 660 for 2022. While the basic engine and chassis specifications were unchanged, the Tiger Sport 660 received more agile steering geometry, a taller seat, and a larger, 17-liter fuel tank. Long-distance touring capability was enhanced by integrated pannier mounts and a twin headlight fairing with height-adjustable screen. Designed to entice a new generation of riders to look at the Triumph brand, the Tiger Sport 660 was suitable for commuting, everyday trips, or longer journeys. It was the perfect all-rounder.

From 2022 Tiger Sport 660 (Differing from the Trident 660)

Wheelbase
55.8 inches (1,418mm)
Wet weight
453 pounds (206kg)
Colors
Blue/Black, Red/Graphite, Graphite/Black

Tiger 1200 GT, GT Pro, GT Explorer, Rally Pro, Rally Explorer

After only four years, the best-selling Tiger 1200 family was completely revised for 2022. In one of the most comprehensive updates in Triumph's history, the three-cylinder engine received a larger bore and shorter stroke, with a new T-plane crankshaft delivering engine firing pulses at 180, 270, and 270 degrees. This provided the improved low-end tractability of a twin with the benefits of a triple's top-end power. While the capacity was reduced slightly, the compression ratio was increased and peak power was up by 8 horsepower over the previous generation. The cylinder head, gearbox and clutch, shaft drive, and bevel box

The Tiger Sport 660 beautifully integrated practicality with modern adventure sports design. Triumph Motorcycles

An all-new Tiger 1200 was available for 2022. Shown here, in front, are the Rally Pro and Rally Explorer, with the GT Explorer and GT Pro behind. Triumph Motorcycles

CHAPTER TEN

2022–2023 Tiger 1200 GT, GT Pro, GT Explorer, Rally Pro, Rally Explorer

(Differing from 2021)

Bore
90.0mm
Stroke
60.7mm
Capacity
1,160cc
Horsepower
148 at 9,000rpm
Compression ratio
13.2:1
Frame
Tubular steel frame, with forged aluminum outriggers. Fabricated, bolt-on aluminum rear subframe.
Swingarm
Twin sided "Tri-Link" aluminum swingarm with twin aluminum torque arms.
Front fork
Showa 49mm USD fork with semi-active damping.
Rear suspension
Showa Monoshock with semi-active damping.
Brakes
Brembo M430 Stylema Monobloc radial calipers, twin 320mm floating discs; Brembo single-piston caliper, single 282mm disc
Wheels
Cast alloy 19x3.0in and 18x4.25in rear (GT, GT Pro, GT Explorer); Spoked 21x2.5in and 18x4.25in (Rally Pro, Rally Explorer)
Tires
120/70R19 front and 150/70R18 rear (90/90-21 front Rally Pro, Rally Explorer)
Wheelbase
61.4 inches (1,560 mm)
Wet weight
528 pounds (240kg) GT; 539 pounds (245kg) GT Pro; 561 pounds (255kg) GT Explorer; 548 pounds (249kg) Rally Pro; 574 pounds (261kg) Rally Explorer
Colors
White (GT); White, Black, Blue (GT Pro, GT Explorer); White, Black, Khaki (Rally Pro, Rally Explorer)

were all new, with every component designed to be lighter. As the new engine was much more compact, the riding dynamics were transformed.

The new frame featured a bolt-on aluminum rear subframe and weighed 5.4 kilograms less than the previous design. Additional weight savings came from the new aluminum fuel tank and tri-link swingarm, which helped make the new generation more than 25 kilograms lighter than before. The five-model range included the road-focused GT and the higher-spec GT Pro, both of which featuring cast wheels and Showa semi-active suspension. They were joined by the Rally Pro with tubeless wire wheels and longer-travel Showa suspension. Completing the lineup were the Rally Explorer and GT Explorer, both with 30-liter fuel tanks. For 2024, the Tiger 1200 received a new Active Preload Reduction feature, reducing the rear suspension preload as the bike slowed and allowing the seat height to be reduced.

2022 Special and Limited Editions

Tiger 900 Bond Edition

Following up on the excitement generated by the Tiger 900 Rally Pro's stunt sequences in *No Time to Die*, Triumph launched a Tiger 900 Bond Edition in 2022. This followed the earlier Scrambler 1200 Bond Edition and included a Matte Sapphire Black paint scheme with blacked-out details. Limited to 250 examples, each was individually numbered.

Bonneville Gold Line Editions

A new range of eight Bonneville Gold Line Editions appeared in the wake of the successful Street Twin Gold Line Edition of 2021. Showcasing the hand-painted gold lining skills of Triumph's paint shop, these new editions were available for only one year. Each

The Bond Special Editions continued in 2022 with the Tiger 900 Bond Edition. *Triumph Motorcycles*

The Bonneville Gold Line Editions featured hand-painted gold lining on the fuel tank. *Triumph Motorcycles*

2010–2024

The T100 Gold Line was one of eight Bonneville Gold Line Special Editions available in 2022. Triumph Motorcycles

The Speedmaster Gold Line was silver and black, available with an optional short front fender. Triumph Motorcycles

The high-spec Scrambler 1200 XE Gold Line Edition featured a distinctive Baja Orange and Silver color scheme. Triumph Motorcycles

2022 Bonneville Gold Line Editions

Bonneville T100 Gold Line Edition
Silver Ice tank with Competition Green infill edged with hand-painted gold lining, Silver Ice fenders and side panels with Competition Green side panel stripes, white and gold Bonneville T100 logo and hand-painted gold lining, matching Silver Ice flyscreen optional.

Street Scrambler Gold Line Edition
Matte Pacific Blue tank with Graphite stripe, gold Triumph tank logos, hand-painted gold lining alongside the tank stripe and around the brushed foil kneepads, Matte Jet Black front and rear fenders and side panel with new gold Street Scrambler logo, matching Matte Pacific Blue flyscreen and high-level fender optional.

Bonneville Speedmaster Gold Line Edition
Silver Ice tank with Sapphire Black twin stripe design, and brushed foil kneepads, edged with hand-painted gold lining. Sapphire Black headlight bowl, fenders, and side panels with unique new gold and silver Bonneville Speedmaster logos. Matching Sapphire Black short front fender optional.

Bonneville Bobber Gold Line Edition
Carnival Red tank and fenders, gold Triumph tank logos, Sapphire Black twin stripe design and brushed foil kneepads, edged with hand-painted gold lining. Sapphire Black side panels with gold and silver Bonneville Bobber logo and hand-painted gold lining. Matching Carnival Red short front fender optional.

Bonneville T120 Gold Line Edition
Silver Ice tank with Competition Green infill edged with hand-painted gold lining and "gold line" logo. Silver Ice fenders and side panels with Competition Green side panel stripes, white and gold Bonneville T120 logo and hand-painted gold lining. Matching Silver Ice flyscreen optional

Bonneville T120 Black Gold Line Edition
Matte Sapphire Black tank, front and rear fenders, headlight bowl and side panels, Matte Silver Ice tank infill edged with hand-painted gold lining and "gold line" logo. Matte Silver Ice side panel stripe graphics with black and gold Bonneville T120 Black logos and hand-painted gold lining. Matching Matte Sapphire Black flyscreen optional.

Scrambler 1200 XC Gold Line Edition
Carnival Red and Storm Gray tank with Aluminum Silver stripe, brushed foil kneepads, hand-painted gold lining, and "gold line" logo. Jet Black side panel and headlight bowl.

Scrambler 1200 XE Gold Line Edition
Baja Orange and Silver Ice tank with Pure White stripe, brushed foil kneepads, hand painted gold lining and "gold line" logo. Jet Black side panel and headlight bowl

287

CHAPTER TEN

The 221 Special Edition Rocket 3 R celebrated the three-cylinder engine's incredible torque. *Triumph Motorcycles*

The Street Twin EC1 Special Edition took its inspiration from the vibrant custom-classic culture of London's East End. *Triumph Motorcycles*

The Thruxton RS Ton Up Special Edition offered an homage to the legendary Ton Up (100-mile-per-hour) boys of the 1950s and 1960s. Here it is posing at the equally legendary Ace Café in London. *Triumph Motorcycles*

represented a different model in the Modern Classic range, and all received bespoke, high-quality individual colors and detailing. The Gold Line Edition designs started with a two-color base scheme, on top of which the gold line was carefully applied by hand with a special soft-bristled sword-liner brush. The paints used for gold lining were formulated by mixing in a cellulose lacquer, and each Gold Line Edition was initialed by the artist.

221 Special Edition Rocket 3 R and Rocket 3 GT

Another one-year-only offering for 2022 was the Special Edition with the 221 Special Edition Rocket 3 R and Rocket 3 GT, celebrating the Rocket 3's class-leading 221 Newton meters (163 pound-feet) of peak torque. The 221's fuel and front fender received a distinctive Red Hopper color scheme and the kneepad got distinctive 221 graphics. The 221 tank graphics detailed the Rocket 3's power, torque, engine size, and bore and stroke. Other bespoke detailing included Sapphire Black fender brackets, headlight bowls, flyscreen, side panels, rear bodywork, and radiator cowls. Along with the Showa suspension, Brembo brakes and 20-spoke cast-aluminum wheels, the twin headlight arrangement, sculpted triple-header exhaust, single-sided swingarm, and hidden folding pillion footrests were all shared with the standard Rocket 3.

Street Twin EC1 Special Edition

Another one-year-only model, the Street Twin EC1 Special Edition took its inspiration from the vibrant custom-classic culture of London's East End—particularly the historic streets of the London postcode EC1. The best-selling Modern Classic, the Street Twin, had proved a popular basis for customization, and the EC1 was ostensibly a Street Twin factory custom. The EC1 paint scheme and premium custom-style detailing included a Matte Aluminum Silver and Matte Silver Ice fuel tank design with hand-painted silver coach lining and dedicated EC1 graphics. The custom theme continued with the Matte Silver Ice side panels that included a Street Twin Limited Edition graphic; both front and rear fenders were Matte Aluminum Silver. An optional Matte Silver Ice flyscreen complemented the EC1 design scheme. Offsetting the silver theme were black ten-spoke wheels, a black headlamp bowl, black-finished mirrors, and black signature-shaped engine covers. Minimalist turn signal indicators, bench seat, and a compact LED taillight completed the EC1's clean silhouette.

Thruxton RS Ton Up Special Edition

The Thruxton RS Ton Up Special Edition was another model only available for one year. It was a homage to the legendary Ton Up (100-mile-per-hour) boys of the 1950s and 1960s—the original café racers—and Malcolm Uphill's first-ever production 100-mile-per-hour lap of the Isle of Man TT in 1969. The Ton Up's classically inspired color scheme included an Aegean Blue fuel tank with Jet Black kneepad graphics and edged with hand-painted silver coach lining. This was paired with a Fusion White seat cowl and front fender, both adorned with Carnival Red "100" graphics. The seat cowl featured additional black hand-painted coach lining.

The Jet Black side panels included a Thruxton RS Ton Up logo and matched the black headlight bowl and rear fender. Other custom-inspired details included Matte Aluminum Silver fork protectors and blacked-out wheels, engine covers, and RSU springs. An optional Aegean Blue cockpit fairing was available. Other equipment was shared with the Thruxton RS, including the high-spec Showa fork, Öhlins shock absorbers, Brembo front brakes, Monza-style fuel filler cap, twin-brushed stainless-steel upswept silencers, single bullet seat, and black bar-end mirrors.

Speed Twin Breitling Limited Edition

In 2022, Triumph partnered with leading watch manufacturer Breitling with the Speed Twin Breitling Limited Edition and Breitling Top Time Chronograph. The unique Polychromatic Blue paint scheme was inspired by the 1951 Thunderbird 6T, driven by Marlon Brando in the iconic 1953 biker movie *The Wild One*. The Polychromatic Blue featured on the rare 1970s blue-dialed Breitling Top Time (Ref. 815). Stylists used a spectrograph to match the color from an original 1950s Triumph factory paint sample book and a mint-condition color chip of the original Thunderbird paint option. The Polychromatic Blue was then custom-mixed, hand-masked, and painted with a matching Jet Black stripe and Breitling script. This unique limited-edition scheme was complemented by hand-painted detailing applied by lead Triumph paint team artist Gary Devine.

The Breitling Special Edition was fitted with higher-specification, fully adjustable Öhlins piggy-back twin rear suspension units with gloss black springs, billet-machined aluminum clutch and alternator embellishers with a dark anodized finish, along with machined Breitling branding. A perforated black leather seat with contrast Gray stitching and bespoke Breitling Speed Twin instrument faces styled from the Breitling Top Time Triumph watch dial completed the premium detailing. To celebrate the Speed Twin's 270-degree parallel twin engine, the Breitling Edition Speed Twin was limited to 270 numbered examples worldwide and owners were offered the opportunity to purchase a Breitling Top Time Triumph limited-edition chronograph with a personalized matching edition number.

Triumph and Gibson 1959 Legends Custom Collaboration

Another collaboration this year was the 1959 Legends Custom Collaboration with Gibson Guitars. Created to support the Distinguished Gentleman's Ride (DGR), a global charity for men's mental health and prostate cancer awareness, the 1959 Legends custom edition was inspired by the shared historical significance of the iconic 1959 Les Paul Standard and 1959 Bonneville T120. The result of this partnership was a custom version of the modern generation of each of these legends: the Gibson Les Paul Standard Reissue and Triumph Bonneville T120 1959 Legends Custom Edition. The T120 1959 Legends Custom Edition featured unique Gibson design detailing, including a hand-painted sunburst paint scheme with a Jet Black painted guitar neck and headstock shape. The 1959 Legends guitar and motorcycle were offered together as a reward to the highest fundraiser for the 2022 Distinguished Gentleman's Ride.

The blue colors on the Speed Twin Breitling Limited Edition tank were matched to the original 1951 Thunderbird 6T. *Triumph Motorcycles*

The Gibson Les Paul Standard Reissue guitar and Triumph Bonneville T120 1959 Legends Custom Edition were created to support the 2022 Distinguished Gentleman's Ride. They were offered together as a reward to the highest fundraiser. *Triumph Motorcycles*

CHAPTER TEN

Spanish star Pedro Acosta dominated the 2023 Moto2 World Championship on the Triumph-powered KTM. *Triumph Motorcycles*

Acosta was presented with a custom Street Triple RS 765 in Valencia at the final round of the 2023 Moto2 season. *Triumph Motorcycles*

In April 2023 Iván Cervantes set a new Guinness World Record on a Tiger 1200 GT Explorer at the Nardò Technical Center in Italy, traveling more than 4,012 kilometers in twenty-four hours. *Triumph Motorcycles*

2023

During the 2022 Moto2 season, Triumph announced several developments for the 765cc Moto2 three-cylinder engine. These included a new cylinder head with a higher compression ratio, a new camshaft profile and longer valves to increase valve lift, and revised valve springs. To maintain the engine's impressive reliability record, improvements focused on new pistons, con rods, and crankshaft to cope with the higher piston pressure of 90 BAR (up from 85 BAR). This resulted in an increase of 5 horsepower and 400rpm. Pedro Acosta, riding for the Red Bull KTM Ajo Team with a Kalex frame, won the Moto2 World Championship and the Triumph Triple Trophy. This was the first time the Trophy was won by the world champion. For this year, Triumph signed a new contract to continue as Exclusive Engine Supplier for the FIM Moto2 World Championship for another five seasons, from 2025 to 2029. A new race gearbox was under development, projected to debut for the 2025 season. At the end of the 2023 season, London's Royal Automobile Club awarded its Torrens Trophy to Triumph Motorcycles for its work as the exclusive engine supplier to the Moto2 World Championship since 2019. The Trophy has been awarded to Britain's highest achievers in motorcycling and motorcycle racing—riders, engineers, manufacturers, and important personalities—since 1979. In the 2023 FIM Supersport World Championship, Harry Truelove and Niki Tuuli joined the PTR Triumph team. Tuuli eventually finished eighth overall.

On April 30, Iván Cervantes officially claimed the Guinness World Records title for the "greatest distance on a motorcycle in 24 hours (individual)." Riding a Tiger 1200 GT Explorer on the High-Speed Ring at the Nardò Technical Center in Italy, he traveled more than 4,012 kilometers in a twenty-four-hour period, beating the previous record by more than 600 kilometers. Triumph also confirmed plans to compete in the FIM Motocross World Championship and AMA SuperMotocross World Championship in 2024.

Many updates to the existing model range extended to colors only. This included the Modern Classics, Roadster, Rocket 3, and Tiger 900 and 850 Sport. In the Modern Classics, the Street Twin became the Speed Twin 900 and the Street Scrambler became the Scrambler 900.

Street Triple R, Street Triple RS, Street Triple Moto2 Edition

The Street Triple 765 lineup was significantly updated for 2023, now headed by a limited-run Street Triple 765 Moto2 Edition. This was the highest-specification and most sporting Street Triple yet, and the closest to a Moto2 racer for the road. The engine incorporated performance updates derived directly from the Moto2 racing program, including new pistons, con rods and

Model New Colors 2023

Speed Twin 900
Matte Silver Ice
Speed Twin 1200
Matte Baja Orange
Scrambler 900
Carnival Red and Jet Black
Scrambler 1200 XE and Scrambler 1200 XC
Carnival Red and Jet Black
Bonneville T100
Meriden Blue and Tangerine
Bonneville Bobber
Red Hopper
Bonneville T120
Aegean Blue and Fusion White
Bonneville Speedmaster
Cordovan Red
Bonneville T120 Black
Sapphire Black and Matte Sapphire Black
Thruxton RS
Competition Green and Silver Ice
Speed Triple 1200 RS
Matte Baja Orange
Street Triple 765 RS
Carbon Black
Street Triple 765 R
Matte Carbon Black
Trident 660
Matte Baja Orange and Matte Storm Gray
Rocket 3 R
Matte Silver Ice, Sapphire Black
Rocket 3 GT
Carnival Red and Sapphire Black, Sapphire Black
Tiger 900 GT and Tiger 900 GT Pro
Caspian Blue and Matte Graphite
Tiger 900 Rally and Tiger 900 Rally Pro
Sandstorm
Tiger 850 Sport
Graphite & Baja Orange

a Showa 41mm upside-down, separate-function, big-piston fork and a Showa Monoshock. The Street Triple RS and Moto2 Edition featured higher-specification Brembo Stylema 4-piston radial Monobloc front calipers with twin 310mm floating discs, while the Street Triple R brakes included front Brembo M4.32 4-piston radial Monobloc calipers. The more commanding riding position of the R and RS included 12mm-wider handlebars, while the Moto2 Edition was more race inspired, with lower and clip-on handlebars that were further forward. Both the RS and the Moto2 Edition featured revised geometry, with a steeper rake and a raised rear. Technological aids included optimized cornering ABS and cornering traction control and four riding modes with more dynamic throttle maps. All models received sharper, more aggressive styling, with the Moto2 Edition in carbon fiber with official branding. The Moto2 Edition was available in two color schemes, with 765 of each produced.

Many models received new colors for 2023, and the Street Scrambler became the Scrambler 900. Triumph Motorcycles

The Street Triple Moto2 Edition came with front and rear Öhlins suspension. Triumph Motorcycles

wrist pins, matched to higher-compression optimized combustion chambers. New valves and camshafts provided increased valve lift, and the gearbox included shorter gear ratios to improve acceleration. With a new, freer-flowing exhaust system, the power was increased for all three versions. The gullwing swingarm was designed to provide excellent torsional stiffness with lateral flexibility. This maximized high-speed stability and was complemented by specific suspension for each model. The Street Triple 765 Moto2 Edition included a 43mm upside-down Öhlins front fork and an Öhlins rear shock. Street Triple RS was equipped with a higher-specification Showa 41mm upside-down big piston fork and an Öhlins piggyback reservoir rear shock, while the Street Triple R was fitted with

CHAPTER TEN

Although the Street Triple RS shared its high-performance three-cylinder engine with the Moto2 Edition, the chassis specification wasn't as high. *Triumph Motorcycles*

From 2023 Street Triple Moto2 Edition, Street Triple RS, Street Triple R
(Differing from 2022)

Horsepower
128.2 at 12,000rpm (Moto2 and RS); 118.4 at 12,500rpm (R)
Compression ratio
13.25:1
Front suspension
Öhlins NIX30 (Moto2)
Wheelbase
55 inches (1,397mm) (Moto2); 55.1 inches (1,399mm) (RS); 55.2 inches (1,402mm) (R)
Wet weight
414 pounds (188kg) (Moto2 and RS); 416 pounds (189kg) (R)
Colors
Yellow, White (Moto2); Silver, Red, Yellow (RS); Silver White (R)

Chrome Collection
Following the success of the 2022 Gold Line Limited Editions, the Chrome Collection was offered as a one-year-only limited edition. Inspired by the chrome-plated fuel tank that first appeared on the 1937 Speed Twin and continued through to the custom Tritons of the 1960s, the Chrome Collection was available in ten specific variants, encompassing the Bonneville and Rocket 3 lineups. Each of the ten new limited editions featured a unique Chrome Edition scheme, chosen specifically to reflect each model's heritage. Triumph's expert hand-crafted chrome finish was at the forefront, the chrome fuel tank complemented by individual, bespoke detailing.

All the new Street Triple 765s, including the Street Triple R here, received new, sharper, more aggressive styling. *Triumph Motorcycles*

2023 Chrome Collection

Rocket 3 R Chrome Edition
Chrome fuel tank with Jet Black accent, Jet Black flyscreen, headlight bowls, front fender, radiator cowls, side panels, and rear bodywork.

Rocket 3 GT Chrome Edition
Chrome fuel tank with Diablo Red accent, Jet Black headlight bowls, fly screen, front fender, radiator cowls, side panels, and rear bodywork.

Bonneville T120 Chrome Edition
Chrome tank with Meriden Blue painted accents, Jet Black fenders, headlight bowl and side panels.

Bonneville Bobber Chrome Edition
Chrome tank with a Jet Black painted overlay, Triumph triangle tank badging, Jet Black fenders and side panels.

Scrambler 1200 XE Chrome Edition
Chrome fuel tank with Brooklands Green painted tank stripe, brushed aluminum fenders and heat shield, Jet Black headlight bowl and side panels.

Bonneville Speedmaster Chrome Edition
Chrome tank with Diablo Red Surround, Jet Black fenders, side panels, and headlight bowl.

Thruxton RS Chrome Edition
Chrome tank with Jet Black painted seam, Jet Black fenders, side panels, seat cowl, and headlight bowl.

Bonneville T100 Chrome Edition
Cobalt Blue tank with Chrome Edition metal stripe, chrome badges and fuel filler cap, Jet Black fenders and side panels.

Speed Twin 900 Chrome Edition
Red Hopper color scheme with Chrome Edition metal kneepad infills, Jet Black tank stripe, Triumph triangle tank badging with metal detailing, Jet Black fenders and side panels with red and silver graphics.

Scrambler 900 Chrome Edition
Brooklands Green color scheme with Jet Black tank stripe, Chrome Edition metal kneepad infills, Triumph triangle tank badging with chrome detailing, Jet Black fenders, side panels and frame cowl.

Speed Triple 1200 RR Bond Edition

Continuing the Bond partnership, Triumph celebrated sixty years of Bond with the release of an exclusive Speed Triple 1200 RR Bond Edition. This included a three-color Black, Granite Gray, and Storm Gray paint scheme in with the official "60 Years of Bond" commemorative logo on the side of the tank. All twenty-five James Bond movies, in their original title fonts, were featured in the tank top design. The tank was finished with official 007 graphics, hand-painted gold lining, and custom gold logo detailing. To emphasize the Bond association, the cockpit included a distinctive

The Chrome Collection was a one-year-only limited edition offered in 2023. This is the Bonneville T120 Chrome Edition. Triumph Motorcycles

The Thruxton RS Chrome Edition included a heritage-inspired chrome tank with black highlighting. Triumph Motorcycles

CHAPTER TEN

The most exclusive of the Bond Limited Editions was the Speed Triple 1200 RR Bond Edition. This included a sixtieth-anniversary black paint scheme featuring all twenty-five James Bond film titles. *Triumph Motorcycles*

DGR founder Mark Hawwa with the Bonneville T120 Black Distinguished Gentleman's Ride Limited Edition. *Triumph Motorcycles*

James Bond gun barrel design. Limited to only sixty worldwide, each Speed Triple 1200 RR Bond Edition was individually numbered from 1 to 60 on a dedicated handlebar clamp badge.

Bonneville T120 Black DGR Limited Edition

To celebrate ten years of the DGR, for 2023 Triumph released the Bonneville T120 Black DGR Limited Edition. Founded in Sydney, Australia, by Mark Hawwa in 2012, the DGR has successfully united hundreds of thousands of classic- and vintage-style motorcyclists all over the world, raising funds and awareness for men's health. The organization has grown to involve more than 90,000 riders taking to the streets in more than 800 cities. More than $37 million has been raised for prostate cancer research and men's mental health. Triumph became the official motorcycle partner of the DGR in 2014.

Triumph created 250 Bonneville T120 Black Distinguished Gentleman's Ride Limited Edition motorcycles as part of the celebration of the anniversary year. With a DGR metallic black-and-white paint scheme, official DGR branding, a custom logo on the tank and side panels, gold detailing, and distinctive brown seat, each came with a numbered certificate. Number 001 was presented to the highest fundraiser for the 2023 ride.

Tiger 900 Rally and GT Aragón Editions

A year after Iván Cervantes won the demanding Baja Aragón, Triumph launched the Tiger 900 Rally Aragón and the Tiger 900 GT Aragón. Available only in 2023, these special editions featured unique colors and specification. The Tiger 900 Rally Aragón included a triple color of Matte Phantom Black, Matte Graphite, and Crystal White, with Yellow accents and Aragón Edition detailing, along with a bespoke twin-color seat design. The lower-specification GT Aragón shared the Aragón Edition detailing, and twin-color seat design, but included a triple color scheme of Diablo Red, Matte Phantom Black, and Crystal White. Engine protection bars were standard on both Aragón Editions, while the Rally Aragón Edition included fuel tank protection bars.

2024

This year, Dorna introduced a rule that participants should use 40 percent non-fossil E40 fuel for the Moto2 World Championship, with a transition to E100 by 2027. Petronas supplied the fuel and Triumph, working in partnership with Dorna, was involved in the testing program for sustainable fuels. Triumph continued with entries the Supersport World Championship, with Ondrej Vostatek and Thomas Booth-Amos on the PTR Triumph Street Triple RS 765 and Jorge Navarro and John McPhee riding RS 765s for WRP Racing. Alongside the new TF 250c single-cylinder Motocross racer, Triumph introduced an all-new TR Series single-cylinder production engine platform for 2024. The Speed 400 and Scrambler 400 X joined the Modern Classic lineup as entry-level models. These were designed to expand the customer base and build brand loyalty, particularly in emerging markets such as India. The Motocross racing program came to fruition with two entries in the 2024 FIM MX2 class and three in the AMA SuperMotocross World Championship. This comprised Supercross and Pro Motocross, with three season-ending SuperMotocross rounds.

Speed 400 and Scrambler 400 X

The Speed 400 and Scrambler 400 X celebrated Triumph's single-cylinder pedigree. The all-new TR fuel-injected single-cylinder engine included a crankshaft weighted to optimize inertia for low-speed riding. A finger-follower valvetrain provided a low reciprocating mass and included friction-reducing DLC coatings. To improve engine aesthetics, the liquid cooling was concealed and the engine cases included machine-cooling fins. This engine was installed in two specific versions: the street-oriented Speed 400 and off-road Scrambler X. While both these models featured a new frame with bolt-on rear subframe and cast-aluminum swingarm, the Speed 400 and Scrambler 400 X each included a specific chassis setup and associated geometry. Along with a higher seat and more upright riding position, the Scrambler 400 X included a longer wheelbase, longer travel suspension, larger front wheel and front disc, and wider handlebars. Bosch engine management included a ride-by-wire throttle, and electronic aids extended to switchable traction control system and Bosch dual-channel ABS. The Speed 400 roadster joined the established Modern Classic Speed Twin 900 and 1200, while the Scrambler 400 X took its inspiration from the Scrambler 900 and 1200.

Scrambler 1200 X, 1200 XE

For 2024, a lower-spec Scrambler 1200 X replaced the 1200 XC, while the 1200 XE received more higher-specification components. Engine developments included a single 50mm throttle body and revised exhaust headers, with peak power now arriving slightly earlier in the rev range. With narrower handlebars, a lower seat, and shorter swingarm, the 1200 X was more street oriented than the 1200 XE. On both versions Marzocchi suspension replaced the previous Showa, the 1200 X with a 45mm nonadjustable fork and the 1200 XE with a fully adjustable 45mm 1+1 fork. The 1200 XE received twin-spring Marzocchi rear shock absorbers and the latest-generation Brembo Stylema radial Monobloc front brake calipers with 320mm floating discs.

The Tiger 900 Aragón Rally Pro and GT Pro celebrated Iván Cervantes's victory in the 2022 Baja Aragón. *Triumph Motorcycles*

The single-cylinder Speed 400 and Scrambler 400 X were built on a completely new platform. *Triumph Motorcycles*

CHAPTER TEN

For 2024 the Scrambler 1200 XE was updated with new suspension, brakes, and electronics. *Triumph Motorcycles*

Thruxton Final Edition

After a twenty-year production run, the Thruxton finally concluded in 2024. Triumph marked its farewell to this iconic café racer with the Thruxton Final Edition, a limited edition based on the higher-spec Thruxton RS. Exclusive features included Competition Green metallic paint with hand-painted gold lining, Thruxton Final Edition branding, and a Final Edition engine badge.

Tiger 900 GT, GT Pro, Rally Pro

Although only recently introduced, the Tiger 900 was updated for 2024 and the previous four model ranges reduced to three models: the Tiger 900 GT, GT Pro, and Rally Pro. The three-cylinder engine received revised engine components, resulting in a 13 percent power increase. With ergonomics and comfort prioritized, the seat was flatter and roomier, and all models included 20mm of height adjustability. Both Pro versions featured heated seats. Comfort was further improved by a new damped handlebar mounting system, while the handlebar position on the Rally Pro was moved back 15mm to improve bike agility in the standing position when riding off-road. New active safety features across the range included an emergency deceleration warning system, while new bodywork emphasized an aggressive, adventure-focused stance.

As in 2023, many updates to the existing model range extended to colors only. This included the Modern Classics, Roadster, Rocket 3, and Tiger Sport

From 2024 Speed 400 and Scrambler 400 X

Type
Single-cylinder, liquid-cooled, four-valve, DOHC
Bore
89.0mm
Stroke
64.0mm
Capacity
398.15cc
Horsepower
39.5 horsepower at 8,000rpm
Compression ratio
12:1
Fuel System
Bosch electronic fuel injection with electronic throttle control
Gearbox
Six-speed
Frame
Hybrid spine/perimeter, tubular steel, bolt-on rear subframe
Swingarm
Twin-sided, cast aluminum alloy
Front fork
43mm upside-down Big Piston fork
Rear suspension
Gas Monoshock
Brakes
Front 300mm disc 4-piston radial caliper Speed 400 (320mm disc *400 X*); 230mm rear disc
Wheels
17x3.00in and 17x4.00in (Speed 400); 19x2.5in and 17x3.5in (400 X)
Tires
110/70 R17 and 150/60 R17 (Speed 400); 100/90 R19 and 140/80 R17 (400 X)
Wheelbase
54.2 inches (1,377mm) Speed 400; 55.8 inches (1,418mm) (*400 X*)
Dry weight
374 pounds (170kg) Speed 400; 394 pounds (179kg) (*400 X*)
Colors
Red, Blue, Black (Speed 400); Green/White, Red/Black, Black/Silver (400 X)

850 and 660. The Modern Classics were distinguished by twelve new color options.

Bonneville Stealth Editions

In the wake of the successful Gold Line and Chrome Collection Custom editions, eight Bonneville Stealth Editions were offered for 2024. (See page 298.) Available for one year only, each Stealth Edition model featured a unique tank design that showcased its hand-painted finish. The innovative paint technique started with a base layer of mirror-finish metallic Silver

From 2023 Scrambler 1200X, XE
(Differing from 2023)

Horsepower	89 at 7,000rpm
Front fork	45mm Marzocchi Upside Down
Rear suspension	Twin Marzocchi
Front brakes	Twin 310mm discs Nissin twin-piston calipers (X)
Wheelbase	60 inches (1,525mm) (*XE*)
Wet weight	502 pounds (205kg) (*XE*); 506 pounds (207kg) (*XE*)
Colors	Red, Gray, Black (X); Black/Gray, Orange/Black, Black (XE)

Ice followed by a dark-to-light Sapphire Black graphite vignette. A translucent tinted lacquer was applied in multiple layers, building a deep, rich top coat to finish the process. The hand-painted finish provided subtle variations, with each tank being unique: to provide a bespoke custom style, the transition from dark-tinted graphite to vibrant color varied in hue and tone under different conditions.

Daytona 660

After an eight-year absence, the sporting Daytona was reintroduced for 2024. The 660 series was aimed at new, younger riders—especially women—and over the previous three years sales had totaled more than 40,000. The Daytona 660 continued on from the successful Trident 660 and Tiger Sport 660 and targeted sports riders who desired style, fun, and everyday usability at an attainable price point. The three-cylinder engine received a new crankshaft with wider gears, new camshafts and profile, new cylinder head, pistons and wrist pins, and new valve gear. With triple-throttle bodies and a three-into-one exhaust system, the power was increased 17 percent over the Trident 660. The suspension, five-spoke cast-alloy wheels, and brakes were all high specification, set off by the sporting full fairing with twin LED headlights and minimal rear bodywork. Clip-on handlebars mounted above the top-fork triple clamp contributed to a more spacious sporting riding position.

Tiger 1200 GT Pro, GT Explorer, Rally Pro, Rally Explorer

The four variants of the Tiger 1200 family were mildly updated for 2024. A new crankshaft, alternator rotor, and balancer increased engine inertia. This resulted in slightly more power and a smoother engine with more precise low-rev torque delivery. The Explorer's dampened handlebars and risers were now fitted to the GT Pro and Rally Pro, further damping vibration. The seat was flatter and more spacious, while higher footpegs

The Thruxton Final Edition celebrated twenty years of the iconic Thruxton café racer. *Triumph Motorcycles*

From 2024 Tiger 900 GT, GT Pro, Rally Pro
(Differing from 2023)

Bore
78mm
Capacity
888cc
Horsepower
106.5 at 9,500rpm
Compression ratio
13.0:1
Wet weight
481 pounds (219kg) GT; 488 pounds (222kg) GT Pro; 502 pounds (228kg) Rally Pro
Colors
White, Graphite/Black, Red/Black (GT, GT Pro); Black, Gray/Orange, Green/Black (Rally Pro)

After only four years, the Tiger 900 was updated for 2024 with more power and new styling. The 900 GT Pro is on the left and the Rally Pro on the right. *Triumph Motorcycles*

CHAPTER TEN

New colors for 2024 included Carnival Red on several models, including the Speed Twin 1200. Triumph Motorcycles

increased cornering clearance. A new Active Preload Reduction feature reduced rear suspension preload as the Tiger 1200 slowed to a stop, lowering seat height by up to 20mm.

Rocket 3 Storm R and GT

Twenty years after the Rocket III's release, Triumph announced a new generation Rocket 3 Storm for 2024. The Rocket 3 Storm boasted more power, 225Nm of torque at only 4,000rpm, and lighter ten-spoke cast aluminum wheels. These reduced unsprung mass and improved steering response. Following the success of recent blacked-out special editions, the new Rocket 3 Storm featured dark, moody colors, and a black finish. Offered as R and GT versions, with three two-tone color options, the GT had more swept back handlebars, a lower seat, and more forward foot controls.

TF 250-X

At the SuperMotocross World Championship Final held at the Los Angeles Memorial Coliseum in September 2023, Triumph publicly revealed the

The lineup of Bonneville Stealth Editions. The Bonneville T120 Blue Stealth Edition is in the foreground. Triumph Motorcycles

Model New Colors 2024

Speed Twin 900
Carnival Red and Phantom Black, Competition Green and Phantom Black
Speed Twin 1200
Carnival Red and Storm Gray, Matte Ironstone and Matte Storm Gray
Scrambler 900
Cosmic Yellow and Graphite
Scrambler 1200 XE and Scrambler 1200 XC
Matte Sandstorm and Matte Jet Black
Bonneville T100
Competition Green and Ironstone
Bonneville Bobber
Jet Black and Ash Gray
Bonneville T120
Jet Black and Fusion White
Bonneville Speedmaster
Pacific Blue and Silver Ice
Bonneville T120 Black
Graphite and Matte Graphite
Thruxton RS
Jet Black and Silver Ice
Speed Triple 1200 RS
Carnival Red
Trident 660
Jet Black and Triumph Racing Yellow
Rocket 3 R
Sapphire Black and Carnival Red with Silver Ice
Rocket 3 GT
Sapphire Black and Carnival Red with Silver Ice
Tiger Sport 850
Graphite and Jet Black
Tiger Sport 660
Snowdonia White and Jet Black

Bonneville Stealth Editions

Bonneville Speedmaster Red Stealth Edition
Bonneville Bobber Purple Stealth Edition
Bonneville T100 Blue Stealth Edition
Bonneville T120 Blue Stealth Edition
Bonneville T120 Black Stealth Edition (Silver Matte finish)
Speed Twin 1200 Red Stealth Edition
Speed Twin 900 Green Stealth Edition
Scrambler 900 Orange Stealth Edition

production-specification TF 250-X motocross bike. The bikes were ridden by Jeff "Six Time" Stanton, who won his last Championship at the Coliseum, and legendary Motocross racer Ricky Carmichael. The TF 250-X was a ground-up design, including a new engine, chassis, and electronics. Designed for competition, the compact and light four-stroke single included a forged-aluminum piston, titanium valves,

From 2024 Daytona 660

Type
In-line three-cylinder DOHC 4 valves per cylinder, liquid cooled
Bore
74.04mm
Stroke
51.1mm
Capacity
660cc
Horsepower
95 at 11,250rpm
Compression ratio
12.05:1
Fuel System
Multipoint sequential electronic fuel injection with electronic throttle control
Gearbox
Six-speed
Frame
Tubular steel perimeter
Swingarm
Twin-sided, fabricated steel
Front fork
Showa 41mm upside-down, separate-function big piston (SFF-BP) fork
Rear suspension
Showa Monoshock
Brakes
Twin 310mm with 4-piston radial calipers; single-piston caliper 220mm disc
Wheels
17x3.5in front and 17x5.5in rear
Tires
120/70 ZR17 front and 180/55 ZR17 rear
Wheelbase
56.1 inches (1,426mm)
Wet weight
442 pounds (201kg)
Colors
Black, Red, Silver

The fully faired Daytona made a return for 2024 with the middleweight sports Daytona 660. This was the third model in Triumph's successful 660 lineup. *Triumph Motorcycles*

an optional MX Tune Pro app, which enabled riders to select mapping, a real-time engine sensor dashboard, and live diagnostic functionality.

The aluminum chassis featured a lightweight spine frame with twin cradles, and high-spec KYB suspension. This included a 48mm AOS coil fork, forged and machined 7075-T6 aluminum triple clamps, and a three-way piggyback coil rear shock absorber. The

From 2024 Tiger 1200 GT Pro, GT Explorer, Rally Pro, Rally Explorer
(Differing from 2023)

Colors
Red, White, Black (GT Pro, GT Explorer); Sandstorm, Black, Khaki (Rally Pro, Rally Explorer)

diamond-like carbon low friction coatings, lightweight magnesium covers, and a wet multiplate Exedy Belleville Spring clutch. A controlled programmable engine management system was complemented by

From 2024 Rocket 3 Storm R and GT
(Differing from 2023)

Horsepower
180 at 7,000rpm
Wet weight
697 pounds (317kg) R; 704 pounds (320kg) GT
Colors
Red with Black, Blue with Matt Black, Black with Granite

Blacked out features distinguished the Rocket 3 Storm R. *Triumph Motorcycles*

299

The TF 250-X was designed for the 2024 Motocross World Championship MX2 class and AMA SuperMotocross World Championship. *Triumph Motorcycles*

Brembo braking system consisted of a twin 24mm-piston floating front caliper, a single 26mm-piston floating rear caliper, and Galfer front and rear discs. DirtStar 7000 Series aluminum rims and machined aluminum hubs were fitted with Pirelli Scorpion MX32 mid-soft tires.

With the TF 250-X already well developed, Triumph was in an excellent place to commence its motocross racing program in 2024. This included an entry in the 2024 Motocross World Championship FIM MX2 class and the AMA SuperMotocross World Championship. The factory-supported race program was established under the new Triumph Racing banner and set up in partnership with Thierry Chizat-Suzzoni. Vincent Bereni was team manager and, while research and development continued at Hinckley, the MX2 team operated out of Eindhoven in the Netherlands. Clément Desalle joined the team as test rider, and Mikkel Haarup and Camden McLellan were signed as the two MX2 riders for 2024. In the US, Triumph Racing operated in partnership with Bobby Hewitt with Steve "Scuba" Westfall as team manager. AMA legend Dave Arnold was engaged for chassis development, with Dudley Cramond responsible for engine building. The team for 2024 AMA Supercross and Pro Motocross consisted

The TF 250-X underwent an intensive development program during 2023. *Triumph Motorcycles*

From 2024 TF 250-X

Type
Single-cylinder, liquid-cooled, four-valve, DOHC
Bore
78mm
Stroke
52.3mm
Capacity
249.95cc
Compression ratio
14.4:1
Fuel System
Dell'Orto
Gearbox
Five-speed
Frame
Aluminum, spine
Swingarm
Aluminum fabrication
Front fork
KYB coil spring fork
Rear suspension
KYB Coil
Brakes
Brembo Twin Piston, 260mm disc (front); Brembo Single Piston, 220mm disc (rear)
Wheels
21x1.6in and 19x1.85in
Tires
80/100x21 and 100/90x19
Wheelbase
1,492mm
Dry weight
104kg

of Evan Ferry and Jalek Swoll. For the 2025 season, Triumph additionally fielded a new TR 450-X in the SuperMotocross World Championship.

Although Triumph's 120-year history has been marked by ups and downs, few motorcycle manufacturers have has such a significant run. The introduction of Edward Turner's Speed Twin established a blueprint for a parallel twin that ensured Triumph's place as the world's dominant motorcycle manufacturer for several decades. While the parallel twin continues to be important, particularly for the traditional modern classic range, Triumph's triple range has expanded to embrace a wider range of popular adventure bikes. The introduction of the 400 single will allow for a foothold in the Asian market and the new TF 250-X motocross signifies an important move into off-road competition. The electric TE-1 is likely to spawn a range of electric motorcycles, and the commitment to supplying engines for the Moto2 World Championship will ensure that Triumph continues as Britain's premier motorcycle manufacturer for a long time to come.

Index

A

Alcoba, Jeremy, 283
Aldana, Dave, 121, 129
Allen, Johnny, 44, 49, 51, 64, 255, 258
Alves, Jim, 26, 48
AMA Grand National Championship, 106, 110, 121, 129
AMA Pro Flat Track, 252–253
AMA Pro Road Racing Championship, 249
AMA SportBike Championship, 242, 245
AMA SuperMotocross World Championship, 290, 295, 299, 300
Anstey, Bruce, 210
Armstrong Equipment, 160–161
Arnold, Dave, 300
Ascot National half mile, 68

B

Baird, Bill, 87, 99, 106
Bajaj Auto India, 260, 270
Baldoloni, Alex, 242
Belstaff, 224, 226
Benn, Tony, 149
Bereni, Vincent, 300
Bettmann, Siegfried, 6, 8, 9
Bezzecchi, Marco, 270
Big Bear Motorcycle Run, 54, 59, 61, 67, 69–70
Bloor, John, 170, 173–174, 233
Bloor, Nick, 269, 274
Bobber TFC, 272–273
Bol d'Or 24 Hour race, 121, 129
Bonneville Speed Week, 226
Booth-Amos, Thomas, 295
British Supersport Championship, 218, 224, 245, 249
Brown, Don, 102, 138
Brufoldt, Jim, 96
BSACI, 112–113, 121, 127, 137, 138, 144
BSA Group, 36, 51, 59, 63, 108, 127, 133, 137–138, 150
Burnett, Don, 80

C

Cambridge (Minnesota) Enduro, 40
Campbell, Malcom, 18
Canadian Pro Sport Bike Championship, 249
Capri, Matt, 226, 229
Caracchi, Stefano, 223
Carmichael, Ricky, 283, 299
Carpenter, Bob, 243
Castro, Don, 121, 129
Catalina Grand Prix, 32, 48, 51, 59, 61, 62
Cathcart, Alan, 174, 229–230
Cedar, Wilbur, 77, 95
Cervantes, Iván, 283, 290
Chizat-Suzzoni, Thierry, 300
Clarke, Freddie, 22
Clubmans Senior TT, 36
Coates, Ron, 23, 36, 120, 122

Colman, Pete, 95, 121
Copeland, Jock, 158
Cramond, Dudley, 300
Crossley, Don, 23

D

Davies, Chaz, 235
Daytona 100-mile amateur race, 23
Daytona 200, 68, 80, 99, 106, 110, 118, 121, 129, 133, 248–249
Daytona Battle of the Twins Race, 168
Dean, James, 46
Desalle, Clément, 300
Devine, Gary, 289
DiSalvo, Jason, 242–243, 249
Distinguished Gentleman's Ride (DGR), 289
Dudek, Joe, 78
Dylan, Bob, 90–91

E

Earls Court Motorcycle Show, 14, 26, 43, 224
Easter Anglo-American Match races, 129
EICMA Milan Motorcycle Show, 212
Ekins, Bud, 27, 44, 48, 51, 54, 59, 61, 67, 78, 79, 87, 89, 274
Ekins, Donna, 274
Ekins, Susan, 274
Elmore, Buddy, 99, 106, 110
Emde, Don, 129
Eslick, Danny, 249, 283

F

Fédération Internationale de Motocyclisme (FIM), 49, 78, 229–230
Fernandez, Augusto, 283
Fernandez, Raul, 276
Ferry, Evan, 300
Fong, Bobby, 254
Fredericks, Les, 14
Fulton, Walt, 32

G

Gaffers' Gallop, 44
Gardner, Remy, 276
Gere, Richard, 167
German Supersport Championship, 224–225
Giles, John, 48, 87, 252
Gould, Rod, 104
Granfield, Charles, 60, 72
Grant, Mick, 145–146
Greenhorn Enduro, 59, 67

H

Haarup, Mikkel, 300
Hailwood, Mike, 60, 121, 129
Halford, Frank, 8
Hammer, Dick, 99, 106
Hammond, Peter, 48
Hargreaves, Bryan, 36
Harris, Les, 170–171, 173
Hartle, John, 104

Hawley, Don, 48, 62
Hawthorne, Jack, 158, 160
Hawwa, Mark, 294
Heanes, Ken, 56, 87, 252
Hele, Doug, 83, 91, 99, 113, 121, 130, 134, 149
Hennegan, Chris, 219, 227
Herz, Wilhelm, 49
Hewitt, Bobby, 300
Hinckley Triumph models
    America, 212, 214–216, 219, 221–222, 226, 229, 234–235, 237, 243, 246–249, 253, 259
    America LT, 248–249, 253, 259
    Belstaff Bonneville, 224, 226
    Bonneville, 203–204, 206, 212, 214–215, 218–221, 223, 229–230, 235–236, 240–242, 247, 251–252, 255
    Bonneville 50th Anniversary, 229–230
    Bonneville America, 205–206
    Bonneville Black, 223
    Bonneville Bobber, 260, 278, 291, 298
    Bonneville Bobber Black, 264
    Bonneville Bobber Chrome Edition, 293
    Bonneville Bobber Gold Line Edition, 286–288
    Bonneville Newchurch, 255
    Bonneville SE, 218–219, 229–230, 235–236, 240–242
    Bonneville SE (2013 Australia only), 247
    Bonneville Sixty, 235–236
    Bonneville Speedmaster, 264, 278, 291, 298
    Bonneville Speedmaster Chrome Edition, 293
    Bonneville Speedmaster Gold Line Edition, 287
    Bonneville Speed Twin, 278–279
    Bonneville Spirit, 255
    Bonneville Stealth Editions, 298
    Bonneville Street Twin, 278
    Bonneville Street Twin Gold Line, 278
    Bonneville T100, 212, 214–215, 218–221, 223, 229–230, 235–236, 240–242, 247, 251–252, 278, 291, 298
    Bonneville T100 110th Anniversary Limited Edition, 241–242
    Bonneville T100 Centennial, 206
    Bonneville T100 Chrome Edition, 293
    Bonneville T100 Gold Line Edition, 287
    Bonneville T100 SE, 251–252
    Bonneville T100 Steve McQueen Edition, 241–242
    Bonneville T120, 258–259, 278, 291, 298
    Bonneville T120 Ace, 268–269
    Bonneville T120 Black, 258–259, 278, 291, 298
    Bonneville T120 Black DGR Limited Edition, 294
    Bonneville T120 Black Gold Line Edition, 287
    Bonneville T120 Chrome Edition, 293
    Bonneville T120 Diamond Edition, 268–269
    Bonneville T120 Gold Line Edition, 287
    Bonneville T214, 255
    Bud Ekins Bonneville T100 Special Editions, 274

301

INDEX

Bud Ekins Bonneville T120 Special Edition, 274
Daytona 600, 210, 212–213
Daytona 650, 218
Daytona 660, 297
Daytona 675, 219–220, 222–225, 227, 229, 235, 239, 242–245, 249, 254, 259–260
Daytona 675R, 239, 242–245, 249, 254, 259–260
Daytona 675 SE, 224–225, 235
Daytona 750, 175, 176–177
Daytona 900, 178–179, 180–181, 184, 186–187
Daytona 955i, 198, 200–202, 204–205, 210–213, 217, 219–220
Daytona 955i Centennial Edition, 204–205
Daytona 955i SE, 217
Daytona 1000, 175, 176–177
Daytona 1200, 178–179, 180–181, 184, 186–187, 191–193
Daytona 1200SE, 192
Daytona 1200SP, 192–193
Daytona Moto2 765 Limited Edition, 270–271
Daytona SE, 212–213, 227, 229
Daytona Super III, 181, 183, 184, 186–187
Daytona T595, 189–192
Ewan McGregor Bonneville, 224, 226
Gibson 1959 Legends Custom Collaboration, 289
Legend TT, 195, 198–199, 201, 204
Rocket 3 GT, 269–270, 288, 291, 298
Rocket 3 GT Chrome Edition, 293
Rocket 3 GT Triple Black Limited Edition, 280
Rocket 3 R, 269–270, 288, 291, 298
Rocket 3 R Black Limited Edition, 280
Rocket 3 R Chrome Edition, 293
Rocket 3 TFC, 269
Rocket III, 212, 216, 219, 221–222, 226, 229, 234
Rocket III Classic, 219, 221–222, 226, 229
Rocket III Roadster, 234–235, 237, 243, 246, 248–249, 253, 259
Rocket III Touring, 225–226, 229, 234–235, 237, 243, 246, 248–249, 253, 259
Scrambler, 220, 223, 226, 229–230, 235–236, 240–242, 247, 251–252, 255
Scrambler 400 X, 295
Scrambler 900, 290, 291, 298
Scrambler 900 Chrome Edition, 293
Scrambler 1200 Bond Edition, 276
Scrambler 1200 Steve McQueen Edition, 280
Scrambler 1200 X, 295
Scrambler 1200 XC, 267, 280, 291, 298
Scrambler 1200 XC Gold Line Edition, 287
Scrambler 1200 XE, 267, 268, 280, 291, 295, 298
Scrambler 1200 XE Chrome Edition, 293
Scrambler 1200 XE Gold Line Edition, 287
Special Edition Rocket 3 GT, 288
Special Edition Rocket 3 R, 288
Speed Four, 209–210, 212–213, 218
Speed 94, 254–255
Speed 94R, 254–255
Speed 400, 295

Speedmaster, 210–211, 214–216, 219, 221–222, 226, 229, 234–235, 237, 243, 246–249, 253, 259
Speed Triple, 198, 200–202, 204–205, 210–213, 216–217, 219–220, 222–225, 229, 235, 239–241, 245–246, 250, 254–255
Speed Triple 15th Anniversary Special Edition, 229
Speed Triple 750, 186–187
Speed Triple 900, 182–184, 186–187
Speed Triple 1200 RR, 284
Speed Triple 1200 RR Bond Edition, 293–294
Speed Triple 1200 RS, 278, 291, 298
Speed Triple R, 241, 245–246, 250, 254–255, 259–260
Speed Triple RS, 265–266
Speed Triple S, 259–260, 265–266
Speed Triple SE, 212–213, 235, 245–246
Speed Triple T509, 189–192
Speed Twin, 266–267
Speed Twin 900, 290, 291, 298
Speed Twin 900 Chrome Edition, 293
Speed Twin 1200, 291, 298
Speed Twin Breitling Limited Edition, 289
Sprint 900, 180, 184–187
Sprint Executive, 192–193
Sprint GT, 238–239, 242–243, 246, 257
Sprint GT SE, 250, 254
Sprint RS, 200–202, 204–205, 210–211
Sprint Sport, 192–193
Sprint ST, 195–197, 201–203, 205, 210–213, 216–217, 219–220, 222–223, 225, 229, 235, 238–239
Street Cup, 262
Street Scrambler, 261–262, 268, 281
Street Scrambler Gold Line Edition, 287
Street Scrambler Sandstorm Limited Edition, 281
Street Triple, 224, 227, 235, 239–241, 245–246, 250, 254–255, 259–260
Street Triple 765 R, 291
Street Triple 765 RS, 276, 291
Street Triple R, 227, 229, 235, 239–241, 245–246, 250, 254–255, 259–260, 262–264
Street Triple Moto2 Edition, 290–291
Street Triple R, 290–291
Street Triple RS, 271–272, 290–291
Street Triple Rx SE, 254–255, 259–260
Street Triple S, 262–264
Street Twin, 258–259, 268
Street Twin EC1 Special Edition, 288
TE-1, 284
TF 250-X, 298–300
Thruxton, 226–227, 229–230, 235–236, 240–242, 247, 251–252, 255, 258–259
Thruxton 900, 213–214, 218–221, 223
Thruxton Ace, 255
Thruxton Final Edition, 296
Thruxton R, 258–259
Thruxton RS, 273, 291, 298
Thruxton RS Chrome Edition, 293
Thruxton RS Ton Up Special Edition, 288–289
Thruxton SE, 235–236

Thruxton TFC (Triumph Factory Custom), 269
Thunderbird, 185–187, 191–193, 198, 201, 204, 206, 212, 233–234, 237, 243, 246, 248–249, 253, 259
Thunderbird Commander, 248–249, 253, 259
Thunderbird LT, 248–249, 253, 259
Thunderbird Nightstorm, 253, 259
Thunderbird SE, 237
Thunderbird Sport, 191, 193, 195, 198, 201, 212, 214–215
Thunderbird Storm, 237, 243, 246, 248–249, 253, 259
Tiger, 197–198, 201–203, 205, 210–213, 217, 219–220, 222, 225, 229, 235
Tiger 800, 237–238, 240–241, 244, 249–250
Tiger 800XC, 237–238, 240–241, 244, 249–250, 255–256, 257, 265
Tiger 800XCA, 255–256, 257
Tiger 800XC SE, 249–250
Tiger 800XCx, 255–256, 257
Tiger 800XR, 255–256, 257, 265
Tiger 800XRt, 255–256, 257
Tiger 800XRx, 255–256, 257
Tiger 850 Sport, 277, 291
Tiger 900, 179, 181, 186–187, 191–193, 274, 275
Tiger 900 Bond Edition, 286
Tiger 900 GT, 274, 275, 291, 296
Tiger 900 GT Aragón Edition, 294–295
Tiger 900 GT Pro, 274, 275, 291, 296
Tiger 900 Rally, 274, 275, 291
Tiger 900 Rally Aragón Edition, 294–295
Tiger 900 Rally Pro, 274, 275, 277, 283, 291, 296
Tiger 1050, 237–238, 240–241, 244
Tiger 1050 SE, 237–238, 240–241
Tiger 1200, 265
Tiger 1200 Alpine Special Edition, 275
Tiger 1200 Desert Special Edition, 275
Tiger 1200 GT, 285–286, 297
Tiger 1200 GT Explorer, 290, 297
Tiger Explorer, 240–241, 244, 249–250, 255–256
Tiger Explorer Wire-Wheel, 249–250
Tiger Explorer XC, 244, 249–250, 255–256, 257
Tiger Explorer XCa, 257
Tiger Explorer XCx, 257
Tiger Explorer XR, 257
Tiger Explorer XRt, 257
Tiger Explorer XRx, 257
Tiger GT Explorer, 285–286
Tiger GT Pro, 285–286, 297
Tiger Rally Explorer, 285–286
Tiger Rally Pro, 285–286, 297
Tiger SE, 229, 235
Tiger Sport, 244, 249–250, 255–256, 257
Tiger Sport 660, 284, 298
Tiger Sport 850, 298
Trident 660, 277, 291, 298
Trident 750, 174–176, 177–178, 180, 184, 186–187, 191–193

Trident 900, 174–176, 177–178, 180, 184, 186–187, 191–193
Trident 900 Sprint, 177–178
Trophy, 246, 250, 255
Trophy 900, 175–177, 179, 181, 184, 187, 191–193, 198, 201–203
Trophy 1200, 175–177, 179, 181, 184, 187, 191–193, 198, 201–203, 205, 210–211
Trophy SE, 246, 250, 255
TT600, 199–201, 204–205, 210–211
See also Triumph models
Hinckley works, 174, 177, 179–180, 189, 204, 209, 216, 230, 247–248
Holbrook, Claude, 8
Honda, 69, 80, 108, 160
Hooper, Harry, 161
Hopwood, Bert, 13, 51, 72, 83, 87, 91, 95, 99, 102, 113, 119, 127, 130, 137

I

International Six Days Trial (ISDT), 23, 56, 76, 78, 87, 252
Isle of Man Formula 750 TT, 133
Isle of Man Junior TT, 210
Isle of Man Production TT, 104, 115, 121, 129, 134, 137, 145–146
Isle of Man Supersport TT, 249
Isle of Man Tourist Trophy (TT) races, 6, 8, 23, 248

J

Jeffries, Tony, 129, 133, 137
Jofeh, Lionel, 102, 112, 119, 127
Johnson, Bill (of Johnson Motors, Inc.), 18–19, 21, 31, 63, 77
Johnson, Bill (speed record holder), 78
Johnson, Dennis, 153, 154
Johnson Motors, Inc., 18, 20, 29, 32, 62–63, 68, 91, 95, 108
Jones, Brian, 149, 155, 156, 158
Jones, Craig, 218

K

Kalinski, Felix, 132–133, 144
Knievel, Evel, 104

L

Leppan, Bob, 96
Lindsay, Bob, 161, 162
Lopez, David, 219, 234
Lyons, Ernie, 22

M

Mallenotti, Michele, 226
Mangham, J. H. "Stormy," 49
Mann, Dick, 121, 129, 137
Manx Grand Prix, 22–23
Manzi, Stefano, 283
Markstaller, Matt, 243
Martin, Bill, 68
Marubeni/Suzuki, 160–161
Marvin, Lee, 62
Martin, Guy, 256–257
Maudes Trophy, 17–18
McConnell, Billy, 245

McCormack, Denis, 32, 63, 77, 102, 127, 132
McCoy, Gary, 229
McCoy, John, 130–132, 153
McGregor, Ewan, 224, 226, 227
McLellan, Camden, 300
McQueen, Steve, 46, 79, 89, 220, 223, 226, 241-242, 251, 252, 274, 280, 281
Meriden Cooperative, 153–155, 158–159, 165, 233
Mettam, Stephen, 138
Miller, Earl, 77, 102
Minonno, Jon, 168
Mockett, John, 189, 204
Moto2 World Championship, 263, 270, 276, 290, 295, 300
Moto Guzzi, 153
Moulton, Wayne, 162, 165
Mulder, Eddie, 85, 99, 106, 133

N

National Championship Dirt Track Race, 90
National Enduro Championship, 87, 90, 99, 106
National Exhibition Centre (NEC) Show, 162, 168, 174
Nelson, John, 95, 155
Nixon, Gary, 90, 99, 106, 110, 118, 121, 129
NORRA Mexican 1000, 268
North, Rob, 129
North West 200 production race, 121
Norton-Triumph Inc. (NTI), 149, 154
Norton-Villiers, 108, 133, 137
Norton Villiers Triumph (NVT), 137, 148, 149, 151, 153–154

O

Ogle Design, 102
Öhlins, 233

P

Paasch, Brandon, 276, 283
Page, Val, 9, 12–13
Peplow, Roy, 252
Phillips, Jimmy, 32
Pickrell, Ray, 129, 133–134
Poore, Dennis, 133, 137, 144, 149, 151, 153
Prentice, Tim, 234
Price, Brenda, 154, 160

R

Racing Spares, 170–171
Reilly, Brian, 156
Ricardo, Harry, 8
Rice, Jim, 121, 129
Richards, Glen, 224, 242
Riedmann, Kenny, 254
Robinson, Brandon, 253
Robinson, Geoffrey, 158, 161
Rocket 3 Storm R and GT, 298
Romero, Gene, 106, 121, 129, 137
Rosamond, John, 161, 165
Routt, Sonny, 120, 122
Royal Automobile Club (London), 290

S

Sanders, Nick, 189
Sandgren, Bob, 59, 61

Sangster, Jack, 9, 12, 18, 32, 72
Savatgy, Joey, 300
Sayer, Ray, 252
Schulte, Mauritz, 6, 8
Scott, Bill, 87
Scott, Gary, 137
Shoemaker, Jake, 253
Shorey, Dan, 60
Smart, Paul, 129
Smith, Buck, 67
Smith, Bud, 90
Smith, Kyle, 276
Soomer, Hannes, 283
Speed Triple Challenge series, 184
Springsteen, Bruce, 185
Stange, Roger, 144, 154
Stanton, Jeff "Six Time," 298
Sturgeon, Harry, 87–88, 91, 95, 99, 102
Supersport World Championship, 283, 295
Superstock 600 European Championships, 223
Suzuki, 160–161
Swoll, Jalek, 300

T

Tait, Percy, 51, 104, 106, 117, 121, 129, 137
Taylor, Robert, 19
Thornton, Peter, 112–113, 119, 121, 127
Thruxton 500 race, 73, 87, 117
Thruxton Cup Challenge, 227
Thruxton Nine Hour race, 45, 51
Tode, Arne, 225
TriCor, 23, 32, 62–63, 68, 108
Triumph
    Hinckley era/works, 174, 177, 179–180, 189, 204, 209, 216, 230, 247–248
    Meriden Cooperative, 153–155, 158–159, 165, 233
    serial numbers, 115
Triumph models
    3T 350, 19–20, 24
    3TA Twenty-One, 68, 71–72, 75–76, 78, 80–81, 87, 90–91, 94, 100–101
    3T DeLuxe, 21, 24, 29
    5TA Speed Twin, 67–68, 71–72, 75–76, 78, 80–81, 87, 90–91, 94, 100–101
    5T Speed Twin, 13–15, 17–21, 24–25, 29, 31–32, 35–39, 43, 45, 53–54, 57, 60, 87
    6/1 twin, 9
    6T Thunderbird, 29–32, 34, 36–38, 41, 45, 53–54, 57, 60, 64, 67, 70, 74, 78, 85–88, 90, 93–94, 98, 100
    250 single, 13
    Bandit, 127
    "Blackbird," 38
    Bonneville Executive Touring, 162–163
    Bonneville TT Special, 99, 150
    Cardinal, 149
    Grand Prix (T100R) 500, 22–24
    Gyronaut X-1, 96
    Harris T140 Bonneville, 170–171
    Model H, 8

INDEX

Model P, 8
Model R Fast Roadster, 8
Ricardo, 8
Royal Wedding T140, 27
T15 Terrier 150, 36, 38, 40, 43–44, 48, 55–56
T20B Super Cub 200, 107, 111–112
T20B Tiger Cub, 101, 107
T20C Tiger Cub 200, 59, 61–62, 68
T20CA Tiger Cub, 61–62, 68
T20J Junior Cub, 61–62, 68, 72, 77
T20M Mountain Cub, 101, 107
T20SC, 80–81, 87, 91, 95
T20SH Sports Home, 80–81, 87, 91, 95, 101
T20SL Scrambler Light, 77
T20SM Mountain Cub 200, 91, 95, 101
T20SR, 80–81, 87, 91, 95
T20S Scrambler 200, 72, 77, 87
T20SS Street Scrambler, 80–81, 87, 91, 95
T20 Tiger Cub 200, 40, 43–44, 48, 55–56, 59, 61–62, 68, 72, 77, 80–81, 87, 91, 95
T20S Tiger Cub, 68
T20T Trials, 77, 87
T20W Woods 200, 72
T21 "Twenty-One," 56, 61
T25SS Blazer, 132
T25T Trail Blazer 250, 132
T90 Tiger 90, 87, 90–91, 94, 100–101, 104, 106–107, 110–111
T100 Tiger 500, 15, 17, 18–19, 19–20, 24–25, 31–32, 36–37, 41, 43, 44–45, 52–53, 57, 60, 64, 67
T100A Tiger 100, 71–72, 75–76
T100C, 35, 38–39, 41, 100–101
T100C 500, 104, 106–107, 110–111, 118–119, 122, 125, 131
T100D Daytona, 147
T100R Daytona Super Sport, 104, 106–107, 110–111, 122, 125, 131, 134–135, 137, 142, 144, 147
T100SC, 78, 80–81, 87, 90–91, 94
T100SR, 78, 80–81, 87, 90–91, 94
T100S/RR, 80
T100SS Tiger, 78, 80–81, 87, 90–91, 94
T100S Tiger 100, 104, 106–107, 110–111, 118–119, 122, 125
T100 Tiger, 15–17, 18–20, 24–25, 29, 31–32, 34–39, 41–45, 52–53, 57, 60, 64, 67, 87
T100 Tiger 100, 100–101
T100T Tiger Daytona, 104, 106–108, 110–111, 118–119, 122, 125
T110 Tiger, 40–41, 44–45, 52–53, 57, 60, 64, 67, 70, 74
T120 Bonneville, 63–64, 69–70, 77–78, 84–85, 92–93, 95–96, 98, 103–104, 108–109
T120 Bonneville TR7A, 69–70, 76
T120 Bonneville TR7B, 69–70, 76
T120C Bonneville 650, 72–73, 77–78, 84–85, 92–93, 95–96, 98, 115, 117, 119–120, 122

T120C TT Bonneville, 84–85, 92–93, 95–96, 98, 103–104
T120R Bonneville 650, 72–74, 77–78, 84–85, 92–93, 95–96, 98, 103–104, 108–109, 115, 117, 119–120, 122, 128, 130–131, 134–135, 137, 140–142, 226
T120RV Bonneville, 134–135, 137, 140–142
T120V Bonneville, 146–147, 151
T140AV Bonneville Executive Touring, 165
T140D Bonneville Special, 159–163, 165
T140E Bonneville, 158–163, 165, 167
T140ES Bonneville Electro, 162–163, 165, 167
T140EX Executive, 167
T140J Bonneville Silver Jubilee Limited Edition 750, 156–157
T140LE Bonneville Royal Limited Edition, 165, 167
T140 TSS 750, 165, 168
T140 TSX 4, 168
T140 TSX 8, 168
T140 TSX Custom 750, 162, 165, 168
T140V Bonneville, 137, 140–142, 146–147, 151, 154–155, 158
T150 Trident 750, 72, 91, 102, 108, 113–115, 119, 128, 133–134
T150V Trident 750, 133–134, 137, 144–146, 149–151
T160 Trident 750, 149–151
T2000 Phoenix, 170
Thruxton Bonneville, 94
Tiger 70, 11, 13
Tiger 80, 11, 13
Tiger 85, 21
Tiger 90, 11, 14
Tina scooter, 77
TR5AC, 76
TR5AR, 76
TR5MX Avenger 500, 147–148
TR5 Trophy 500, 25–27, 29, 35–36, 36–40, 43, 48, 54–55, 57–58, 60–61, 67
TR5T Trophy Trail 500 (Adventurer), 142, 144, 147
TR6A Trophy, 70–71, 74
TR6B Trophy (Trophybird), 70–71, 74
TR6C Trophy 650, 75, 88, 90, 98, 100, 104, 109–110, 117–118, 122, 128, 130–131, 134–135, 137
TR6R (SR) Trophy 650, 75, 88, 90, 93–94, 98, 100, 104, 109–110
TR6R Tiger, 117–118, 122, 128, 130–131, 134–135, 137, 140–142
TR6RV Tiger, 134–135, 137, 140–142
TR6SC Desert Sled Trophy 650, 88, 90, 93–94
TR6SC Trophy 650, 78, 85–87, 88, 90, 98, 100
TR6SR Trophy, 78, 85–87, 88, 90, 93–94, 98, 100
TR6SS Trophy, 78, 85–87, 88, 90, 93–94, 98, 100
TR6 Tiger, 117–118, 122
TR6 Trophy 650 (Trophybird), 51, 54–55, 57–61, 67, 79, 93–94, 98, 100, 104, 109–110, 242
TR7A/B, 70

TR7RVS Tiger 750, 162–163, 165, 167
TR7RV Tiger 650, 137, 140–142
TR7RV Tiger 750, 137, 140–142, 146–147, 151, 154–155, 158–163, 165, 167
TR7T Tiger Trail 750, 165, 167
TR20 Trials, 80–81, 87, 91, 95
TR25W Trophy 250, 108, 111–112, 118–119, 125
TR65 Thunderbird 650, 165, 167–168
TRW, 20–21
TRX75 Hurricane 750, 138–139
TS8-1, 165
TS20 Scrambler, 80–81, 91, 95
See also Hinckley Triumph models
Triumph Motorcycles America (TMA), 154, 155, 158, 160, 165, 170
Triumph-Norton Inc. (TNI), 137
Truelove, Harry, 290
Turner, Bobby, 31, 32
Turner, Edward
  death of, 137
  Grand Prix model and, 22–23
  Japanese competition, 69
  pictured, 12, 21, 51, 127
  profile of, 12
  retirement from Chief Executive position, 87
  Speed Twin and, 13–14, 300
  Thunderbird model and, 29
  Tigress scooter and, 59–60
  Triumph under BSA and, 32, 36
  US distributors and, 20, 32
Turner, Eric, 72, 77, 95, 119, 127
Tuuli, Niki, 290

U
Uphill, Malcolm, 115, 117

V
Vale, Henry, 25
Vetter, Craig, 138–139
Vigil, Ernie, 267–268
Vostatek, Ondrej, 295

W
Westfall, Steve "Scuba," 300
White, Roger, 61
Whitworth, David, 22
Wickes, Jack, 113, 127, 149–150
Wicksteed, Ivan, 14
Williams, Les, 134
Wilson, Jack, 49, 155
Wilsmore, Mark, 269
Winslow, Marius, 14
Wood, Stuart, 260
World Supersport Championships, 223, 229, 235, 242
World Supersport Next Generation, 283
Wright, Russell, 49
Wynn, Keenan, 62